P9-DMV-314

The Story of Science

Aristotle Leads the Way

An underwater encounter between an archaeologist and a sphinx near Alexandria (see page 127)

The Story of Science

Aristotle Leads the Way

Joy Hakim

Published in Association with the National Science Teachers Association

Smithsonian Books
Washington and New York

Published by Smithsonian Books

Carolyn Gleason	Editorial and Production Director

Produced by American Historical Publications

Byron Hollinshead	President
Sabine Russ	Managing Editor and Picture Editor
Lorraine Hopping Egan	Editor
Kate Davis	Copy Editor
Marleen Adlerblum	Designer and Illustrator
Lisa Savage	Administrative Assistant

ISBN: 1-58834-160-7
Library of Congress Control Number: 2004107324

British Library Cataloguing-in-Publication Data available

Manufactured in the United States of America
12 11 10 09   3 4 5 6

Grateful acknowledgment is made to publishers who have granted permission to quote from their copyrighted works (see page 276).

This book is for Stephen and Henry Hollinshead
and their grandfather too.

A Writer's Reasons

"The sight of stars always sets me dreaming," Vincent van Gogh wrote to his brother, Theo. Somehow it consoles me to think of that tormented painter finding repose by looking heavenward. So, in a notebook I keep (don't all writers have them?), I put Vincent's words about stargazing right next to those of Huck Finn. Huck says, "We had the sky up there, all speckled with stars, and we used to lay on our backs and look up at them, and discuss about whether they was made or only just happened."

In 1888, Vincent van Gogh painted *Starry Night over the Rhone*.

Funny thing: the Greeks asked that question, Mark Twain asked it, and we're still asking. The big questions don't seem to go away. As to stargazing? That's how science got started. Where *did* those stars come from? What are they made of? And where are they going? Those are questions for all of us, not just for the astrophysicists in our midst.

That's a conviction that led me to write these books (this one is the first of a projected six). The first three books in the series tell the story of physics and chemistry, from ancient Sumer to today's string theory. Which means they deal with the very big (the universe), the very small (atoms and particles), and the in-between as well. It is, I believe, an enthralling tale, and one that is basic to our human heritage.

I'm convinced, and I hope to convince you, that science is not just for scientists. In the twentieth century, we compartmentalized knowledge; in the information age, that doesn't make sense. Today, you can be a hermit on a mountain peak and still have access to the world's learning. For scholarship to be so available, so democratic, is unprecedented in world history. To use that opportunity well, we all need to be generalists first (then we can find specialties). And no field of knowledge is as basic or as creative as science. That ragtag notebook of mine—cluttered with thoughts from poets, artists, and philosophers—has helped me realize that the human quest to understand the universe underlies almost all other creativity. I have more to say on this subject, but books should speak for themselves. Read on and see if I persuade you.

Iron atoms recorded by an electron microscope (see page 90)

Joy Hakim

There's More to This Story

No one writes a book like this without a lot of help. From the beginning, scientists, educators, friends, and my wonderfully supportive family have been enthusiastic and generous with their time. Maybe they sensed that a project like this needed many heads to make it happen.

When I asked questions of Hans Christian von Baeyer, he said helping with books intended for a general audience that includes young readers might be one of the most important things he could do. Hans, who is Chancellor Professor of Physics at the College of William & Mary, and whose own books are terrific, welcomed me to his classroom and helped get me started. Neil de Grasse Tyson, the Frederick P. Rose Director of the Hayden Planetarium of the American Museum of Natural History, read the draft manuscript and made a number of useful suggestions. Physicists Rocky Kolb and Chris Quigg read early versions of the manuscript and sent cogent comments. John Hubisz, professor of physics at North Carolina State University and former president of the American Association of Physics Teachers, read much of the manuscript and offered specific and wise thoughts. Roland Otto, head of the Center for Science & Engineering Education at Lawrence Berkeley National Laboratory, read a first draft, made helpful suggestions, found some errors, and informed me. Gerry Wheeler, executive director of the National Science Teachers Association and the author of an excellent physics text, encouraged, read some chapters, and also answered queries. Richard Schwartz, a mathematician who is also an artist, at the Institute for Advanced Study at Princeton University, was exact and clear in answering questions on mathematics and very generous with his time and knowledge. Thanks to David Beacom at the NSTA, Juliana Texley read the manuscript and gave an informed, detailed, and supportive critique. Texley is a master teacher of science and lead reviewer for *NSTA Recommends*. Tanya Didascalou Waterman, who is a friend as well as a great physics teacher, answered scientific questions and sent thoughts on her native Greek language.

Ruth Wattenberg, editor of *American Educator*, was one of a host of educators who actively encouraged. An article in that publication, previewing these books, generated an enthusiastic response from teachers. Tom Adams, in the California Department of Education, a visionary educator, was generous with good advice. I am especially grateful to Frederick Seitz, professor and former president of Rockefeller University, and one of America's most distinguished scientists. His personal encouragement and support has meant a great deal. As president of the Richard Lounsbery Foundation, Dr. Seitz was responsible for a substantial grant that helped bring the books to fruition. I've also had support and financial assistance from Doug MacIver and the Talent Development Middle School project at Johns Hopkins University. Working with teachers and students in urban classrooms, the JHU writer/educators produced brilliant and innovative teaching materials to coordinate with *A History of US*. I'm thrilled that Maria Garriott and Cora Teter of TDMS are developing teacher/student materials to accompany the science books.

I'm constantly awed by the teachers I meet. Again and again, teachers like John Holland in Yorktown Heights, N.Y., have encouraged me in this project. I wish there were room to thank you all, but what a gift we have in so many of you who teach our children.

Students have also read and commented on this manuscript, especially Ben Brown in California, Natalie and Sam Johnson in Colorado, and Madelynn von Baeyer (Hans's daughter) in Virginia.

I'd like to think that these books are without errors, but James McPherson (who teaches history at Princeton) once told me that no books dealing with history are errorless. Any that you might find are my responsibility (so let me know, and they'll get corrected in future editions).

Now, as to Those Who Did a Lot of the Work...

These books are a coproduction. Byron Hollinshead, who brought my history books into being and made them a marriage of pictures and content, has done that and much more with these books. He's been a guide and advisor whose good taste and sound advice always seem to be exactly what is needed. Sabine Russ is responsible for the pictures and for a whole lot of the creative and detailed work that goes into putting a book together. Sabine has been quite simply indispensable. Lorri Egan, as editor, has brought expertise in science writing along with intelligence, wit, and astonishing dedication. Kate Davis, the copy editor, has refused to let me get away with sloppy thinking (or writing). I've known lots of editors, but no one as dedicated as Kate. The beautiful design is the work of Marleen Adlerblum, who, with patience and ingenuity, dealt with the endless complexities of layouts and revisions. (I keep changing words.) Lisa Savage capably assisted with picture research and with many other aspects of the project. Don Fehr, director of Smithsonian Books, was enthusiastic about this project from the time that he learned about it, and he has been able to spread his enthusiasm throughout that remarkable institution.

Contents

Key for Feature Sections

Science ∞ *Math* *Language Arts*  *Technology and Engineering* *Geography* *Philosophy*

1 Birthing a Universe

In the beginning God created the heaven and the earth. And the earth was without form, and void; and darkness was upon the face of the deep. And the Spirit of God moved upon the face of the waters. And God said, Let there be light: and there was light.

—Genesis 1:1–3, the first book of the Hebrew Bible and the Christian Old Testament (King James Version), and a holy foundation for the Islamic Qur'an (Koran)

For the forming of the earth they said, "Earth."
It arose suddenly, just like a cloud, like a mist, now forming, unfolding.

—The Popol Vuh, the sacred book of the Quiché (a Mayan people), ancient oral stories first written down in the sixteenth century

Some foolish men declare that a Creator made the world. The doctrine that the world was created is ill-advised, and should be rejected. . . . Know that the world is uncreated, as time itself is, without beginning and end.

—The *Mahabharata*, a sacred Hindu epic poem composed in India between 400 B.C.E. and 400 C.E.

The universe was called *an-ki*, which meant "sky-earth."

The wise ones, who were deep thinkers and observers, said the sky was like an upside-down soup dish—solid and bowl shaped. Some called it a heavenly vault. Perhaps it was fashioned of tin, although more than a few thought it was made of the beautiful blue gemstone lapis lazuli. Others said there were three layers of translucent crystal. Stars, which were celestial fires, shone through holes in that glorious roof. Water was stored between the layers, they said. When a hole got unplugged, it rained.

And the Earth? It was a flat disk set in a surrounding ocean.

Below Earth was the vast underworld. Each night it was visited by the Sun. Once a month the Moon made the same journey to the lower region.

Like old friends, these same stars (left) paraded across the ancient sky each clear night. Nights were darker then, free of electric lights and smog, so the points of light in the sky loomed bright and dense—like speckled stone rather than polka dots. Despite the overwhelming scale, the regularity of the nightly routine was comforting. The ancients even found the shapes of familiar animals in the unchanging star patterns. Yet there were unsettling questions too: How did the stars get there? Why did five of them—out of thousands—follow different paths? What made the stars return night after night? Would it always be so?

CELESTIAL means "having to do with the sky or heavens."

From where did life come? From the ocean—which was eternal; out of it had come everything living.

These were the thoughts of the Sumerians, who, 5,000 years ago, created what may have been the world's first great civilization. They built city-states with temples and schools in the land that is now Iraq.

They believed themselves to be modern people—and they were. The Earth was ancient—and so were humans. For millennia they had moved with the seasons, the animals, and nature's crops. Then perhaps someone spilled seeds and noticed that, later, grains grew where those seeds fell. It would have taken a genius, someone with a scientific mind, to make that connection between seeds and soil and time and growth. But the connection was made, and humans began to plant crops and settle in organized communities.

Long before the Sumerians appeared, people had farmed, herded animals, and lived inside walled villages. The oasis town of Jericho, between Israel and Jordan, is 10,000 years old. (The story of Jericho and of Joshua, who made its walls tumble, is told in the Bible.) But the Sumerian cities were different from other ancient settlements. In the old cities almost everyone farmed and herded. In Sumer, work was divided in a new specialized way. A few Sumerians were priests or ruling officials, others were craftsmen, traders,

The quotation from the Bible that starts this chapter is in a translation called the King James Version. James I (1566–1625), who was king of England when the first English settlers arrived at Jamestown in 1607, was awkward and irritable and not very popular. But he authorized a new version of the Bible, which has become one of the great literary works of all time.

The *Mahabharata* is the longest epic poem in any language. This great Hindu work begins with a rivalry between cousins, tells of civil war, and reaches a climax with a ferocious battle. Underneath the action are serious thoughts on religion, ethics, government, and philosophy. Written in Sanskrit (the classic language of Hindu India), it has some 88,000 verses in its "short" form.

B.C. or B.C.E.?

The letters B.C.E. after a date stand for "before the Common Era," or before the year 1. The letters C.E. mean "Common Era," or from the year 1 to now. After the year 1000, we drop C.E. and just use the date.

Some people say B.C. ("before Christ") and A.D. (*anno Domini*, "in the year of the Lord"). B.C.E. and C.E. are more current and include both Christians and non-Christians.

Keep in mind that for B.C.E. dates, bigger numbers are earlier years: 300 B.C.E. is longer ago than 100 B.C.E. For C.E. dates, the opposite is true. The year 1800 is longer ago than 1900. We're about to go on a journey where keeping time and civilizations straight is important. Timelines help.

1200 B.C.E. 1000 B.C.E. 800 B.C.E. 600 B.C.E. 400 B.C.E. 200 B.C.E. 1 B.C.E. | 1 C.E. 200 C.E. 400 C.E. 600 C.E. 800 C.E. 1000 C.E. 1200 C.E.

BEFORE THE COMMON ERA — THE COMMON ERA — TO THE PRESENT

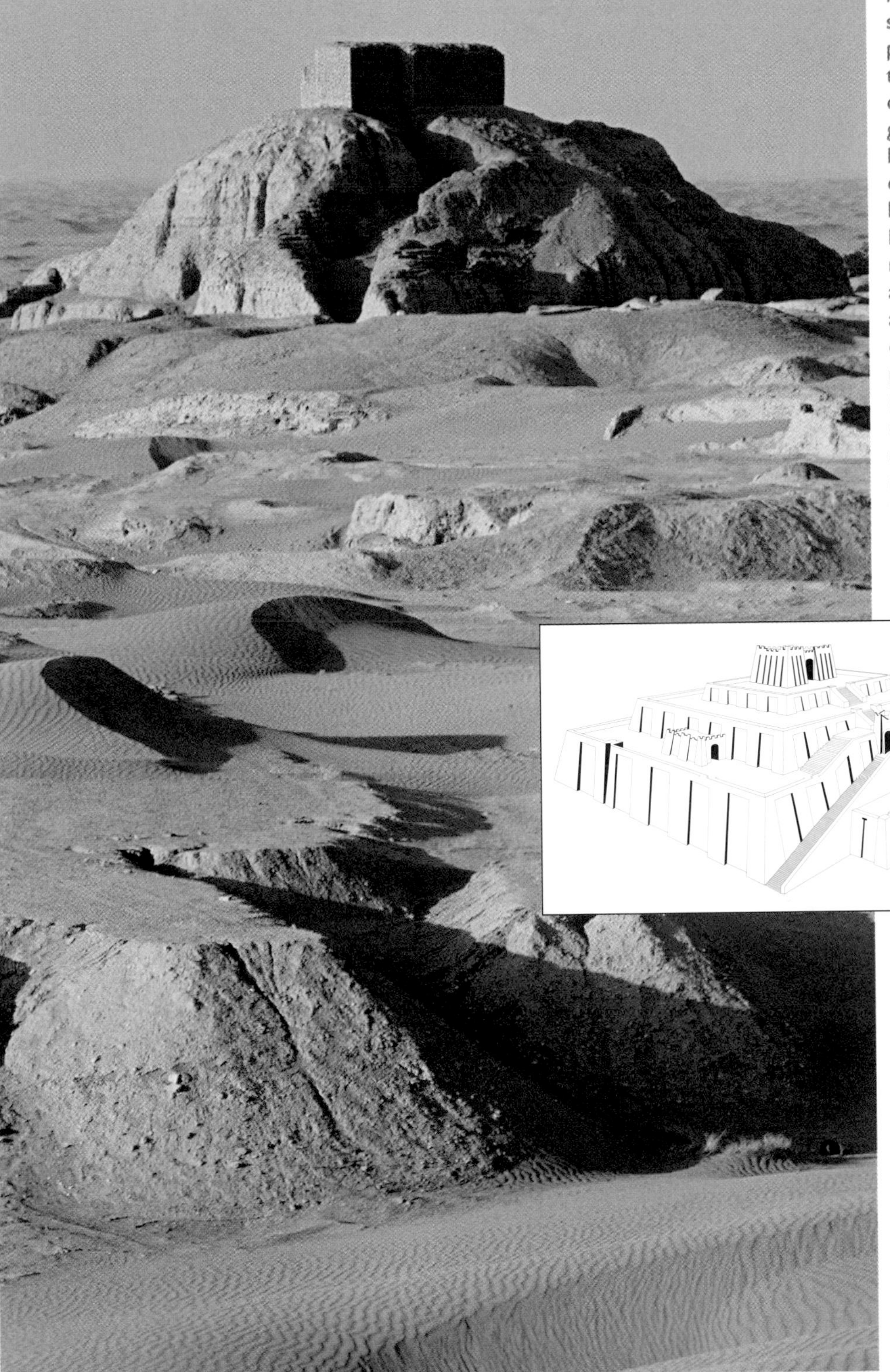

In Mesopotamian societies, priests in pyramid-like terraced temples, called ziggurats, guarded and kept knowledge secret, especially scientific knowledge. Knowledge gave rulers power. In ancient China, astronomy was considered a royal privilege. Ordinary persons who owned or used astronomical instruments faced harsh penalties. Does knowledge still give people power?

The Ziggurat of Ur, from ca. 2100 B.C.E., sits on an enormous mud-brick platform surrounded by desert. Parts of the temple were reconstructed in the twentieth century. The drawing (adapted from the British Museum) shows how the complete ziggurat might have looked.

Ancient Civilizations of Mesopotamia

Reading about the civilizations of Mesopotamia can be confusing, especially because they were all roughly in the same place. So here's something to help keep them straight. Mesopotamia ("between rivers") is the ancient land between the Tigris and Euphrates Rivers in what is now Iraq. Archaeologists have evidence of settlements there from the fifth millennium B.C.E., which is about 7,000 years ago. Sumer is the name of the southern Mesopotamian civilization of city-states that reached a pinnacle at about 3000 B.C.E. One thousand years later, Sumer was in decline and was absorbed by Babylonia and Assyria. Babylonia flourished from about 1750 B.C.E. until its decline in the sixth century B.C.E. It finally fell to the Persians in 539 B.C.E.

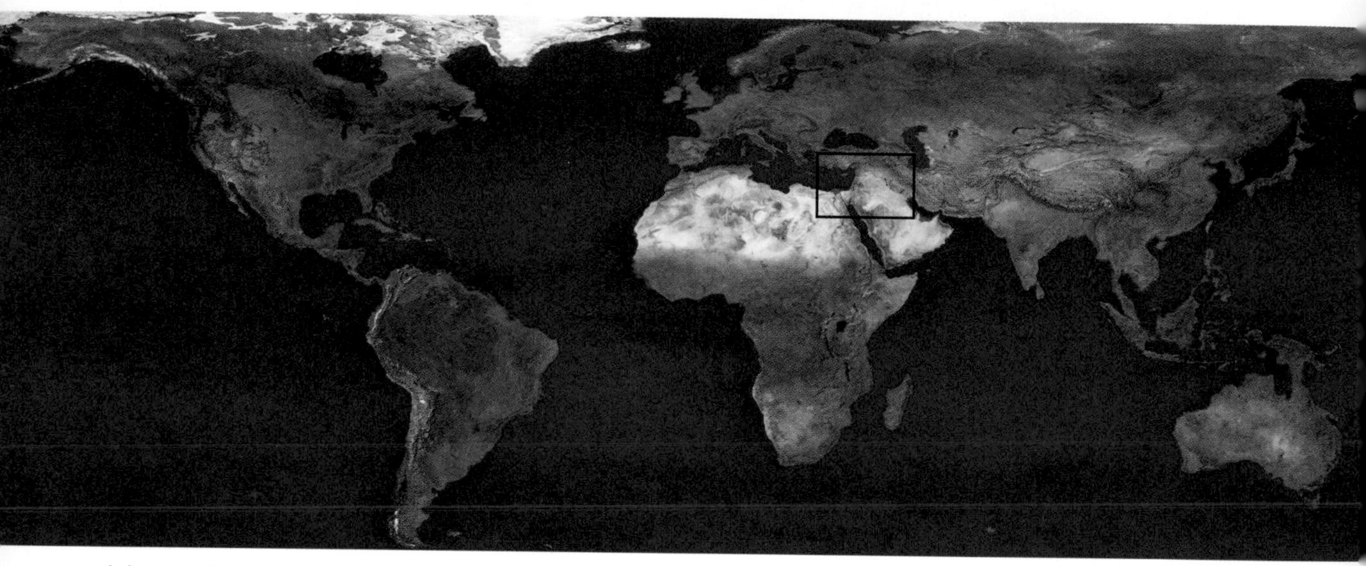

A *lamassu* (meaning "protective spirit") sports a human head, a bull's and lion's body, wings, and five legs. Such limestone beasts guarded important doorways in Assyrian palaces.

laborers, or farmers. The cities were well-to-do, with serving and governing classes. Farms were productive; water was channeled for irrigation and personal use; and trade brought ideas and goods from far places.

Dividing up the work meant that some of the city dwellers were stuck with dreary jobs, but others had time to study, plan, invent, and think.

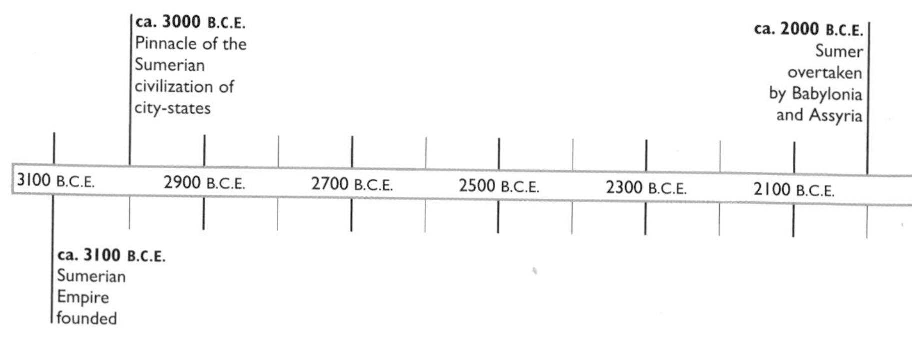

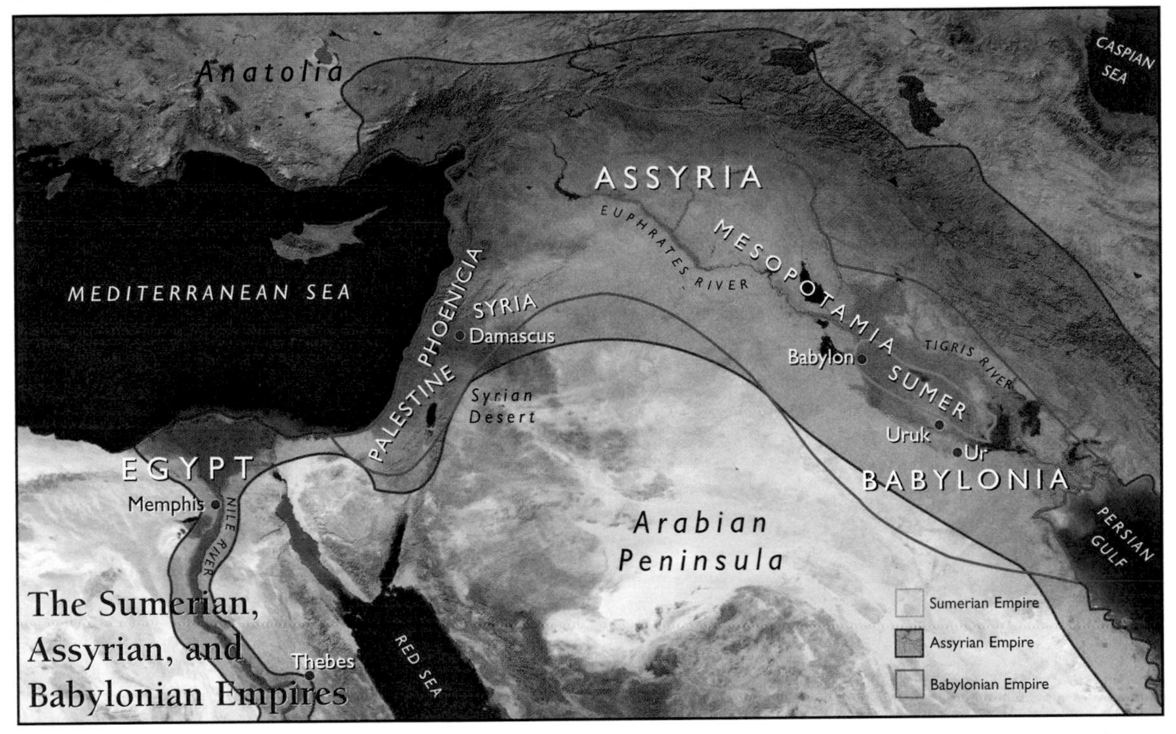

The Sumerian, Assyrian, and Babylonian Empires

The Sumerian Empire, a southern Mesopotamian civilization of city-states, existed from ca. 3100–ca. 2000 B.C.E. until it was eventually taken over by Babylonia and Assyria. The Babylonian (ca. 1750–539 B.C.E.) and Assyrian (ca. 950–612 B.C.E.) Empires overlapped for a period of years, during which ruthless kings and hostile armies struggled for power in southern Mesopotamia. The Assyrian Empire peaked when its borders extended to the Mediterranean, but it was eventually conquered by the Babylonians in 612 B.C.E. The Babylonian Empire remained strong until it fell to Cyrus the Great and his Persian army in 539 B.C.E.

They thought about laws and numbers and art and government and architecture. They learned to use the wheel. They taught themselves writing and may even have invented it. They studied the stars, watched the Moon, and used its recurring cycles to make a lunar (moon-based)

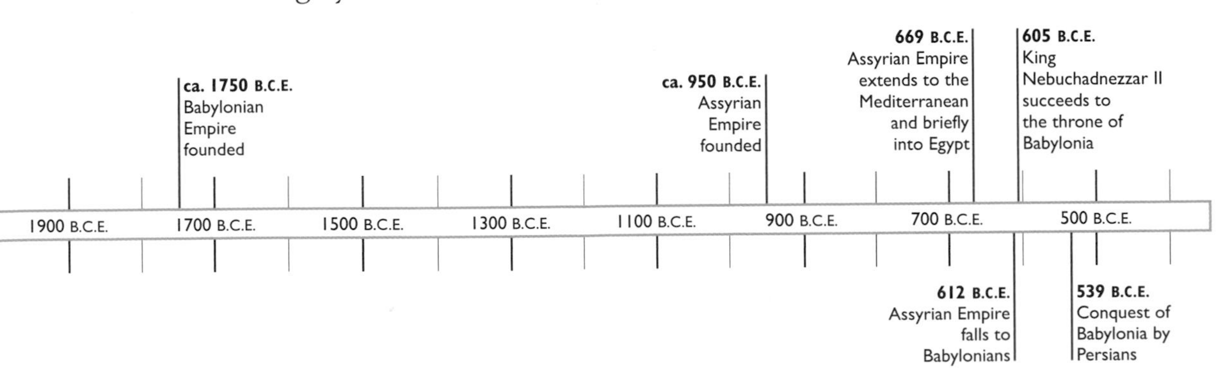

This Sumerian game board was found in the royal graves of Ur. It is about 4,500 years old and made of wood inlaid with shell, red limestone, and lapis lazuli. Players cast dice or threw sticks before racing their counters from one end of the board to the other. The fate of the dotted pawns was poetically compared to wins and losses of food, drink, and love. The rosette spaces were lucky.

IS IT IN THE STARS?

Clearly, the Sun—bringing light and warmth—affects life on Earth. So too does the Moon, whose presence causes ocean tides. Even our celestial neighbors Mars and Venus tug on Earth while passing by.

But do the positions of stars and planets also influence future events on Earth or the personalities of its inhabitants? That idea, believed by many, led to astrology, which turned out to be a bit like fortune-telling and not science at all. That's why it's called pseudoscience. *Pseudo* (SOO-doh) means "false" in Greek.

calendar that had months with names. The Sumerians worshiped a moon god; her name was Nanna.

It takes 12 cycles of the Moon to hold the four seasons, which led naturally to the idea of a 12-month year. Everyone could see those seasons repeat themselves like a rolling wheel. "Is that Nanna's mystery, or can it be explained?" the Sumerians must have asked themselves.

Just 800 kilometers (about 500 miles) away from Sumer's fertile, river-washed lands, the Egyptians lived in another river region. Theirs was an already-ancient society that put people in layered categories from slaves to godlike royalty. They too studied the sky and learned to predict the seasons. But the Egyptians did it by paying attention to the Sun, not the Moon. They worshiped sun gods and created a solar (sun-based) calendar. It was better than the lunar calendar. (Why? See chapter 3.)

On the tomb of Harkhebi, a third-century B.C.E. astronomer, it is written:

> *Wise in sacred writing . . . clear-eyed in observing the stars, among which there is no erring... who observes the culmination of every star in the sky... who divides the hours of the day and night without error... knowledgeable in everything which is seen in the sky...*

A magnificent Egyptian pendant from 1350 B.C.E. Two baboons carrying moon discs on their heads worship the sun god in the form of a scarab (beetle). The beetle holds up a solar disc, and all three sit inside a celestial boat.

"Who divides the hours of the day and night without error... " Everyone understood that it was knowledge from observing the heavens that helped regulate days, that informed farmers when to plant, and that gave constancy to festivals and ceremonial days.

Those long-ago people had minds just as good as ours. And, since there were no lights at night except for the Moon and stars and planets, they gazed at those heavenly beacons and knew the bright ones as well as you know the houses in your

Something that is CONSTANT doesn't change. Holding a festival on the same day each year is an example of CONSTANCY. In science, a constant is an exact and unchanging number, like the speed of light in a vacuum.

The epic of Gilgamesh was found on 12 big clay tablets by archaeologists who dug up the library of Assurbanipal, a seventh-century B.C.E. king. The story is much older; some say it was written in ca. 2000 B.C.E. Mighty Gilgamesh was an arrogant, part-immortal king; the gods sent a wild man to tame him. What happened is told in a story filled with friendship and heroism. Read it and you might cry at the end.

An ancient terra-cotta (right) shows the demon genie and guardian of the cedar forests Humbaba, whose head was cut off by Gilgamesh and his friend Enkidu.

A seventh-century B.C.E. Assyrian stone bas-relief from the palace of Assurbanipal in Nineveh, Mesopotamia, shows the king in his chariot inspecting booty after a battle. The panel probably belonged to a group of reliefs that decorated the king's private apartments. It depicted all the triumphs in war and sport he was most proud of.

Assyria's King Assurbanipal (ah-ser-BAH-ni-pahl), who certainly was proud of his abilities, spoke these words in about 650 B.C.E.:

"I learned the hidden wisdom, the whole art of the scribes and astrologers,
I can interpret omens of heaven and earth,
I participate in the councils of experts,
I can discuss treatises of divining with the skilled seers,
I can find difficult reciprocals and products which are not easy to calculate,
I can read elaborate texts in which the Sumerian and Akkadian are obscure,
I can decipher stone inscriptions which date from before the flood."

neighborhood. When they looked at the stars, they must have wondered, What are they made of? and How and why do they travel across the sky every night?

Those are the questions of science, but science was still an infant, just taking baby steps out of the cradle. So, while the stargazers took notes, kept records, and made useful observations, they couldn't answer the big questions, the ones we humans would keep asking through the ages: How were the heavens formed? How does the universe work?

For answers, the ancients turned to storytellers or priests. Out of a sincere but misguided belief that stars and planets rule human activity came a false science, a pseudoscience, called astrology. Fanciful stories of the stars became explanations for the unexplainable. As to how the world began? The best stories about that, which were told and told again, are part of our heritage. We call them creation myths. Every culture seems to have them.

2 Telling It Like They Thought It Was: Myths of Creation

At first within the darkness veiled in darkness,
Chaos unknowable, the All lay hid.
Till straightway from the formless void made manifest
By the great power of heat was born the germ.

—Rig-Veda, a sacred Hindu hymn from ca.1500 B.C.E. It is the oldest known religious text in any Indo-European language.

When all the stars were ready to be placed in the sky, First Woman said, "I will use these to write the laws that are to govern mankind for all time. These laws cannot be written on the water as that is always changing its form, nor can they be written in the sand as the wind would soon erase them, but if they are written in the stars they can be read and remembered forever."

—Diné (Navajo) creation story, as told by George Johnson in *Fire in the Mind*

The sky is like a hen's egg and is as round as a crossbow pellet; the earth is like the yolk of the egg, lying alone at the center.

—Zhang Heng (78–139 C.E.), Chinese astronomer, *Hun-i chu*

Once, or so it was said, there was only Chaos and two emperors: Hu (of the Northern Sea) and Shu (of the Southern Sea). Hu-shu (the names together mean "lightning") looked at Chaos, whose name was Hundun, and saw that he was unformed. So, for seven days, the two emperors zapped him with thunderbolts. They made seven openings in Hundun's great body: holes for seeing, hearing, speaking, breathing, eating, reproducing, and eliminating. When lightning pierced Chaos, life began; or so Chinese parents told their children.

CHAOS (KAY-ahs) means "confusion" or "disorder." *Merriam-Webster's Collegiate Dictionary* says, "the confused unorganized state of primordial matter before the creation of distinct forms."

In Greece, in ancient days, it was the words of the poet Hesiod (HEE-see-uhd) that parents spoke to explain the gods

A 525 B.C.E. frieze from Delphi, Greece, depicts the battle of the gods against the Giants. (Giants, with a capital *G*, are different from Titans, who were gigantic in a lowercase way.) The god Apollo and the goddess Artemis (far left) are approaching the shielded Giants, one of whom has been slayed. Because Giants couldn't be killed by divine hands, the god Zeus, with the help of a mortal woman, had a son named Heracles, who did most of the killing. The Giants eventually lost and ended up buried beneath volcanoes.

and the world's beginnings. Hesiod, who lived 28 centuries ago, said that at first there was only Darkness and Chaos.

Then Love appeared; she brought light.

And then Hesiod wrote,

Earth, the beautiful, rose up.
...And fair Earth first bore
The starry Heaven, equal to herself.

Ouranos (UR-ah-nohs) is a Greek word for "heaven." It became *Uranus* in Latin and, later, the name of the seventh planet of the solar system.

The **Cyclopes** (sy-KLOH-peez) were terrifying creatures, each with one round eye. The singular is **Cyclops** (SY-klops). Our word *cycle* comes from the same root. Hesiod says the Cyclopes were mighty blacksmiths imprisoned by the gods. Zeus freed them, and, in thanks, they gave him the thunderbolts they made on their forge. The poet Homer described them as lawless barbarians.

Mother Earth was called Gaea (JEE-uh), and Father Heaven, Ouranos. From them came the monstrous one-eyed Cyclopes, the powerful giants called Titans, the gods who lived on Mount Olympus, and—after some warfare between the monsters and gods—ordinary people.

The Egyptians said the world began with the ocean, which was called Nun. When Nut-the-sky-goddess pulled a hill out of Nun-the-ocean, it was the Earth's beginning. The creation of that first hill was soon commemorated with a pyramid, and another, and another.

In ancient Egypt, creation myths appeared in many variations. In this mural, found in the 3,000-year-old tomb of a Pharaoh, the air god, Shu, holds aloft the sky goddess, Nut, separating her from Geb, god of the Earth.

In India, poets spoke of a world held in the branches of a great tree, and from that World Tree image grew stories of the Tree of Life and the Tree of Knowledge.

The Chinese believed the world was flat and covered by a heavenly canopy studded with stars.

Ancient Peruvians thought the universe was a huge box with a ridge for a roof where the great god lived.

Hesiod and the others were attempting to explain the universe—its mystery and its power—in the days before there was science to help with the explaining.

A drawing of the Hindu universe depicts the dome-shaped Earth supported by six elephants that are standing on a giant tortoise resting on a snake.

How do we tell myth from science? Why aren't origin stories science? It's not because the myths and stories are wrong.

DEFINING WORDS AS SCIENTISTS DO

observation: To a scientist, observation means more than just looking at something. Scientists are careful observers who study an object or event in detail and record those details thoroughly and precisely to make a scientific record. *As Greek thinkers observed a lunar eclipse, they noted and drew pictures of the changing appearance of the moon.*

hypothesis: A hypothesis is a possible and reasonable explanation for a set of observations or facts. Once you have a hypothesis, you can begin testing it, but it is not yet accepted as fact. *The Greek idea of a round Earth began as a hypothesis.*

theory: To a nonscientist, a theory and a hypothesis are the same thing. They are ideas that attempt to explain things. *The American Heritage Dictionary* says that a theory is an assumption, a conjecture, a speculation. But to a scientist, a theory is a *well-tested* explanation of observations or facts. Brockhampton's *Dictionary of Science* defines *theory* as "in science, a set of ideas, concepts, principles, or methods used to explain a wide set of observed facts." A scientific theory is verified—checked and well tested. British physicist Stephen Hawking says a theory "must make definite predictions about the results of future observations." *When Greek thinkers observed the curved shadow of Earth on the Moon during an eclipse, their round-Earth hypothesis began to seem like a sound theory.*

fact: Facts are information tested and shown to be accurate by competent observers of the same event or phenomenon. *When astronauts took pictures of the spherical Earth, the fact of its round shape could be seen clearly.*

Accepted science sometimes turns out to be wrong.

Myths are imaginative attempts to explain what seems unexplainable. They appeal to emotions. No one can prove a myth, and that's the measure: Science is about proof. But the ancient myths were important; they were part of a process that would stretch minds. Those myths got people thinking, as good science fiction usually does.

Real science starts with a question. Scientists are searchers, looking for solutions. So once a question is asked, the hunt is on for an answer. In the language of science, an untested answer or an idea or a possibility is called a *hypothesis* (hy-PAHTH-i-sis). The next step is to test the idea. The goal is to find mathematical and experimental proofs that confirm or reject the hypothesis. If a hypothesis survives the tests, it becomes part of a *theory*. (Albert Einstein's Theory of Relativity has passed all its tests.)

Doing science is a bit like making your way through a jungle when your eyesight is limited. Some paths are straight and easy, some are baffling. Will we ever see the whole picture? Perhaps not. That doesn't mean that what we are

Elaborately painted burial chambers in pyramids and royal tombs have given us valuable clues about the lives and beliefs of the ancient Egyptians. The sarcophagus hall of Pharaoh Seti I (who reigned ca. 1302–ca. 1290 B.C.E.) is decorated with astronomical scenes and the voyages of the sun god, Ra. Every evening, Ra is swallowed by the sky goddess, Nut, and travels through her body. Nut gives birth to Ra again each morning, and he passes alongside her body during the day. The detail on the opposite page shows the Northern constellations above the falcon-headed Ra during his daytime journey.

Why do hurricanes and spiral galaxies look similar? Compare this satellite image (right) of a giant hurricane named Floyd (September 1999) with the Hubble space telescope image of a majestic spiral galaxy (opposite page) named NGC4414. Check out the pinwheel shape, the saucer-flat profile, the rotating clouds, the concentration of energy and matter near the center. Those common bonds are no accident. Hurricanes are clouds of moisture. Galaxies, a trillion times larger, are mostly clouds of gas (even though the stars in them grab center stage). Both water and gas are fluids, substances that flow according to the universal laws of science—in this case, the physics of fluid dynamics.

doing is wrong; it just means that even though we can map that jungle with astonishing detail, our knowledge of the whole may always be incomplete. In I Corinthians 13 (part of the New Testament), Paul says, "For we know in part, and we prophesy in part."

Considering that can make you humble. We're just human beings trying our best to understand the wonder and vastness of the world around us. We've been asking questions and searching for answers for as long as we've walked the Earth. That's a good part of what makes us human.

We've learned that science is a detective story, a sleuthing adventure, a step-by-step course, a never-ending process that builds on itself, providing answers with ever greater precision. Wrong turns are sometimes taken—and eventually found out—and then there are new hypotheses, new experiments, new formulas, and new

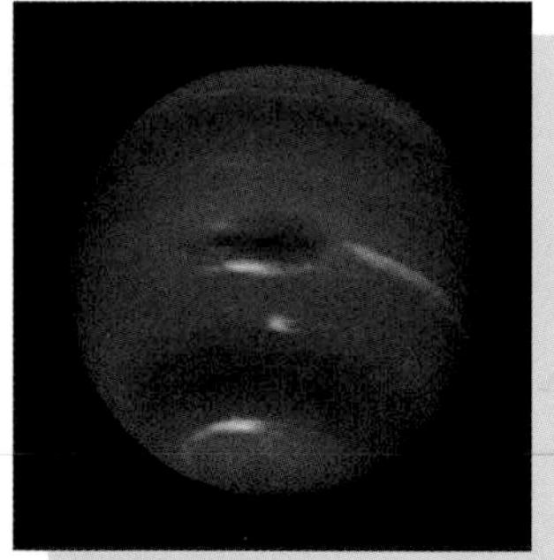

GUESS WHAT?

A science or math prediction is not a fortune-teller's hazy gaze into the future. It's not even a guess. It's a calculated foretelling of events and outcomes—eclipses, earthquakes, experiment results—based on facts.

The planet Neptune was a math prediction before it was a discovery in 1846. Uranus was not orbiting according to the path carefully plotted by mathematicians. Something big, with a serious gravitational pull, was throwing it off. Astronomers predicted the size and location of the object and pointed their telescopes there. Sure enough, they found the dim pinprick that is our eighth planet.

SPIRAL BOUND

The philosopher Immanuel Kant, back in the eighteenth century (that was Ben Franklin's century), understood that the spiral shape found in a hurricane can also be found in the "systems of stars gathered together in a common plane, like those of the Milky Way." Kant was doing what scientific thinkers have done throughout history. He was using what he knew in order to speculate (or hypothesize) about something not yet seen.

Twentieth-century astronomers confirmed the spiral shape of many star systems. Euclid, a math professor in about 300 B.C.E., put the awe of it all in written words: "The laws of nature are but the mathematical thoughts of God."

theories. So, in brief, being a scientific thinker means staying awake and keeping your mind open.

We humans have always wanted to know about ourselves and about the world we inhabit. We've asked two basic questions since the beginning of recorded time. One is about

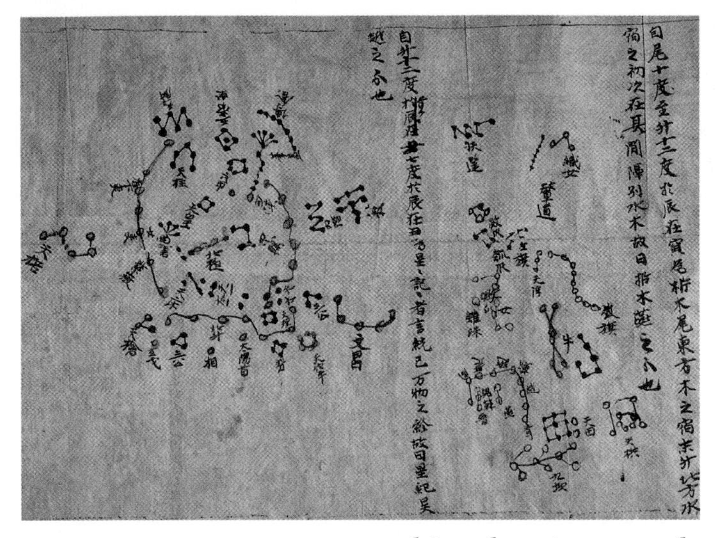

This is a detail from the oldest known portable star map. It is part of a long Chinese scroll from the T'ang dynasty (618–907 C.E.), which shows 12 sections of the night sky seen from the Northern Hemisphere. To determine the season, astronomers observed which direction the handle of the Plough constellation was pointing in the early evening. North Americans call that same group of stars the Big Dipper. Can you find it?

the very big: What is this universe of ours all about? Which leads to other questions: What are the stars made of? How many are there? Are they static, or do they change? What is Earth's place in the universe?

The other basic question is about the very small: What is life? Which leads to: What's an atom? Are there still smaller particles? What makes us tick? And (a big question): Is there anything that ties us to the stars and to all of creation?

The stars have always been part of the quest; you can't understand our place in the universe without understanding them. So while most long-ago peoples confused myth with reality, a few looked at the heavens and kept careful records. Chinese sages charted the paths of the stars and planets, identified 300 constellations, and understood that eclipses of the Sun and Moon come in 18-year cycles. They recorded their findings on bones and turtle shells and passed them on to their heirs.

Where did the world come from? The ancient Chinese scholars didn't worry much about that. For them, the world was just there. The Chinese were more concerned with the "how" of nature than the "why." To make use of the "how'" they needed to measure and weigh and compare and understand numbers.

Numbers have been the surprise of science. The universe seems to work in ways that can be expressed precisely with numbers, ratios, and mathematical equations. A seventeenth-century Italian professor of mathematics named Galileo Galilei wrote, "[The universe] cannot be read until we have learned the language and become familiar with the characters in which it is written. It is written in mathematical language." The Greek philosopher Plato had already scooped him on that idea. Some 21 centuries earlier, Plato had said,

Mathematical beauty? Most of us agree that flowers are beautiful, and the number and spacing of flower petals are mathematical. (More about this in chapters 9 and 25.) Art and architecture have always borrowed from nature. Especially during the Gothic period, European architects became fascinated with numbers and geometry as keys to understanding nature. Consider this gorgeous stained-glass rose window from the thirteenth century. It's the north window of the great Notre Dame Cathedral in Paris, France. Each area of the rose contains smaller geometric shapes. In the petals surrounding the center are figures of prophets, priests, kings, and judges.

"The world is God's epistle to mankind....It was written in mathematical letters." The ancients saw math as divine and beautiful. We know that arithmetic's numbers, geometry's shapes, and algebra's ratios are human made, but amazingly they explain much of the universe. Are they beautiful? Yes, they abound in art, music, and nature. As for equations, those with beauty are usually simple.

An EPISTLE is a formal letter written for the purpose of teaching an important idea or belief.

Einstein's greatest equation, $E=mc^2$, ties E (energy) to m (mass) through an unchanging number, c (the speed of light in a vacuum). That equation implies that mass and energy are actually just forms of the same thing. And that unity can be expressed with an awesomely simple formula.

Is that what scientists mean by "mathematical beauty"? Or is it the numerical spacing of a rose's petals, or the precise spiral pattern in the seeds of a sunflower, or the curve of an elephant's tusk? They are not random

Geometry is Born

We think the Egyptians invented geometry, which is about shapes, space, and measuring. They needed it. They had a difficult river to deal with. Each year the Nile overflowed its banks. That flooding deposited rich soil on the farmlands it touched. It also washed out all boundary markers. You can imagine angry farmers fighting over whose land was whose. It didn't make for good neighbors. The Egyptians took this so seriously that in their Book of the Dead (a religious tome), dead souls had to swear to the gods that they had not cheated their neighbors out of their land.

For the rulers of Egypt there was something even more important than dead souls—taxes. The pharaohs collected taxes based on land ownership. They needed to know who owned which parcel of land.

To solve the problem, surveyors reset the boundaries after each Nile flood. They usually worked in teams of three and were known as rope stretchers because of their long, knotted ropes. They learned to measure plots of land by dividing them into rectangles and triangles. And that, we believe, is how geometry was born.

Herodotus, a Greek historian, wrote about Egyptian mathematics (which was history to him). This is what he said:

> *Sesostris [king of Egypt] made a division of the soil of Egypt among the inhabitants, assigning square plots of ground of equal size to all, and obtaining his chief revenue from the rent which the holders were required to pay him year by year. If the river carried away any portion of a man's lot, he appeared before the king and related what had happened; upon which the king sent persons to examine and determine by measurement the exact extent of the loss; and thenceforth only such a rent was demanded of him as was proportionate to the reduced size of his land. From this practice, I think, geometry first came to be known in Egypt, whence it passed into Greece.*

Egyptian mathematicians were famous in their time. Of course, they influenced the other Mediterranean peoples, especially the Greeks. But the Egyptians never took geometry beyond everyday useful functions. It was the Greeks who made it abstract, something to do for its own sake. They thought of geometry as the science of reasoning.

A surveyor carrying a knotted cord is measuring a field in the presence of a farm couple. The wall painting was found in the tomb of Mennah, a Pharaoh's scribe and estate inspector in ca. 1400 B.C.E.

happenstance; they are mathematically predictable. (Note: It's the average pattern of a large sample of roses or sunflowers or elephant tusks that is predictable. You can never be sure how any single one may turn out.)

That marriage of numbers and nature is something that anyone, even those of us without sophisticated math skills, can learn to see. It is an essential part of the scientific story.

So here we are, heading way back in time and about to watch a drama unfold. It is the human attempt to understand our universe. It is the great thinking quest that has distinguished our species.

Observation, experimentation, and much headwork will be needed to begin to make sense of the world around us. The early astronomers were up to the challenge: in charting patterns in the sky, they achieved amazing insights. Among other things, a few learned to foretell the blotting out of the Sun or the Moon that is an eclipse. Which means they came to realize that there's a pattern to it.

Some of those ancient sky watchers used their knowledge wisely to help their communities function more efficiently; some just used it to gain power.

Almost all—the wise and the wicked, the learned and the pretenders—shared a common belief. They thought the heavens were controlled by gods whose reasons could not be known. And then, astonishingly, some thinkers off in a corner of the Mediterranean world began to question that idea. They began the process that would separate science from mythology. But the big idea of science—that the universe is governed by mathematical laws that can be understood by humans—well, that idea would take time to be born.

This small marble sculpture of a stargazer is believed to be 4,800 years old and comes from the Cyclades, Islands in the southern Aegean Sea. Angular-shaped heads and eyes staring at the sky are two of the most distinctive features of artifacts from the Cycladic culture of the early Bronze Age.

3 Making Days: Were the Calendar Makers Lunatics or Just Moonstruck?

When there were all those gods
administering to panthers,
jumping over mountains,
and lighting stars and comets and a moon,
what was their one Belief?
what was their joining thing?

—Gwendolyn Brooks (1917–2000), American poet, *In the Mecca*

It is the very error of the Moon;
She comes more near the earth than she was wont,
And makes men mad.

—William Shakespeare (1564–1616), English playwright and poet, *Othello*

Perhaps it was the Moon that got people thinking about keeping a written record of time. The silvery Moon goes from new moon (invisible) to full moon (bright) to new moon again in a 29.5-day pattern that any sky watcher can follow. Keep track of that pattern, with its changing phases, and you will have useful information. You'll know the nights that will be dark and the nights that you can hunt. You can make predictions.

The ancients, who were struggling to understand the ways of the universe, thought that the Moon and planets affect the way people on Earth behave. They believed the position of the planets could determine your luck. (That's what astrology is all about.) They were wrong, but they didn't know it, and it gave them a reason to study the stars carefully.

Some 3,000 years ago, scribes in Uruk, Mesopotamia, inscribed an astrological calendar on this clay tablet, using cuneiform writing (see pages 32 and 33).

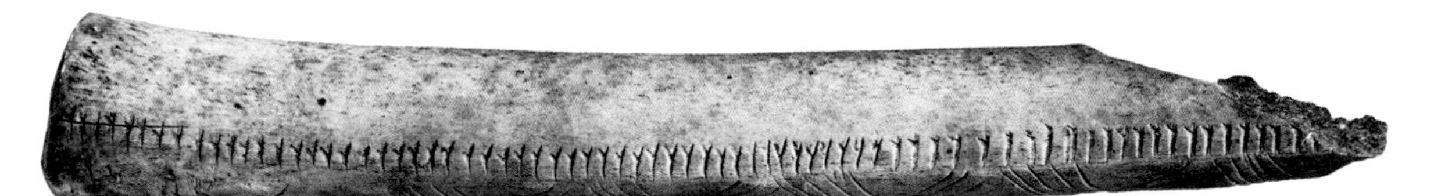

This eagle bone was found in a cave in Le Placard, France, and is believed to be at least 13,000 years old. Could the row of tiny markings be accidental scratches? Not a chance, say experts on Stone Age artifacts. The grooves were chiseled by human hand, perhaps to record lunar phases. Other bones found elsewhere show similar patterns. Were they early calendars?

Looking at the Moon, it seemed to them a flat dish that wasn't very big. Did anyone realize then that the Moon has no light of its own, that moonlight is actually reflected sunlight? It's not likely.

Did the keen observers understand that we on Earth see only one side of the Moon? There was a clue: the shadows on the Moon's "face" never change. The thinkers may have asked, "Why don't we see the other side?" It was a puzzle that needed solving.

The other big puzzle was this: "Why don't we see a full moon all the time? What causes those changing phases?" That would be tough to figure out. It would take many, many centuries to do it.

So while no one understood the reasons behind the action in the heavens, everyone could see the nightly moon show and note its repeating pattern. And that gave ancient peoples a way to measure the passing of time.

This is important. If you use a lunar calendar and start it so that the first day of the year is planting time, then 3 years later you are planting a month too soon, and by the time a decade has passed you are planting in midwinter. After 33 years, the first day of the year is back where it is supposed to be, having traveled through the entire solar year.

—Isaac Asimov, American science writer

How did they record those measurements? An intriguing bone from a baboon's thigh, found in South Africa's Lebombo Mountains near Swaziland, has 29 grooved notches. (Landlocked Swaziland is south of Mozambique, not far from Africa's east coast.) The bone is some kind of counting device and is about 37,000 years old, from the Paleolithic period. Could those bone markings be a record of the Moon's journey? Was this an early calendar?

We don't know, but we do know it was the Moon that the first calendar makers in Mesopotamia and China and America used to record time. The Babylonians of Mesopotamia found that twelve 29.5-day moon cycles

WHAT MAKES THE SEASONS?

It's the tilt of the Earth. That's the short answer to remember, whenever anyone asks. Seasons are not about being closer or farther from the Sun, a common mistake. They're not about planet Earth receiving stronger or weaker sunlight. They're about the tilt. (It's worth repeating.)

Shine a flashlight on a globe (or any tilted ball), and you can see what a huge difference direct or indirect rays make. The tilt causes the Sun's rays to strike the Earth head-on in places (like the equator, where it's always warm) and at an angle elsewhere. There's more tilt talk coming up.

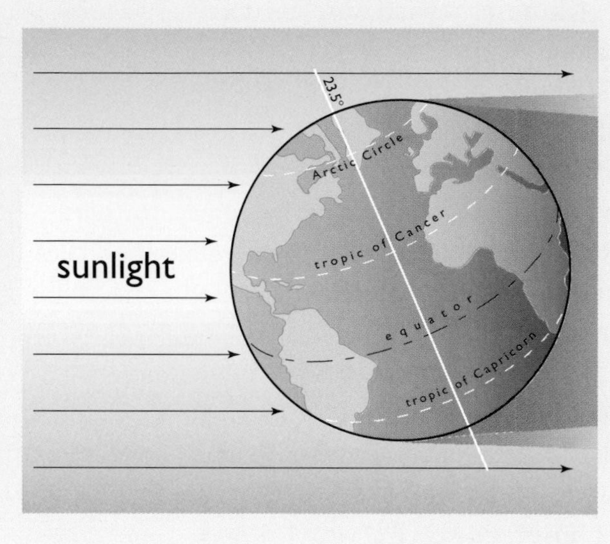

(or months) make a year of 354 days. Unfortunately for the calendar makers, the Moon's year doesn't keep pace with the Sun's year, which is about 11 days longer. And it's the Sun's year that determines the seasons. In an agricultural society, where the main purpose for having calendars was to tell when to plant and when to harvest, that was a problem. So extra makeup days were added to the Babylonian calendar to make it come out right with the seasons.

Even though moon calendars didn't work well, people stuck with them. The Moon had a mystical attraction, a religious aura, and a romantic one, too. Poets sang of the Moon, dancers cavorted by moonlight, and priests often prayed to a moon god. As to those in love? In ninth-century Japan, the poet Ono no Komachi wrote:

> *This night of no moon*
> *There is no way to meet him*
> *...My heart is consumed by fire.*

The Chinese, under Emperor Yao (way back in ca. 2357 B.C.E.), established a lunar calendar and found it soon fell behind the seasons. So every 19 years, they adjusted the calendar by adding seven months.

Above, a modern Jewish calendar records the period of the Omer, the 49 days from the second day of Passover to the day before Shabuoth. Since the calendar is lunar, the dates change from year to year. Below, the Gezer Calendar (ca. 925 B.C.E.) is an agricultural calendar carved on limestone with Hebrew inscriptions about seasons and planting and harvesttimes.

Chinese, Jewish, and Islamic calendars are still lunar. The Islamic year has 12 months of alternately 29 and 30 days. There is no attempt to mesh with the seasons, and so Ramadan, Islam's holy month of fasting, comes at different times in different years. (Imagine Christmas moving from winter to spring to summer to fall before getting back to winter.)

Egypt was the first civilization, as far as we know, to look to the Sun for its calendar. How did they manage to think differently from other civilizations?

Perhaps they noticed that the brightest star in the sky, Sirius (also known as the Dog Star), seasonally disappears into the heavens. Then, when it is seen again at dawn in the Egyptian sky, it is in a direct line with the rising Sun. What made Sirius's reappearance seem magical was that each year it came at the start of the annual flooding of Egypt's great river Nile. It became Egypt's New Year's Day, the first day of the month of Thoth, in late August. By recording Sirius's arrival year after year, Egyptian astronomers came to realize that the Sun's year is 365.25 days. So they made a calendar that divided the year into 12 months of 30 days each, with 5 extra days that became the birthdays of the major gods. Every few years they added another special day to take care of that extra quarter day. Our modern calendar, with its leap year, is a grandchild of Egypt's innovation.

The calendar is intolerable to all wisdom, the horror of all astronomy, and a laughing-stock from a mathematician's point of view.

—Roger Bacon, English philosopher, from an appeal to Pope Clement IV

The Egyptians, and a few others, understood that the cycle of seasons and the length of days are determined in a ballet danced by Sun and Earth. They didn't understand that it was Earth, spinning on its axis, that was doing the pirouettes. Those who lived in the Northern Hemisphere (Europe, North America, Asia) just knew that on June 21 (June 20 on certain leap years) the Sun appears at its apex (highest point) in the sky, and the day lasts the longest. That day became known as the summer solstice (SUHL-stis), the first day of summer. In the Southern Hemisphere (Africa, South America, Australia), June 21 is the first day of winter.

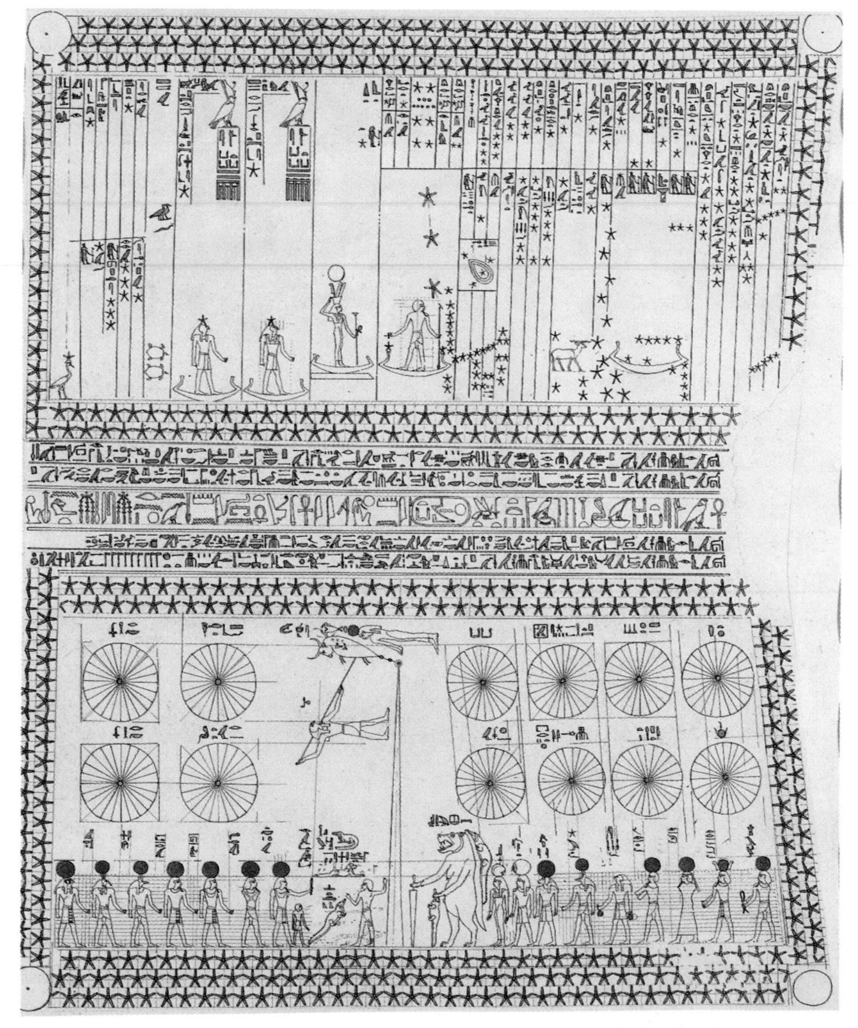

This drawing of an unfinished ceiling painting comes from the tomb of Senmut, who was a close advisor to Hatshepsut, the only woman who held the title of Pharaoh. (They lived in the fifteenth century B.C.E.) The astronomical scenes describe celestial objects in the skies—night (upper half) and day (lower half). The upper half tracks the changing position of the constellation Orion as it travels the night sky. Orion's distinctive trio of stars (representing his belt) loom above Herf-Haf, the celestial ferryman, who is always looking backward. Isis (with the solar disk above her head) and other deities accompany him in separate boats. Also present are the planets Mars, Venus, Jupiter, and Saturn. In the lower half, deities along the bottom are carrying solid solar disks on their heads toward the center. The 12 large, spoked circles mark the months of the lunar calendar and feast days.

After June 21, things begin to change. In the Northern Hemisphere, the Sun peaks at a slightly lower point in the sky each day, reaching its lowest point above the horizon on about December 21. That's the winter solstice, the shortest day of the year. (But not in Southern Hemisphere cities like Buenos Aires, Argentina, where December days are long and hot.) On two days, called equinoxes (EE-kwuh-noks-uhz)—March 20 or 21 and September 22 or 23—day and night are about the same length.

It took sophisticated sky watching and centuries of record keeping to predict and chart the equinoxes and the solstices, but it happened in cultures around the globe.

Day to day, the Sun rises at different times and in different spots along the eastern horizon—and this photograph proves it. Actually, it's a series of photos of the same piece of sky, taken at the same time each morning over the course of a year and combined into one image. The extreme points show the Sun's morning position on solstice days in June (left) and December (right). (Because the photo location is in the Northern Hemisphere, the summer solstice is June 20 or 21; in the Southern Hemisphere, those are winter solstice dates.)

This annual figure-eight pattern has a name—the analemma. If the Earth's orbit were a circle (it's not—it's an ellipse), and if the Earth weren't tilted, the analemma wouldn't exist. All these morning suns would appear in the same spot in the sky every day.

WHY DOES THE MOON DAZZLE, THEN DISAPPEAR?

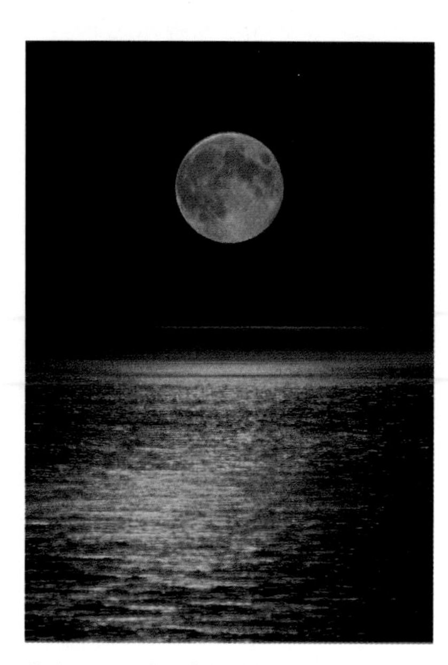

Only one side of the Moon—this blotchy visage—faces Earth all the time, no matter what the phase. Turn the page upside down, and you'll see what people in Australia and Argentina see. Either way, a full moon can only appear directly opposite the Sun, because it reflects the Sun's rays fully. As the Sun sets in the west, the full moon rises in the east. In the morning, the reverse happens—the Moon sets in the west, and the Sun rises in the east. That makes the full moon the only phase we see all night long. Moon phases are easy to figure out if, when you look at one, you ask yourself: "What time is it?" Then: "Where is the Sun?"

For a long time, no one understood that the phases are caused by the Moon's journey around Earth at the same time as Earth is moving around the Sun. The Moon has no illumination of its own. When its face looks bright, it is reflecting the Sun's rays toward us. About 7 percent of the sunlight that hits the Moon bounces our way. Where the Moon is on a given night—in relation to Earth and the Sun—determines what we see.

When the Moon is between the Sun and the Earth, its illuminated surface is turned toward the Sun but away from us on Earth, and the Moon appears to be invisible. It's a new moon, and there's not much to see. As that sphere moves around the Earth, its bright side begins to come into view. Starting as a fingernail-shaped sliver, it grows fatter each night, staying visible in the sky later and later. About a week after a new moon, we on Earth can see half of a disk (called a quarter moon because the Moon is one-fourth of the way in its orbit). A week later, at full moon, the side of the Moon that faces the Sun is also facing Earth: so the full moon is a bright dish that shines all night, and we are dazzled.

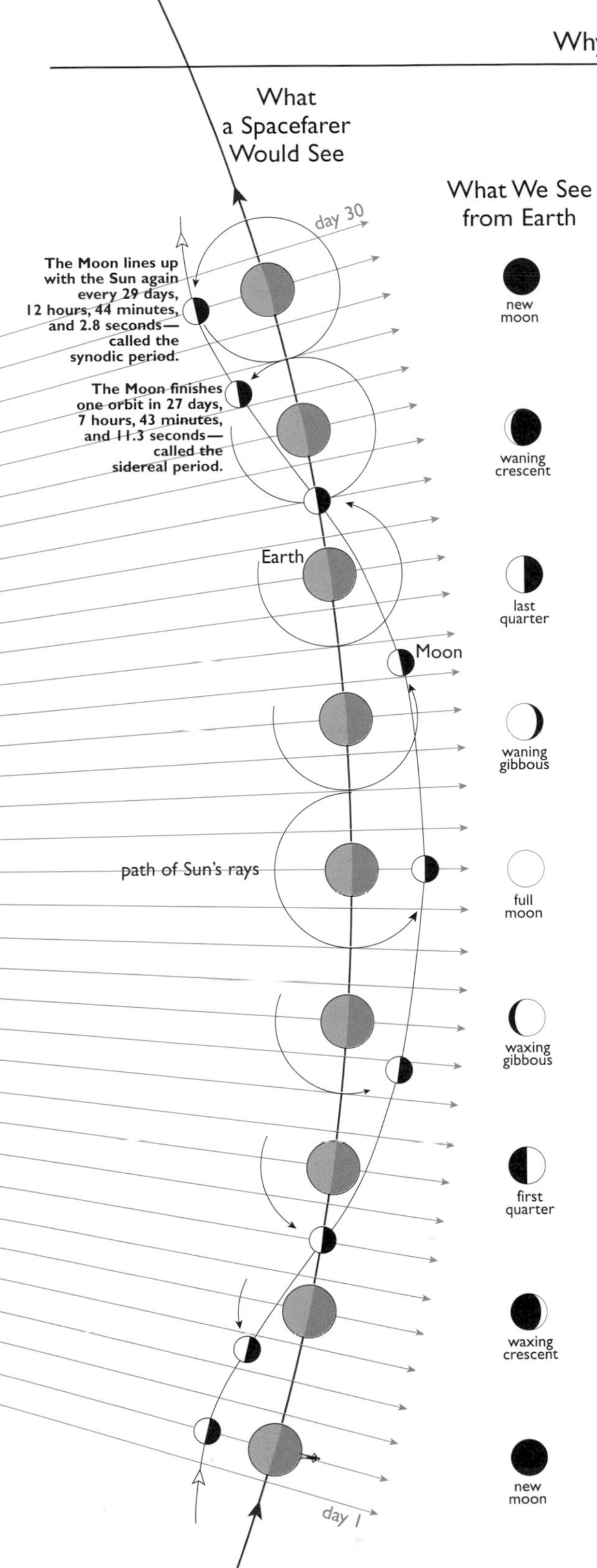

Science books often talk about the "eight phases of the Moon" (left), but as the photo series shows (right), there can be many more. From the upper left sliver to the bottom right new moon, the Moon changes appearance so gradually that it's not always easy to tell when it is truly "full" or "gibbous" or "quarter" or "crescent."

GIBBOUS is from the Latin word for "hump." It's the not-quite-full-moon phase that looks like a hump or bulge. A WAXING moon is growing bigger, from new moon to full moon, and a WANING one is shrinking to invisibility again.

Starting at the bottom of the diagram (left), imagine yourself standing on the blue Earth and looking at the orbiting Moon. The moon phases ("What We See from Earth") show what the Moon looks like in the sky. Notice that to a spacefarer, the Moon always appears half-lit on whatever side is facing the Sun.

(diagram not to scale)

Archaeologists call the building complex (above) "the palace," although it probably served as an administrative center in the ancient Mayan city of Palenque (now in southern Mexico). The tower rising above the ruins is an astronomical observatory. Originally, the top didn't have a roof. Today, visitors can still watch the Sun set directly over the Temple of Inscriptions, a building farther down the hill.

Palenque (pah-LENG-kay), in Mexico, was a Mayan astronomical center (from 600 to 800 C.E.), with ceremonial buildings laid out on an exact line so that the Sun's rays on the solstice pierced its inner recesses. (It made for drama as well as good astronomy.) The same thing was true at Stonehenge, huge circles of stones raised in England from about 3000 to about 1500 B.C.E., and it was also true in thousands of other European sites, which may have been centers of sun-based worship.

Why were the solstices so important? Did the ancients worry that the Sun might decide not to repeat its journey? Suppose winter lasted forever? When the winter solstice arrived (and the Sun stopped sinking in the sky), it was a time to celebrate and give thanks to the gods. Ah, if only we could waft ourselves to ancient Stonehenge, join in, and find

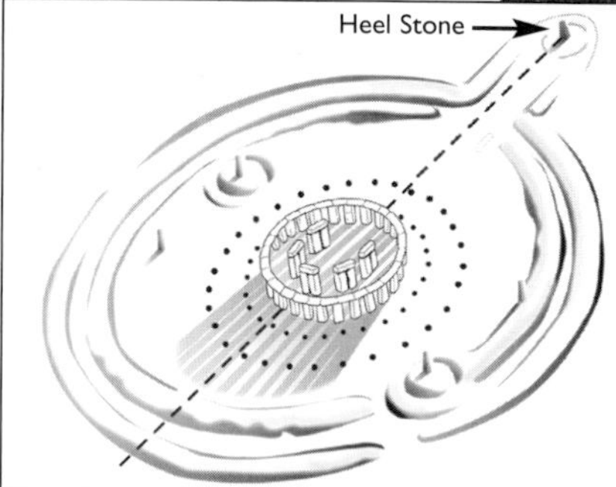

Stonehenge was laid out to greet the summer solstice. On June 20 or 21, the longest day of the year in the Northern Hemisphere, you can stand in the center of the stone circles and watch the Sun appear to "rest" on top of the giant Heel Stone.

out what people were thinking. (You can visit Stonehenge today, see some astonishing ruins, and wonder.)

In the land of the Maya, each god was associated with a star. By studying the position of stars, the high priests were expected to foretell which gods would be in control at a given time. Thus, they could tell (with a helpful god wielding power) when there would be widespread good fortune or (with a devil dominating) when to prepare for disaster. It was not science; the focus was on magic and mysticism. Yet out of the Maya's careful observations of the heavens came one of the most sophisticated calendars the world has known.

Actually, three calendars worked in unison: a religious moon-based calendar of 260 days, a practical sun-based calendar that was 365 days long, and a third calendar—the Long Count—that began in 3114 B.C.E. and will end (and begin again) in 2012. Making those three calendars work in harmony was like synchronizing three different-sized but coordinated chain wheels on a bike. The Maya did it.

But none of the ancient calendars got things quite right, so there was fiddling and fine-tuning in the centuries that followed. A calendar is just a human way of organizing time. Most of us do it in two ways—with clock time and calendar time. Clock time chases itself in circles, with morning and evening following each other in endless repetitions. But calendar time is linear. It's a timeline marking past, present, and future.

These Mayan glyphs were carved into a limestone lintel (a beam across the top of a door, window, or fireplace). They add up to the Mayan equivalent of the date February 11, 526 C.E. Some glyphs represent days and months. The gods' heads, seen in profile, stand for numbers.

SYNCHRONIZE means to "keep in time with." *Chronos* is Greek for "time."

NAMING DAYS AFTER—WHO ELSE?—THE GODS

CELESTIAL BODY/ GOD	DAY NAMES			
	BABYLONIAN	ROMAN	SAXON	ENGLISH
Sun	Shamash	Sol	Sunnan (or Sun)	Sunday
Moon	Sin	Luna	Monan (or Moon)	Monday
Mars	Nergal	Mars	Tiw	Tuesday
Mercury	Nabu (or Nebo)	Mercurius	Woden	Wednesday
Jupiter	Marduk	Jupiter	Thor	Thursday
Venus	Ishtar	Venus	Frigg	Friday
Saturn	Ninurta	Saturnus	Saeturn (or Saturn)	Saturday

Ra, the sun god of ancient Egypt, had disguises. This colorful mural (ca. 1300 B.C.E.), which was found in the tomb of Senedjem at Deir el-Medina (western Thebes), combines three myths in one image. Starting from the left, the Sun rises between two sycamore trees, appears as a bull calf, and then appears as Ra, who carries on his falcon head the Sun encircled by a cobra.

In ancient days, clock time and calendar time were mostly in the hands of the elite—priests, astronomers, and mathematicians. The Egyptian seers settled on a 24-hour day, perhaps because they thought the sun god, Ra, visited the 12 constellations during the day and the 12 regions of the underworld during the night.

The Babylonian astronomers gave us our week. They noticed five unusual "stars" (they were really planets) along with the Sun and the Moon, added five and two, and came up with a seven-day week. One day was named for each of those heavenly bodies. The Greeks didn't bother with a week, but for centuries, the Romans had one with eight days.

As to hours, they were usually measured with sundials, but they don't work well at night. Still, those dials led to complaints about time's pressures. Plautus, a Roman comic poet who lived in about 200 B.C.E., wrote this:

The gods confound the man who first found out
How to distinguish hours! Confound him, too,
Who in this place set up a sundial,
To cut and hack my days so wretchedly
Into small portions.

MUSINGS IN TIME

The solar year in the twenty-first century lasts 365 days, 5 hours, 48 minutes, and 46.5 seconds. The year has slowed 10 seconds since 1 C.E. The decrease in the length of the year is due to the gradual slowing of the Earth's rotation on its axis, an average of half a second per century. Until the mid-twentieth century, that slowdown was suspected but couldn't be accurately measured. Then scientists came up with a new definition of the second, based on the frequency of a cesium atom (instead of the rotation of Earth). Microchip-sized atomic clocks—using cesium atoms probed by microwaves in a vacuum—now give an accuracy equal to 100-quadrillionths of a second.

What would Plautus say about today's atomic clocks that measure nanoseconds, which cut and hack our days into ever smaller portions? Meanwhile, the profound among us ask, "What is time?"

We now know that time isn't what it appears to be. It's a relative thing. That means it is slightly different for me sitting at my desk than for an astronaut soaring through space. And it would be very different for a traveler zooming at close to the speed of a light beam. As to the calendar? There is no reason to believe that the calendars we use today will be around forever. Imagine yourself as a space voyager. Will you use a calendar tied to our Sun if you're heading beyond our solar system?

PET STARS

The Babylonians noticed that familiar groups of stars, like visiting relatives, drop by during the same season in the same place in the sky each year, and then they disappear. They gave pet names—the names of animals—to 12 star groups, one for each month. This annual parade of celestial animals was the inspiration for our 12 zodiac constellations.

An early fifteenth-century ceiling fresco (above) in the Old Sacristy of San Lorenzo in Florence, Italy, depicts a lively zodiac. The Sun peeks between Gemini and Cancer, and a majestic lion, Leo, approaches on the left.

TAKE A NUMBER, THEN WRITE IT DOWN

How did people write letters in ancient Mesopotamia? They pressed their words into clay tablets, like the one above, which is a king's letter to a high official. The tablet below lists numbers of goats and sheep.

W**hen did people** first begin to count? Was it when they considered their fingers and toes? Or when they counted their children? Or their sheep?

Whenever it was, counting was the beginning of arithmetic (about adding, subtracting, multiplying, and dividing), and arithmetic, along with geometry (about shapes, space, and measuring) is the foundation of mathematics.

At first, in order to count, people kept tally sticks, made notches on bones, piled up pebbles, or put knots in a rope. But with the advent of cities and large-scale trading, bookkeeping systems and written records were needed. So the first writing had to do with recording business transactions; later, writing that expressed language was developed.

The concept of number came slowly everywhere. Often the symbol for six when it stood for fish was different from six when it stood for sacks of wheat. The abstract idea of number, as separate from the objects counted, took thousands of years to evolve.

GETTING IT WRITE

Egyptian picture-based writing is called *hieroglyphic*. It was usually cut into stone. We have examples of hieroglyphic writing from about 3000 B.C.E. But those hieroglyphs were difficult and slow to draw and cut. They would end up mostly on tombs and fancy plaques (which is where we use Roman numerals today). For everyday figuring, the Egyptians began to use a kind of shorthand called hieratic script; eventually, it didn't look at all like its hieroglyphic origin.

Egyptian scribes did their writing on papyrus, a paper made from a tall, grassy reed abundant along the Nile River. Thin strips sliced from the roots of the plant were put next to each other with edges overlapping. One layer of strips was laid over another at right angles, and then the whole fabric was smoothed, pressed, and polished. Finally, 20 or more papyrus sheets might be pasted side by side to create a long scroll. Papyrus, like most paper, isn't long lasting. We don't have many examples of ancient Egyptian hieratic writing.

The Greeks did their everyday writing on slate tablets coated with a kind of wax that they scratched with a marker called a stylus. To erase, they just smoothed the wax. But wax is ephemeral (look it up!).

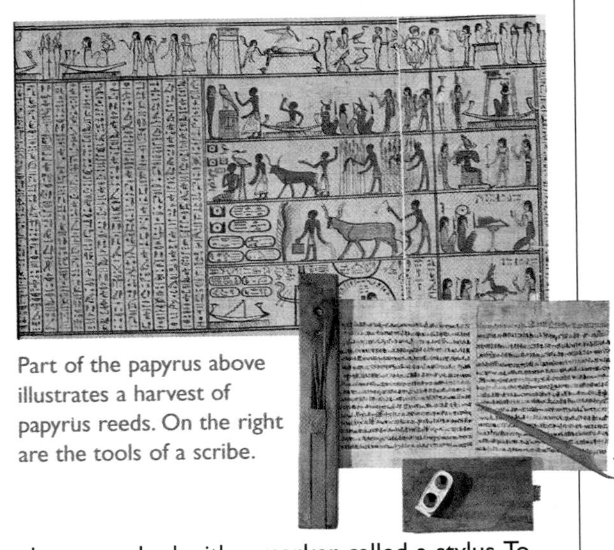

Part of the papyrus above illustrates a harvest of papyrus reeds. On the right are the tools of a scribe.

In India's river-based towns in the Indus Valley, a culture arose with writing and sophistication that seems to have rivaled that of Babylonia and Egypt. We have yet to decipher clay tablets found there, so we don't know details.

Going from scratch marks and pebbles to a language of mathematics was like turning on a light in a dark room. First came words for numbers and then written symbols.

In Mesopotamia, there was an abundance of clay soil and river reeds. Clay could be turned into writing tablets, and the sun would bake it hard. Reeds, cut to have triangular tips, became writing instruments. When pressed into soft clay, those reed tips left wedge-shaped marks. From the shape of those marks comes our name for Sumerian writing, *cuneiform* (kyoo-NEE-uh-form), derived from the Latin *cuneus*, meaning "wedge."

Thousands of cuneiform tablets have survived time. Most of them deal with daily needs, like beer, bread, and barley. I'm looking at a tablet with symbols that denote the daily barley ration for a worker; other symbols show the amount of barley needed for 1,800 portions. (Was someone running a big business? Or planning a city's rations?)

We know that, because of those long-lasting tablets, the Sumerians did complicated arithmetic problems, often dealing with ratios. Those who practiced the art of mathematics seem to have been an elite group of insiders (a guild), who kept their skills a closely guarded secret.

4 Ionia? What's Ionia?

Though leaves are many, the root is one.
—William Butler Yeats (1865–1939), Irish poet, "The Coming of Wisdom with Time"

We dance round in a ring and suppose,
But the Secret sits in the middle and knows.
—Robert Frost (1874–1963), American poet, "The Secret Sits"

There are more things in heaven and earth, Horatio,
Than are dreamt of in your philosophy.
—William Shakespeare (1564–1616), English playwright and poet, *Hamlet*

Does where we live affect the way we are? Along the Aegean Sea, Mother Earth is no shy flower. There, though spectacularly beautiful, she is also boisterous and hard to tame. Her steep mountains glisten white, showing off the stones that form them. Gray-green olive trees and scruffy bushes cling to hillsides. Sapphire waters surround raw land. The sun-filled azure sky is no less intense. This is no easy place to live. Crops grow poorly. Earthquakes and volcanoes terrify and leave scars. The sea swallows its victims.

AZURE is sky-blue. It comes from a Persian word for the gorgeous blue mineral lapis lazuli. (The *lazuli* part is what morphed into "azure.")

The Greek-speaking people, who settled sea-washed Turkey, Greece, and the islands between, needed sharp minds to survive, and they developed them. About 3,000 years ago, some of them began using their minds in new ways. It was a time when superstition and fear guided most thought and action. Rulers who claimed godly powers told subjects how to think. But in Ionia, on the Turkish coast, there were no god-kings. There, cities thrived. Traders and travelers brought new ideas. Ordinary people were free to think for themselves. A few, who examined the world with clear, unfrightened minds, drew conclusions from what they observed.

A Greek vase from ca. 500 B.C.E. shows a sacrificial bull being led to an altar surrounded by priests and gods.

Elsewhere, the ways of the world were seen as mysterious.

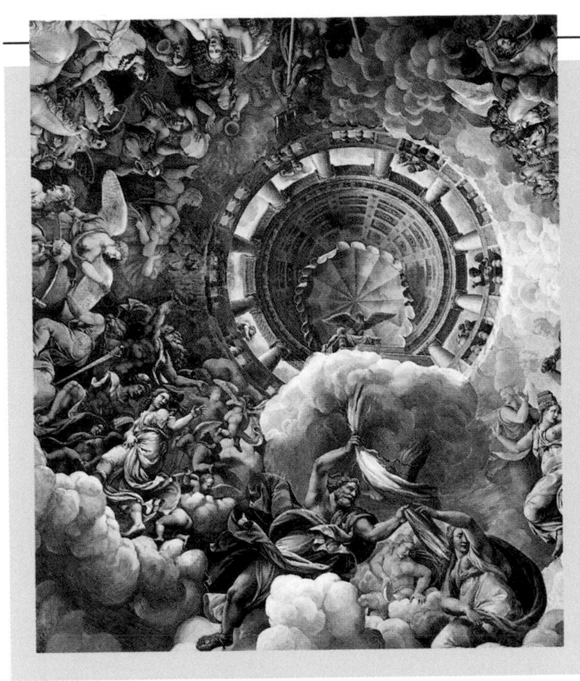

THE OLYMPUS GAMES

Keeping up with all the Greek gods is a task; there were so many of them. Those gods, who were said to live on Mount Olympus, had superpowers and they were immortal. Otherwise they behaved like ordinary people: they could be nasty and troublesome, or helpful and generous.

Perhaps it was because the tales grew so bizarre that thinking people stopped taking the gods seriously. Eventually they became part of the culture, as characters from good stories usually do. Today no one worships the Greek gods, but their stories—the Greek myths—are still among the best you will ever read.

There's turmoil on Mount Olympus in this 1535 Italian ceiling fresco from the Palazzo del Tè in Mantua. Note Jupiter hurling thunderbolts in the foreground.

Gods seemed impulsive and baffling. People brought sacrifices to those gods: they brought their most precious possessions, sometimes their own children. They were trying to bribe unknowable forces, not grasping that there are laws of nature that *can* be understood.

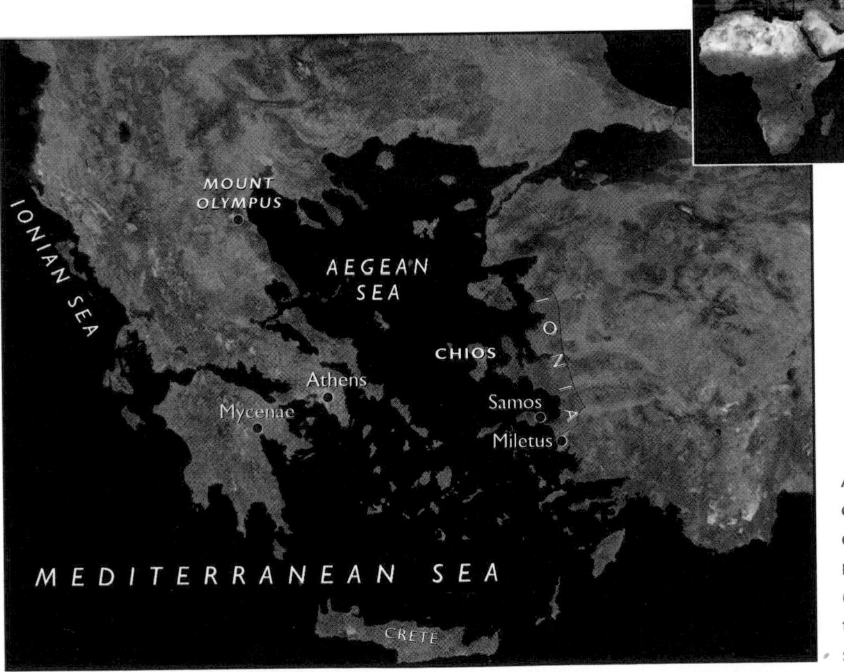

Ancient Ionia is usually thought of as western Turkey—to the east of today's Greece. The modern Ionian Sea, however (and this gets confusing), is on the other side, the western side of Greece.

"But," writes astronomer Carl Sagan, "in Ionia, a new concept developed, one of the great ideas of the human species. The universe is knowable, the ancient Ionians argued. . . . There are regularities in Nature that permit its secrets to be uncovered. Nature is not entirely unpredictable; there are rules even she must obey."

Thales (THAY-leez), a sixth-century B.C.E. Ionian living in the prosperous port of Miletus, is said to be the world's first philosopher-scientist-mathematician; the first to look for explanations in observed facts, not myths; the first scientist to leave his name on his ideas.

Named poets had come before him. Homer, the blind Ionian bard, had written his epics, the *Iliad* and the *Odyssey*, in the eighth century B.C.E. The story he told of the Trojan War had happened 400 years before that. Earlier, only kings and gods had been celebrated by name; now poets were singing of heroes and ordinary people and of themselves.

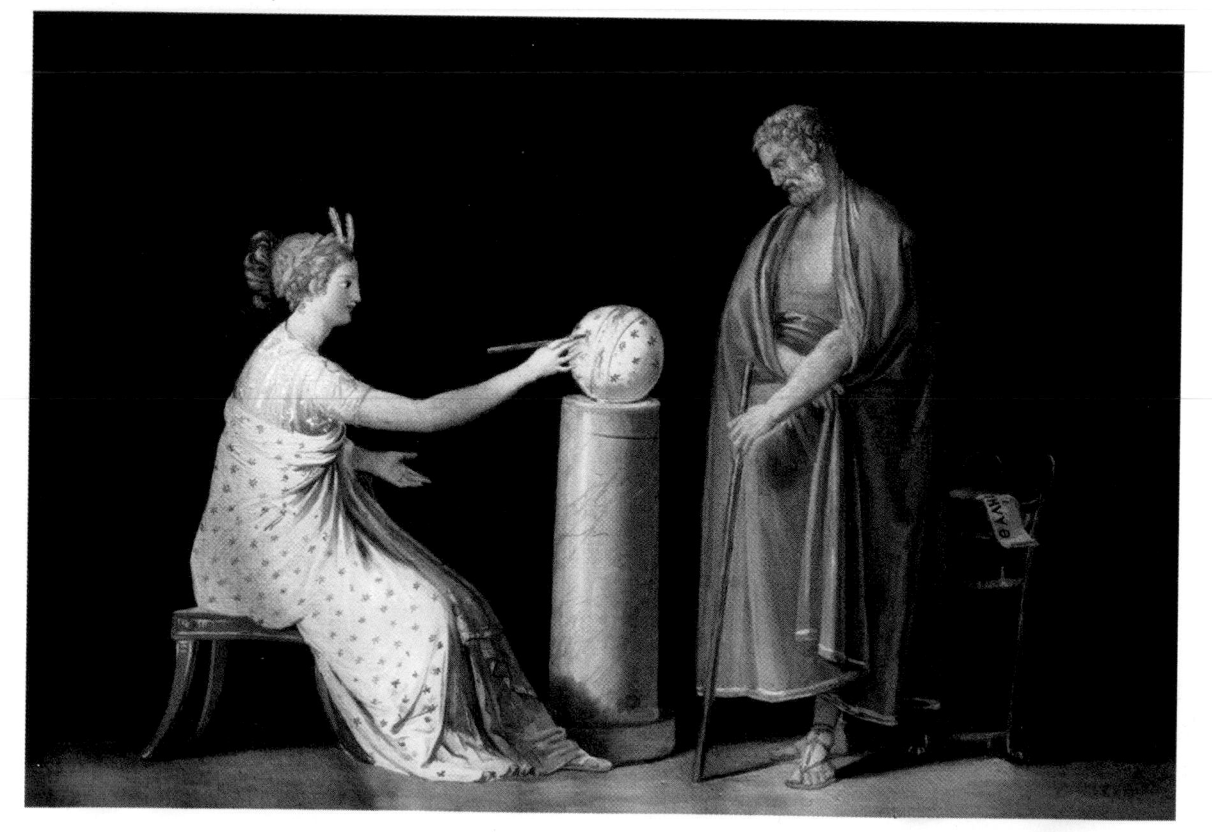

This painting of Urania, the muse of astronomy, and Thales is thought to be a work by Antonio Canova (1757–1822), the Italian master of neoclassical ("new classic style") sculpture.

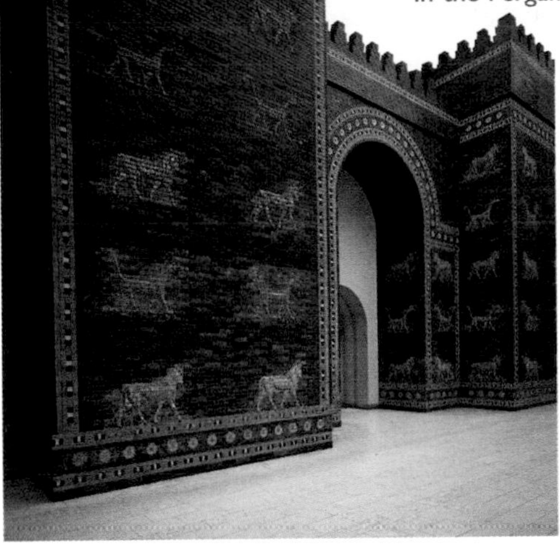
The city of Babylon once stood behind this glazed-brick gate, built in 575 B.C.E. by Nebuchadnezzar II. The king dedicated it to Ishtar, the goddess of love and protector of the army. The ruins of the Ishtar Gate were found in 1902. It now stands, restored, in the Pergamon Museum in Berlin, Germany.

In another Mediterranean land, Hebrew poets spoke of strife and love and human foibles. Something new—individualism—was appearing on the human stage. Ionia wasn't the only place it turned up, but it was a crossroads, a good place for cross-fertilization. The Aegean Sea, which licked Ionia, also washed the sands of Mycenae (my-SEE-nee). That spectacular Bronze Age city had disappeared by Thales' time, but tales of its splendor remained. Thales inherited the wisdom of these Mycenaeans, the Minoans, and other peoples who were ancients to him—such as the Amorites, the Hurrians, the Kassites, the Hittites, and the Israelites.

Foibles are small oddities of character.

Each of these civilizations had its own set of stories about the origin of the universe. Thales may have known of them all, and others too. We're told by a later, fifth-century C.E. Greek named Proclus that Thales traveled widely. The Mediterranean Sea was a highway that led south to Egypt and to ports east and west. A road went from Thales' home in Miletus to the fabled city of Babylon, where the savvy ruler Nebuchadnezzar (neb-uh-kuhd-NEZ-uhr) had a stable of astronomers at work. (Babylonia had succeeded Sumer as the leading culture in Mesopotamia.)

From Babylon, traders and scholars went to India and Mongolia and China, learned of those cultures, and brought home goods that dazzled. Back in Ionia, they were joined by soldiers, returned from the Persian Wars, who had tales to

FABLED KINGS

Have you heard the story of King Minos? There really was a Minos: he lived on the fabulous island of Crete. And there was an Agamemnon too. He was king of Mycenae on the Greek mainland. You can visit both places today and see splendid ruins and opulent treasures.

This gold mask from a tomb in Mycenae is called the "Mask of Agamemnon"—although there are doubts about the face's identity.

FERMENT, in everyday language, is unrest or turmoil. Scientifically speaking, fermentation is a chemical change in which complex organic compounds are turned into simple substances. For example, yeast or bacteria change sugar (complex) to alcohol and carbon dioxide (simple).

tell of that, to them, exotic culture. The soldiers bragged of their army's engineers, who had built marvels. Some were now working in the Ionian ports.

Given all this intellectual ferment, Thales may have been exasperated by the fickle gods worshiped by most people of his time. The god-stories had gone too far. Perhaps he realized the importance of a fresh start that turned away from myths toward observation and thought. He would use no gods to explain nature. Rather, he would use his senses and his intelligence and teach others to do the same. It was the beginning of the scientific approach.

We don't know much about Thales as a person, except what others tell us. They speak of a many-sided genius who was a lawgiver, a civil engineer, an astronomer, a mathematician, and a teacher. It is said that he predicted the solar eclipse of 585 B.C.E., that he changed the direction of the Halys River, and that he figured the height of a pyramid by measuring the length of its shadow. (To do that, he probably just waited until the time of day when the shadow of a stick is the same length as the stick itself. When that happened, he knew the pyramid's shadow would be the same length as the pyramid's height. It was easy to measure the shadow.) Thales' mind had made a connection and then taken a creative leap and come up with something very useful.

Perhaps most important to people of his time, Thales worked out a way to use triangles to tell how far a ship is from shore. (See pages 42–43.) For seafaring people, that was an enormous achievement. Thales seems to have been the first to take mathematical ideas and proceed step-by-step with them in a chain of logic, leading from an axiom (a generally accepted rule) to a direct proof.

English words with **HELIO** usually have something to do with the Sun: **HELIOSPHERE** (the sun's atmosphere), **HELIUM** (a gas in that atmosphere), and **HELIOCENTRIC** (the Sun-centered solar system where we reside).
A **PYRE** is a wood fire for burning a sacrificed animal or, in a funeral pyre, a deceased person. It comes from the Greek word for "fire." *Pyric* has to do with burning; *pyrotechnics* are fireworks; and *pyromaniac*? You can figure that one out yourself.

He must have spent most of his time observing and experimenting, which is what he was doing when he rubbed amber with a cloth and noticed that papyrus and other light objects stuck to it. Thales had created static electricity. Although he couldn't explain it, he must have been fascinated. The Greek word for "*amber*," *ēlektron*, became the root for *electricity* and *electronics*.

WORDS FROM THE PAST

Diogenes Laërtius, a third-century C.E. writer, said, "Thales seems by some accounts to have been the first to study astronomy, the first to predict eclipses of the sun and to fix the solstices."

Don't always trust words from the past. Diogenes Laërtius was not right when he said Thales was the first to study astronomy. The Babylonians had studied the stars with skill. They and others (such as the Chinese) could predict eclipses. Diogenes Laërtius didn't know that.

Still, words from the past can tell us important things. Diogenes Laërtius tells us something was new in Ionia. For the first time, he names the discoverers of knowledge—individuals such as Thales. In other cultures, knowledge was the property of the ruling class, and scientists were unnamed and unknown.

And what about that eclipse that made Thales famous? After he predicted it, how could people believe that the Sun is carried across the heavens in a carriage driven by Helios, the sun god? Or that Helios's son Phaëthon borrowed the chariot and drove it too close to Earth, forcing Zeus to terminate him with a thunderbolt? Actually many could and did believe it. Sacrifices to Zeus would burn on pyres for another thousand years. But Thales and a few others rejected the old supernatural religions and their magical incantations. They turned to the natural world for answers and, because of that, are often seen as founders of what we now call "Western civilization."

As to Thales' personality? Plato, a Greek philosopher who lived a few generations later, wrote in a work called *Theaetetus* (thee-AY-ti-tuhs) that Thales, "when star-gazing and looking upward, fell into a well. A clever and pretty maidservant from Thrace scoffed at him for being so eager to know what

THALES' PROPOSITIONS

Thales is credited with five propositions, which were guidelines for the development of geometry.

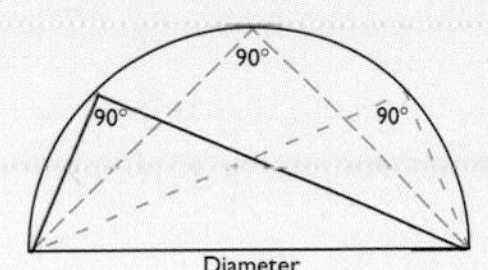

Any angle inscribed within a semicircle is a right angle (90°). That's known as Thales' Theorem.

A circle is bisected by its diameter.

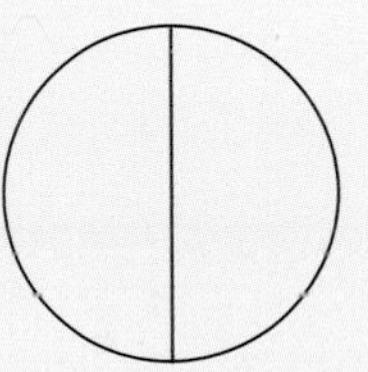

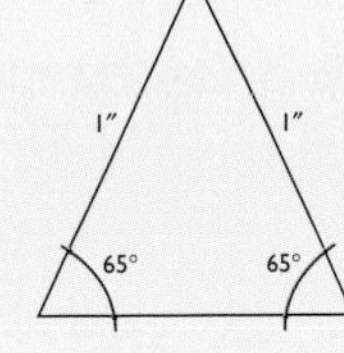

An isosceles triangle has two sides of equal length. The two angles opposite those sides are equal to each other.

When two straight lines cross each other, they form four angles. The angles opposite each other will be equal.

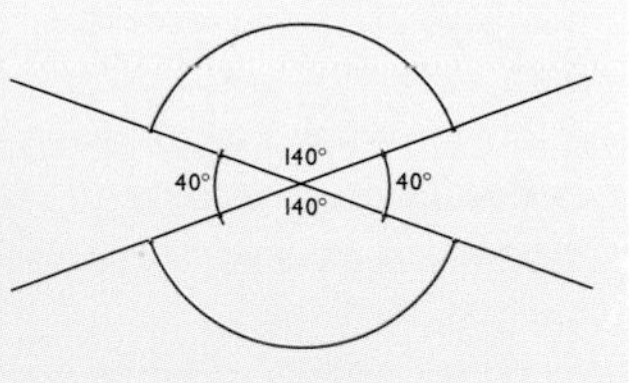

Two triangles are equal in all respects if they have two angles and one side that equal those in the other triangle.

DOES MOTHER MATTER?

The original Indo-European word for both "mother" and "matter" was *måter.* Its descendants include *mutter* (German) and *moeder* (Dutch). Both mother and matter, in ancient minds, were the origin of all things. I think the ancients had it right.

was happening in the heavens, that he did not notice what was in front of him, nay, at his very feet." Thales was so busy thinking about the stars, he didn't notice the well or the pretty girl!

When he was chided for his poverty, Thales decided to do something about it. Aristotle, a fourth-century B.C.E. Greek philosopher, explains in a famous book called *Politics*:

> *According to the story, he knew by his skills in the stars while it was yet winter, that there would be a great harvest of olives in the coming year; so, having a little money, he gave deposits for the use of all the olive-presses in Chios and Miletus, which he hired at a low price because no one bid against him.*
>
> *When the harvest time came, and many presses were wanted all at once, he rented them at any rate which he pleased and made a lot of money. Thus he showed the world that philosophers can easily be rich if they like, but that their ambition is of another sort.*

In other words, money wasn't his goal, although he didn't find it hard to amass. What he really wanted to do was understand the world about him.

The Sumerians and the Egyptians had said the ocean was the source of life. Thales looked for evidence to prove that hypothesis. When he discovered sea-animal fossils far inland, he thought he had it.

Then he took a step in another direction, saying that *everything* is made of water, and he was wrong. But the idea behind this hypothesis, that *all* things in nature share common particles, is one we take seriously. Thales was trying to discover a basic unit or element of life. When he came up with water—which takes three forms (solid, liquid, gas)—it was a reasonable start for a hunt that continues today.

While others trembled at nature's wonders and puzzles, the Ionians asked hard questions. And then they looked to the world around them, not to mythology or to wizardry, to find the answers.

Scientists observe, collect data, record, think, and make deductions. They are always asking questions. To a scientist, there is never a final answer. Albert Einstein said science is "never completely final, always subject to question and doubt." The Ionians knew how to ask questions. So do children. Some adults lose that ability. Remember, scientists are picky about words, so we're repeating these definitions:

A HYPOTHESIS is a possible explanation of observations or facts. A THEORY, to a scientist, is a well-tested explanation, a verified hypothesis. In popular usage, a theory is just an idea.

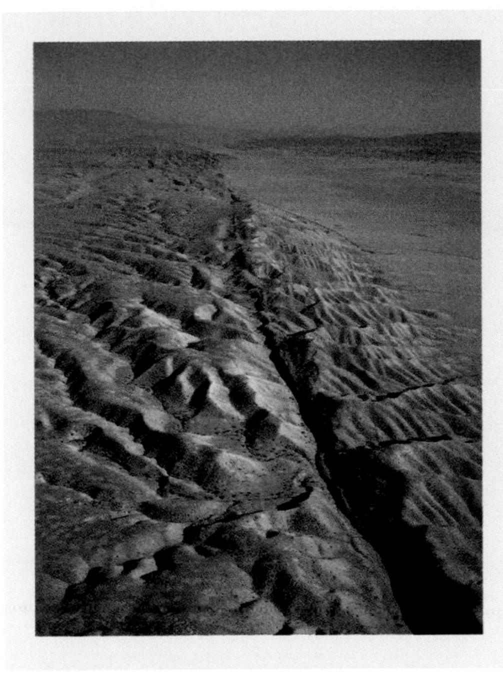

EARTH AFLOAT

Another of Thales' ideas was that the Earth floats on water and that we have earthquakes because of the motion of the liquid. He wasn't entirely wrong: Molten (melted) rock is a layer of liquid, and the Earth's crust floats on it and slides around on it. When two pieces of crust slip past each other—*Wham!* Earthquake.

If Earth were solid rock, we'd have no earthquakes. For more on this, look into the theory of plate tectonics.

The San Andreas Fault is a gash that splits California from bottom to nearly top. The coastal side of the state is part of the Pacific Plate, which is straining hard to slide northwest, in the opposite direction of the continental side. It's not a smooth ride. The jagged edges of the plates are stuck together. They can only move in fits and starts when tremendous pressure builds up at a point of contact—causing an earthquake. (To see how it works, make two fists, interlock your knuckles, and try sliding them past each other.)

Thales asked, "What is the nature of matter?" By that he meant: What are we made of? What is the world made of? Is there one thing that ties everything together?

The world is full of differences, and yet the Ionians believed that underneath all the complexity there is a unity that explains everything. Were they right?

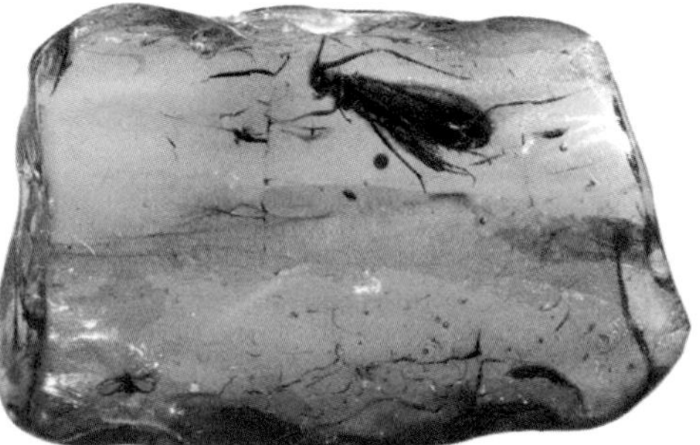

Trees ooze sticky resins that trap and entomb tiny animals. Resin hardens into amber, creating a fossil that's pretty enough to be sold as jewelry—with or without the prehistoric specimen. This insect (above) was entombed some 45 million years ago in the Baltic Sea region of northern Europe.

We're still working on that. Today we think we're close to having an answer. We know of more than 100 elements, and we also know that each is composed of tiny particles called atoms, and that all the atoms in an element are alike. For a long time we thought that was as small as you could get. We now know that atoms are composed of still smaller elementary particles (some of their names are quarks and electrons and photons). Scientists are investigating the idea that under those particles there are still smaller—much smaller—vibrating bits of matter.

Was Thales right? Is there some kind of particle that is basic to all life? Tune in to today's science, and you'll find out where we are in this search: it is heating up.

MEASURING WITH THE MIND

he eye can see a whole lot farther than the hand can measure. Every Ionian, at one time or another, stood on the shores of the Aegean Sea and watched white specks—sailing vessels—slowly approach. If merchants could know how far those ships were, they could predict when they would dock (and be prepared to do business). Soldiers might better prepare to meet enemy ships. But measuring vast distances, especially across the flat and featureless sea, seemed impossible.

Yet even when the speck was too far away to tell friend from foe or merchant ship from warship, Thales found a way to calculate its distance and, from that, when it would arrive. He did it by using his mind to shrink this ocean-sized problem down to a handheld triangle.

We're not sure which mathematical approach Thales used (there's more than one that will work), so here is our best educated guess.

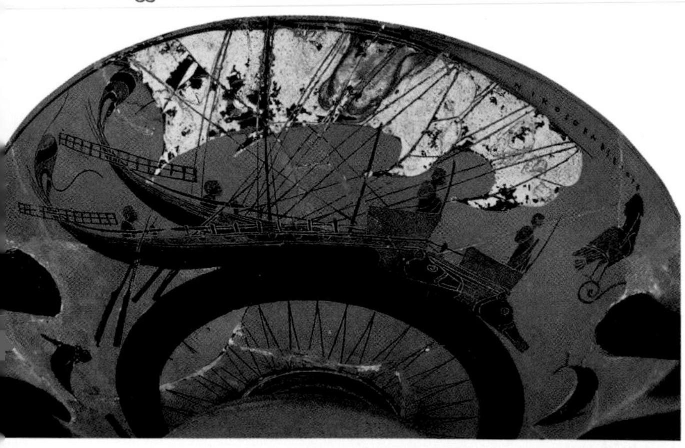

The Greeks liked to see sailing vessels on their drinking vessels. In ca. 520 B.C.E., an artist painted these boats in black silhouette and then etched in tiny details—an art style called black-figured Attic. (Attica is the greater Athens area.). Did the black horizon circling the base suggest a round Earth?

WHAT IS A RATIO?

Ratio is easy. It's just a relationship that can be expressed as a fraction. Put another way, every fraction expresses a ratio. Compare Thales' handheld triangle (in illustration below) with the triangle that has a ship at its tip. The ratio is 1 to 100, which can also be written 1:100. The small triangle is 1/100 of the big triangle.

But the world isn't all triangles. Suppose you want to compare things that are completely different from each other. In that case, you usually call a ratio a rate. Think distance and time, and you have the ratio of kilometers per hour. Think money and time: If you make $10 an hour, how many hours will it take to amass $200? That ratio is simple and constant; you can use it to make predictions—about your future earnings, for instance.

The Mesopotamians were using ratios for similar purposes thousands of years ago. We can see their ratios pressed into clay tablets. As cities grew, it became important to predict future needs; knowing how to use ratios gave them a head start. How much farmland do you need to feed a family of six? How much to feed 1,000 families?

Most ratios aren't that simple. Here is what Ernest Zebrowski Jr. says in a book called *A History of the Circle*. "There is nothing in nature's rule book that requires every ratio to be constant. Most ratios, in fact are not constant. If, for instance, it took 24 rowers to row a galley at 15 miles per hour, this does not mean that 48 rowers would get the boat up to 30 miles per hour and that with 144 rowers, the boat would hit 90 miles per hour. (In fact, this line of reasoning would suggest that the ancients could have broken the sound barrier just by getting together enough rowers.)" A nonconstant ratio, such as boat speed to rowers, is called a variable ratio.

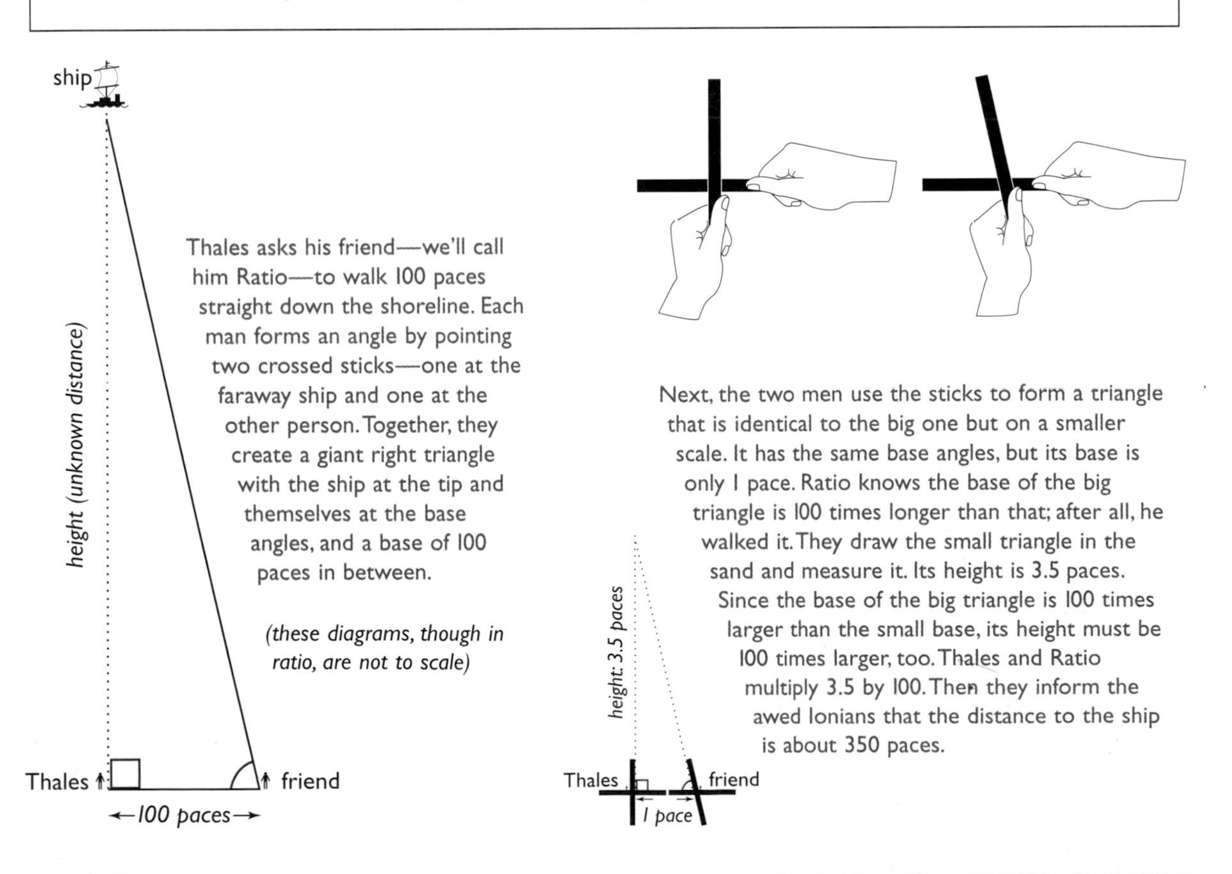

Thales asks his friend—we'll call him Ratio—to walk 100 paces straight down the shoreline. Each man forms an angle by pointing two crossed sticks—one at the faraway ship and one at the other person. Together, they create a giant right triangle with the ship at the tip and themselves at the base angles, and a base of 100 paces in between.

(these diagrams, though in ratio, are not to scale)

Next, the two men use the sticks to form a triangle that is identical to the big one but on a smaller scale. It has the same base angles, but its base is only 1 pace. Ratio knows the base of the big triangle is 100 times longer than that; after all, he walked it. They draw the small triangle in the sand and measure it. Its height is 3.5 paces. Since the base of the big triangle is 100 times larger than the small base, its height must be 100 times larger, too. Thales and Ratio multiply 3.5 by 100. Then they inform the awed Ionians that the distance to the ship is about 350 paces.

5 The "A" Team

Nothing endures but change....
Much learning does not teach understanding....
It is wise to listen, not to me but to the Word, and to confess that all things are one.
—Heraclitus (ca. 535–475 B.C.E.), Greek philosopher, *On the Universe*

Words have a longer life than deeds.
—Pindar (ca. 518–ca. 438 B.C.E.), *Nimean Odes*

Just because they were long ago doesn't mean they were stupid.
—Hans Christian von Baeyer (1938–), American author and professor of physics

Thales did a new kind of thinking, but it might not have gone anywhere if he hadn't been a teacher. His students spread his words and extended them, too.

Anaximander (ca. 611–ca. 547 B.C.E.), who was a pupil of Thales, thought there were many inhabited worlds—including one on the Moon. (Today we know the Moon is uninhabited, but we're taking the idea of life in other solar systems seriously.)

You'll see the abbreviation *ca.* in front of some dates. That stands for *circa*, which means "about." We don't have exact dates for some people and events in ancient history, so by using *ca.*, we're telling you we've considered clues and made an educated guess.

Anaximander (an-ak-suh-MAN-der) said that the first animals on Earth came out of the water and evolved into more complicated forms of life (which is true). He also thought that the first people were aquatic—sort of mermen and mermaids (untrue).

A woodcut from the *Nuremberg Chronicle* (1493), a pictorial history of the Earth, shows Anaximander of ancient Greece in fifteenth-century German clothes.

Don't laugh; he was doing heavy thinking, and he didn't have much to guide him. We have the advantage of generations of other peoples' wisdom and observations. That's called hindsight, and it makes us think we are very smart.

Not much remains of ancient Miletus in present-day Turkey. The ruins are mostly Roman.

Anaximander was one of three thinkers that I call the "A" team. That's because they were there at the beginning, when science first got real. Besides, they had grade-A minds, and their names all happen to begin with *A*.

Like Thales, they came from Miletus and were among the first to put aside superstition and use their minds to try to understand the world about them. They questioned old ideas. That was threatening to the keepers of the old ideas, who were apt to be political or religious leaders or both. Those leaders often made life difficult for the scientists. It was a problem that innovative scientists would face again and again.

Thales and the "A" team were true scientists, which doesn't mean they were always right. But they came up with hypotheses—the best possible explanations they could devise—and then used them as bases for observation and serious thinking. What they *didn't* do, which later scientists would do, was try to *prove* their hypotheses with controlled experiments.

Anaximander is often called the Founder of Astronomy. That's not quite true. People had been looking at the stars from the earliest of times. Some of those people kept careful records and made accurate predictions. The Babylonians and the Chinese were particularly good at it. But Anaximander did do something new. He tried to picture the whole Earth and understand its place in the cosmos. (He drew it on a map, which no longer exists.) He figured out that the Earth's

"A" NAMES

Should you remember the "A" names in this chapter? If you do, your knowledge will be impressive. (The names muddle my head, so I use nicknames to keep them straight. To me, the "A" players are Mander, Menes, and Goras.) Beyond the names, it is more important to get a general picture of the time and place (sixth and fifth centuries B.C.E.) and to understand that scientists build on the work of those who have come before.

Part of learning how to learn is making decisions on what to memorize and what to put in your knowledge background. Ancient Ionia was special; don't forget that. And Thales is someone to remember.

The Greek word *kosmos* derives from "order" or "order of the universe." It came to mean "the universe," which to the Greeks seemed perfectly arranged and orderly. In English, COSMOS means "the universe." Those who use COSMETICS are putting themselves in order. A COSMOPOLITAN person is at ease anywhere. COSMIC phenomenon are universal. And MICROCOSMIC? Clue: The Greek word *mikros* means "little."

surface must be curved (that explains the changing position of stars when one travels). And, in a breakthrough thought, he described the sky as a transparent sphere that moves and carries the Sun and stars; it was not just an arch over the Earth. The idea of spheres began to seep into astronomy; it would hold sway for more than 1,000 years.

A SPHERE is a ball that's perfectly round. We now know that Earth isn't perfectly round—it bulges at the equator and is flattened at the poles. A PHENOMENON is an event or thing that you can see or sense; usually it is amazing or special. The plural is PHENOMENA. Something PHENOMENAL is extraspecial. These words come from a Greek root meaning "to appear."

But Anaximander's Earth wasn't a sphere. (That idea would come a few decades later.) It was a kind of pudgy cylinder with a top that had a north-south curve. Earth stood all by itself in the middle of the universe. Nothing was holding it, said Anaximander. That idea of an unsupported Earth was hard for most people to imagine; it took a big intellectual leap. Once it was made, it wasn't difficult to go from a freestanding cylinder to a free-floating globe.

Anaximander's student, Anaximenes (ca. 570–500 B.C.E.), believed, like Thales, that there is a single element behind all of nature. He thought that element was air. He said that air is made up of tiny particles and is the primary form of matter. (Remember, Thales had said water was the base of all matter.)

So how did Anaximenes (an-ak-SIM-uh-neez) account for the variety of life? He seems to have believed that different mathematical quantities of air's particles produce different forms of matter, and that methods of change, such as condensation, are part of the process. It was good thinking—not right, but not so far off either.

An Artist's Interpretation of Anaximander's Universe

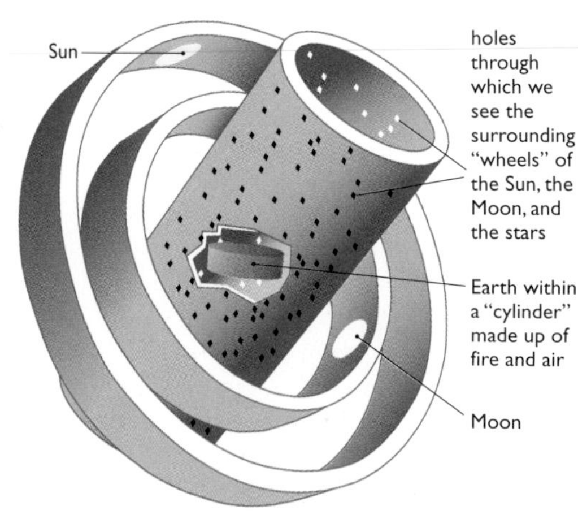

We now know the universe is made of particles and atoms and elements. Each of the elements—like oxygen or gold—has its own kind of atom. Numbers make one atom different from another. It is the number of tiny particles (called protons and electrons) in and around an atom's nucleus that determine what it is. So Anaximenes, who talked of particles and numbers, was heading in the right direction.

He is said to be the first Greek to understand that Mars and Venus are different from the stars. (He was right; they are planets.) He also said that the rainbow is a natural phenomenon and not a goddess.

The Greek myths say that the heavens are held in place by a demigod named Atlas. He was a Titan, condemned by Zeus to that backbreaking job because he took part in the Titans' war against the gods. He stood with his burden on the Atlas Mountains in Africa. Was that a poetic way of saying that those lofty mountains hold up the skies? Some people took it seriously. How else would you explain stars in the heavens if you didn't know about gravity? In the sixteenth century C.E., a Flemish geographer named Mercator (his real name was Gerhard Kremer) put a picture of Atlas on a book of maps. After that, *atlas* came to mean any book of maps.

You're looking at the bottom of a Greek cup. Atlas (left) watches an eagle feast on the liver of his brother Prometheus, who was bound to a mountain by a vengeful Zeus.

But Anaximenes took a backward step when he rejected Anaximander's cylinder and described Earth as a flat disk. Science is like that: it isn't all straight uphill.

When the third member of the "A" team arrived on the scene, power and politics were changing the Mediterranean world. The action had moved to the Greek mainland. Anaxagoras (ca. 500–428 B.C.E.) left Ionia and went off to Athens, which was the up-and-coming city-state. If you've ever wondered where true strength lies, consider Anaxagoras (an-ak-SAG-uh-ruhs). His ideas influenced the most important generation in Greek history, maybe in world history, and that means he influenced you. Anaxagoras is said to have taught Pericles, Euripides (yoo-RIP-i-deez), and Socrates.

Pericles was a great political and military leader who built the Parthenon (one of the world's most famous buildings; it's in Athens) and was a champion of democracy. Read his story, and you won't forget him. Read Euripides' plays, and you won't forget them.

And Socrates? He was said to be the wisest man in Greece. Besides, Socrates was the teacher of Plato, who was the teacher of Aristotle. You'll hear about Socrates, Plato, and Aristotle again in this book, and all your life. Their ideas are a big part of *your* heritage—no matter where you come from. So, was Anaxagoras—the teacher who reached all of them—

WITHIN REASON

The Ionians believed that "reason rules the world." What they meant was that the mind could be used to understand the world about us. That was a huge step. Most Greeks thought there were mysterious forces or fickle gods who determined things like thunder and hurricanes and that there was no rational explanation for nature's events.

The Ionians questioned that idea. They were thinking boldly, and they made some brilliant guesses. But they didn't have the technology to be careful scientific investigators or to prove their conjectures.

After the Persian Wars, Pericles (ca. 495–429 B.C.E.) raised a new Athens from the ashes and led Greece into a golden age of politics, arts, and learning.

important? Did he have power? What do you think?

When Anaxagoras went from Ionia across the Aegean Sea to Athens, he took the idea of other inhabited worlds with him. Anaxagoras said those other worlds "have men on them, and these have houses and canals just as we do." He, too, said that matter originally existed as tiny particles, which he called "seeds." He was prescient, which means "way, way ahead of his time." At the world's beginning, he said, a powerful Mind brought order out of chaos.

In 468 B.C.E. a huge hunk of rock—a meteorite—fell from the sky into the Aigospotamos River. Anaxagoras must have seen it or at least known of it. It probably got his mind churning. Anyway, he came up with ideas about the heavens that seemed wild at the time.

The Moon, said Anaxagoras, is made of ordinary matter (it is), has mountains (it does), and shines because of reflected light (it does). Most people thought the Sun and Moon and planets were gods or made of godly stuff. Anaxagoras, with an astonishing leap of mind, said that the Sun is a fiery stone almost as large as Greece. That thought left most political and priestly leaders aghast. They believed the Sun was divine. They said his idea was so dangerous that Anaxagoras should

Pericles built Athens's most famous structure, the Parthenon, painted here in golden hues in an 1818 watercolor. The religious temple is made entirely of marble.

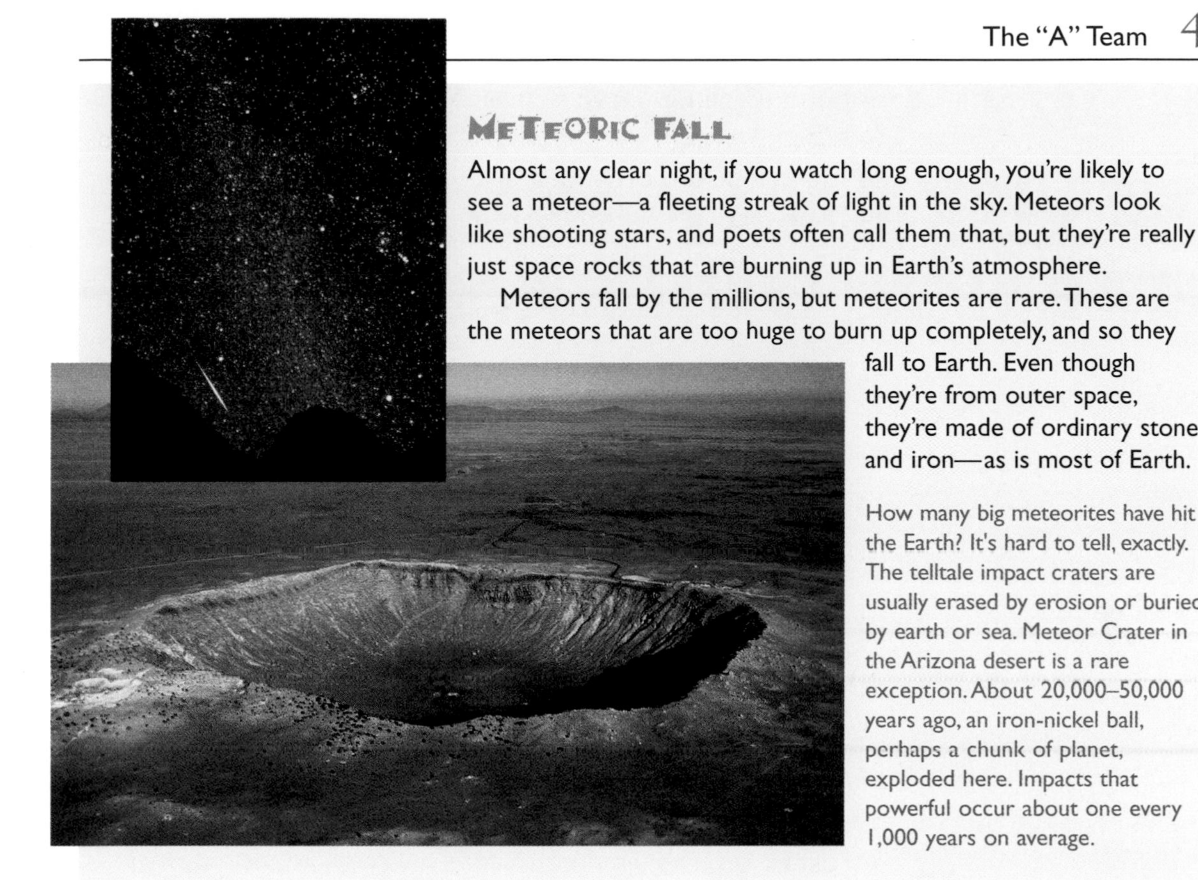

Meteoric Fall

Almost any clear night, if you watch long enough, you're likely to see a meteor—a fleeting streak of light in the sky. Meteors look like shooting stars, and poets often call them that, but they're really just space rocks that are burning up in Earth's atmosphere.

Meteors fall by the millions, but meteorites are rare. These are the meteors that are too huge to burn up completely, and so they fall to Earth. Even though they're from outer space, they're made of ordinary stone and iron—as is most of Earth.

How many big meteorites have hit the Earth? It's hard to tell, exactly. The telltale impact craters are usually erased by erosion or buried by earth or sea. Meteor Crater in the Arizona desert is a rare exception. About 20,000–50,000 years ago, an iron-nickel ball, perhaps a chunk of planet, exploded here. Impacts that powerful occur about one every 1,000 years on average.

be imprisoned or killed for the crime of impiety. Anaxagoras said, "Reason rules the world," which didn't help his case with the authorities.

Pious people are religious. **Impiety** is disrespect for religion. **Surreptitiously** means "secretly."

Anaxagoras was a serious scientist looking for answers in nature, not in magic or in tales of gods. Do you see any danger in that? The political and religious leaders did. Anaxagoras was tried and convicted. Then he got out of town, surreptitiously, and died in exile.

A great scientist sneaking out of town? It wouldn't be the last time. Scientists have been persecuted, jailed, and even killed—and their works have sometimes been destroyed with them. Why? Can science be dangerous?

It was for Anaxagoras. There were those who feared his questioning scientific mind. They dumped him and his ideas. And that helped chill Ionian science. Too bad; it had started so well.

The *Nuremberg Chronicle* (1493) features a woodcut of Anaxagoras too (see also page 44).

More on Numerals, Geometry, and Math's Origins

It was the Babylonians who figured out a way to measure circles and spheres. They divided them into 360 equal parts—today called degrees. Why 360? Probably because 360 equals six 60s, and the Babylonians liked the number 60. It was easy to deal with.

The Babylonians didn't do well with remainders. Divide 10 by 7, and you get 1 r. 3, or $1\frac{3}{7}$. Fractions (another way of thinking of remainders) are the leftovers when you divide. Sixty can be evenly divided by 1, 2, 3, 4, 5, 6, 10, 12, 15, 20, and 30. Using 60, the Babylonians hardly had to worry about remainders.

Those long-ago people, who lived in the land that is now mostly Iraq, actually had two number systems, one with a base of 10, the other with 60. Our 60-minute hour comes to us from Sumer. (To reconnect you with a bit of history, in southern Mesopotamia—the ancient land between the Tigris and Euphrates Rivers—the Sumerians built what may have been the world's first great urban civilization. See page 5.)

BABYLONIAN

This tablet showed ancient Babylonians how to calculate the area of a plot of land. Cuneiform numerals (below) were pressed into the clay with a wedge-shaped tool.

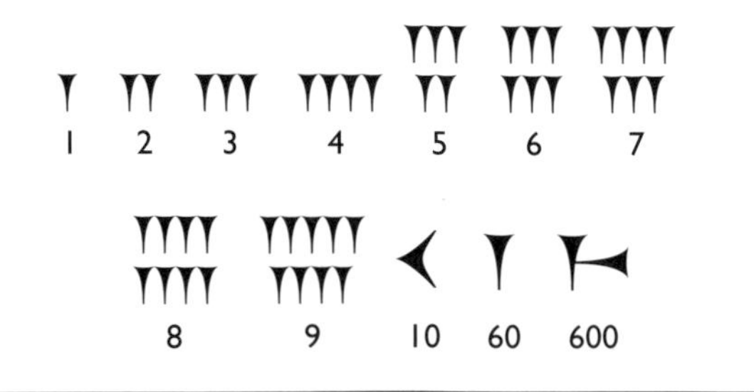

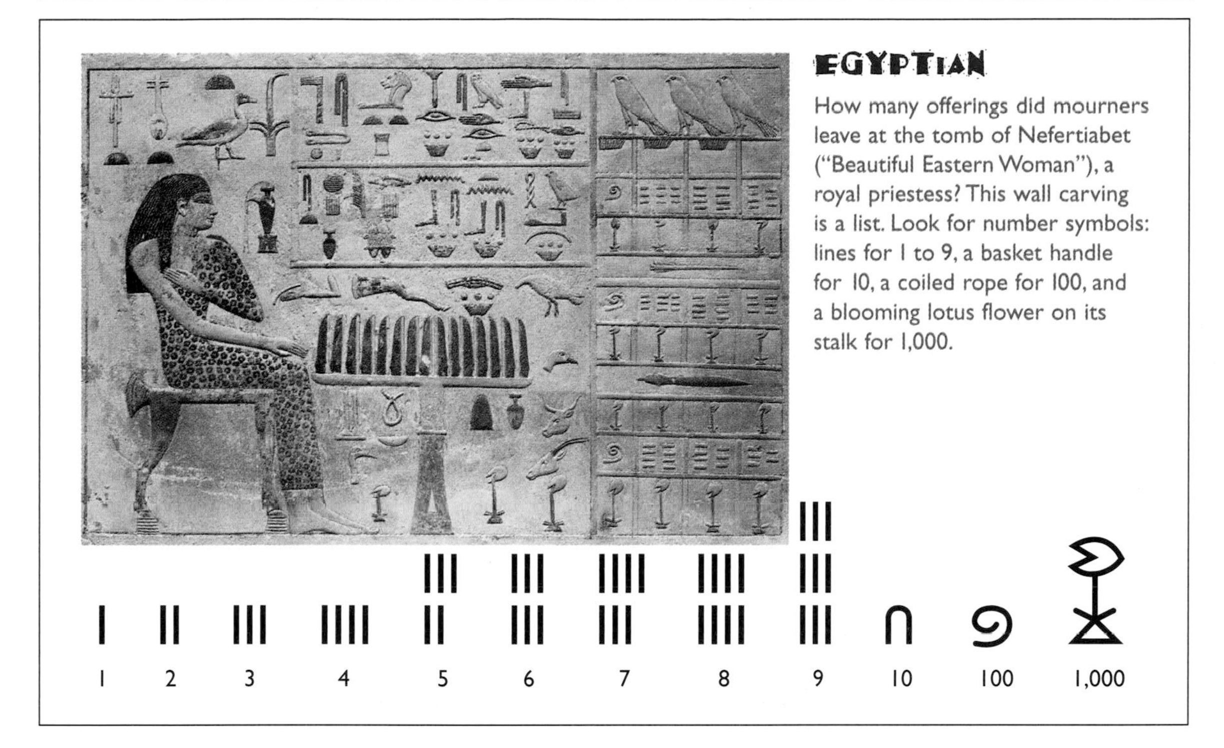

EGYPTIAN

How many offerings did mourners leave at the tomb of Nefertiabet ("Beautiful Eastern Woman"), a royal priestess? This wall carving is a list. Look for number symbols: lines for 1 to 9, a basket handle for 10, a coiled rope for 100, and a blooming lotus flower on its stalk for 1,000.

Meanwhile, the Egyptians (and later the Greeks) used a decimal, or base-10, system. We have examples of Egyptian hieroglyphic numerals from about 3000 B.C.E. The big problem: there was no zero, not even as a placeholder. Without zero to lead you from 9 to 10 or 19 to 20, there is no clear pattern in the number system. You have to memorize symbols for all the key numbers—1, 10, 100, 1,000, etc.—which means a lot of memorizing. (There's more on zero in chapter 25.) Mathematics wasn't for ordinary people.

The ancient Phoenicians used letters from their alphabet as numerals, a practice copied by the Greeks and Hebrews. But if an alphabet letter can also be a numeral, then people might take the letters in your name and read them as a number. Because numbers were thought to have special properties—sometimes magical, sometimes bad—this could get confusing, upsetting, or mystical. (Do you know anyone bothered by number 13?)

The Romans, who came later than the number-conscious Greeks and Hebrews, went back to the primitive idea of strokes: I, II, III. Some say that V, which equals 5, represents the natural V made by a hand when the thumb is extended and the fingers grouped. But some other Roman numerals are letters and were copied from the Greeks: X for 10, L for 50, C for 100, D for 500, and M for 1,000. Combining those symbols with strokes, you get:

VI = 6, IX = 9, XII = 12, XIV = 14

(more about numbers and math on next page)

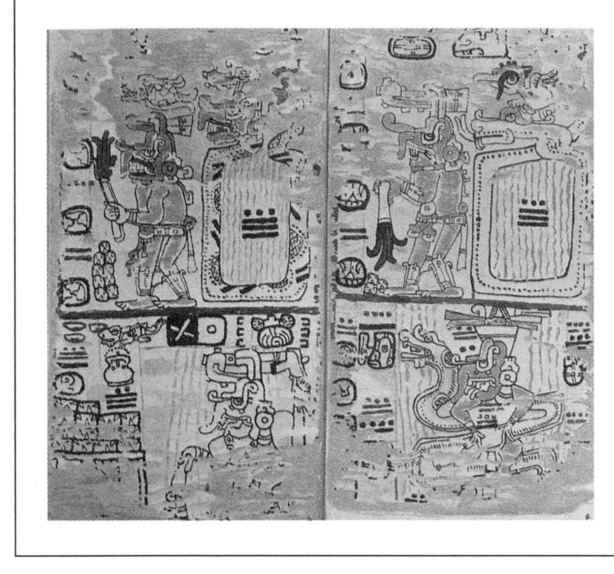

MAYAN

This Mayan codex from ca. 900 C.E. tells of the rain god Chaac. Could the square-shaped area next to him be a body of water? And why is it marked with number 18?

1 2 3 4 5 6 7

8 9 10 11 12 13

14 15 16 17 18 19

(A reminder: When the stroke comes after the letter, it is added, but when the stroke comes before the letter, it is subtracted.) What is XC? CIX? CC? MCCXVII? DLVI? DCCCLII? (Answers on this page.)

The Roman-numeral system held sway in Europe for way over 1,000 years. There wasn't much progress in European mathematics during that millennium. Try adding two big Roman numbers. Add MCXIX and DCLVII, for a start. The Romans produced no mathematical geniuses. Can you understand why?

The Maya, in ancient Mesoamerica, did produce skilled mathematicians. It helped that they used zero, although they didn't understand that it could be more than just a placeholder, that it really is a number. Like the Egyptians, the Maya had two ways of writing numbers: they used glyphs (picture symbols) or dots and bars. Take a look at Mayan glyphs (see page 29)! To do basic arithmetic in ancient America, you had to be very bright and also study long and hard.

Today, the numerals that we use have conquered the planet, in good part because they are so simple to use. How did we get them? Keep reading; chapter 25 has details.

Answers: 90; 109; 200; 1,217; 556; 852

WARNING: THIS IS A BASIC INFORMATION BOX WITH DEFINITIONS OF NUMBER TERMS. (It is important or it wouldn't be here.) HOWEVER, YOU MAY KNOW ALL THIS, IN WHICH CASE, JUST SKIP IT.

The numbers used in counting (1, 2, 3, 4, and so on) are called **natural numbers** (also called **cardinal numbers**). Because there is no last number, natural numbers are an infinite set. When you add zero to this set, you have the set of **whole numbers.** To repeat: Whole numbers are the natural numbers plus zero.

Positive numbers are all those greater than zero.

Negative numbers are all those less than zero. Negative numbers (–1, –2, –3, etc.) are especially

EGYPTIANS HAD MATH TEXTBOOKS, TOO

Alexander Henry Rhind was a young Scotsman who was told to travel to a warm climate because of his health. So he went to Egypt, where he did some studying of ancient artifacts for a book on tombs. In 1858, when he bought an old roll of papyrus, his name headed for the history books. That papyrus, which is about 34 centimeters (about 13 inches) high and some 3.2 meters (about 10.5 feet) long, turned out to be a kind of math textbook in the handwriting of a scribe named Ahmes, who copied it in about 1650 B.C.E. Ahmes tells his readers that the work was compiled 200 years before his time. Today it is known as the Rhind Papyrus and is in the British Museum.

The papyrus (detail shown above) contains 84 problems and solutions. Most are arithmetic, but there is also geometry and even some algebra. At the time the papyrus was discovered, most of what we knew about Egyptian math came from hieroglyphics carved on tombs and monuments. Hieroglyphic digits (1 through 9) are just straight up-and-down strokes. (One stroke equals the number 1, nine strokes equal 9.) Ahmes used hieratic (cursive) writing, meant for business and everyday use, and there was a different symbol for each digit and for multiples of 10. The Rhind Papyrus showed us that the Egyptians had made the important leap to symbolic numbers, a tremendous contribution to the development of the modern number system. We couldn't have learned that fact from the math in Egyptian tombs.

useful in measuring things—temperatures below zero, locations below ground level, and your budget if you're in debt. But the idea of negative numbers seemed weird to most people, so they had to be discovered, and that couldn't happen without zero. For thousands of years, mathematicians (like the Greeks) got along without zero or negative numbers.

Integers are all the positive numbers, zero, and all the negative numbers.

Digits are the numerals *0* through *9*.

Chop a natural number, and what do you get? A **fraction**. One half, ½, is the most common fraction. Some fractions are proper, and some aren't. A **proper fraction** has a **numerator** (on top) less than its **denominator** (below). Example: ³⁄₅.

Decimals were invented in the sixteenth century. Decimals are just another way of writing fractions. They bring sense and precision to number lines. The fraction ½ is .5 as a decimal. Whenever you see a decimal dot, you know that everything to the right of it is between 0 and 1.

Mixed numbers are the sum of an integer and a proper fraction (or decimal). Example: 2½ or 2.5.

Fractions are sometimes called **rational numbers**, because they express a ratio or relationship between the whole of a thing and its parts. Example: ¾ is a ratio. (See page 43 for more on ratios.)

If all this seems boring, well, basics are apt to be. But these are mostly logical, descriptive names for numbers. If you want to think about something illogical, wait until you read about those upsetting critters known as irrational numbers.

6 Elementary Matters: Earth, Air, Fire, and Water, says Empedocles

Some say the world will end in fire,
Some say in ice.
From what I've tasted of desire
I hold with those who favor fire.
But if it had to perish twice,
I think I know enough of hate
To say that for destruction ice
Is also great
And would suffice.

—Robert Frost (1874–1963), American poet, "Fire and Ice"

Thales said life's basic element is water. Anaximenes said it was air. Other Greeks said fire or earth. Empedocles (em-PED-uh-kleez), who lived in the fifth century B.C.E. (about the same time as Anaxagoras), said it was all four of those and that two forces, Love and Strife, guide all growth and action. Love and Strife? Well, scientifically, he was thinking of attraction and repulsion. Yes, the ancient Greeks sometimes reached conclusions that seem strange to us. But they were asking the right questions. We still analyze the world in terms of elements and forces.

That idea of four basic elements—earth, air, fire, and water—was one of the longest lasting and most influential scientific ideas in all of world history. For centuries and centuries and centuries (about 2,300 years), people believed it, although it would turn out to be wrong. Schoolchildren were still being taught about earth, air, fire, and water in the eighteenth century.

Pull and *push* are simple words to describe any force. A magnet both pushes and pulls, depending on which end you use. Gravity is two objects pulling on each other. A nuclear force is a push-pull relationship between parts of an atom. The scientific words for push and pull are ***attraction*** and ***repulsion***. Thales is said to have discovered that a certain ore attracts iron. The ore came from a region in Thessaly (now in central Greece) named Magnesia. From that we get what word?

In the eighteenth century, people still believed in earth, air, fire, and water. A costume design for the ballet *The Elements* (1737), by French composer Jean-Féry Rebel, has them all.

GOOD GUESS BUT NOT QUITE RIGHT

The Greeks were wrong: earth, air, fire, and water aren't elements. But they can be compared to the four states of matter: solid, liquid, gas, and plasma. Plasma? It's an atomic substance packed with charged particles. The Sun's atmosphere is a plasma.

Empedocles was wrong in the elements he chose, but right in his idea—that instead of a world where everything is different and unrelated, there are certain basic substances that combine to make up everything else. We now realize that earth, air, fire, and water aren't basic elements. Today we know of 112 elements (such as iron, gold, calcium, oxygen, and carbon) and we keep discovering more. It was the Ionians who got us searching in the right direction.

Empedocles, who was a philosopher, a scientist, and a poet, was so wise that some people wanted him to be king. He refused, suggesting democracy instead of kingly rule. When he used doctoring skills to cure a plague in Sicily, many thought him a god. Later, as an old man and tired of

The universe is made up of more than 100 elements, from aluminum to zinc. They are the basic forms of matter. Everything else comes when those elements are combined into compounds. Chemistry deals with the elements and their chemical compounds. When we say something is "elemental," we mean it is basic or sometimes just plain simple. "He's in his element" means he's doing the one thing he does best. Elementary school? Why did we choose that name?

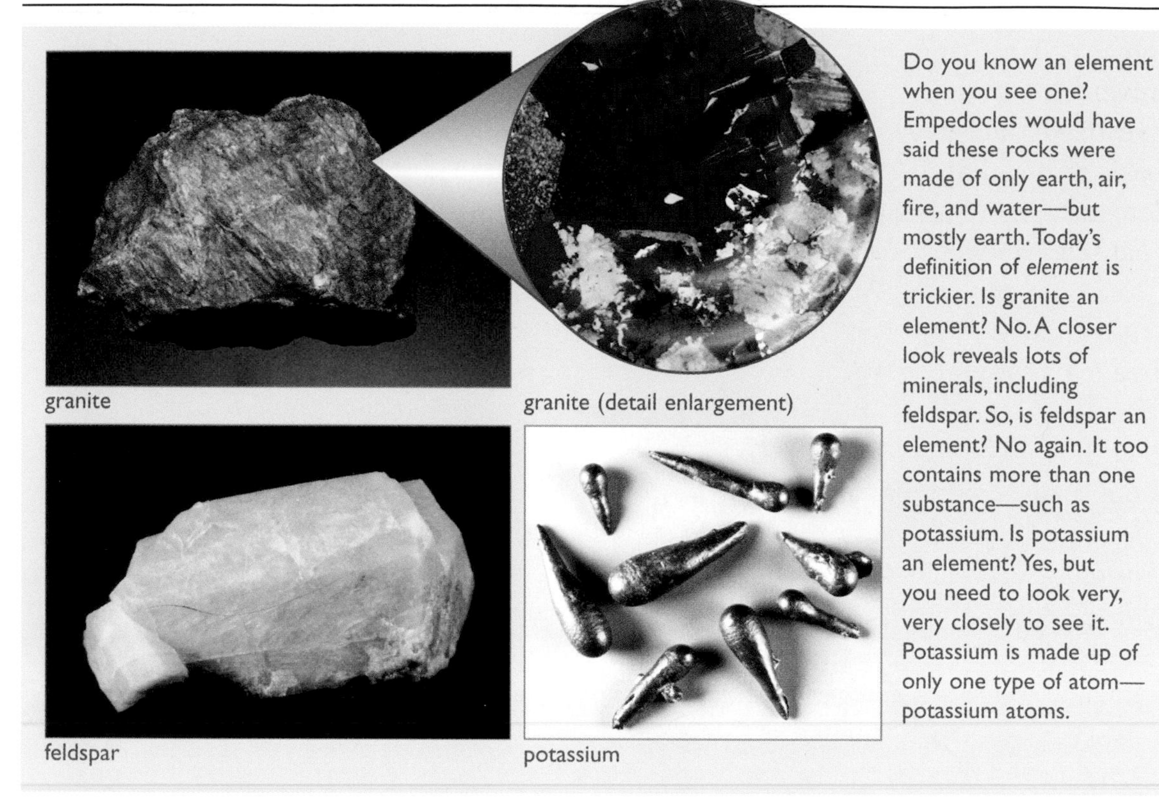

granite

granite (detail enlargement)

feldspar

potassium

Do you know an element when you see one? Empedocles would have said these rocks were made of only earth, air, fire, and water—but mostly earth. Today's definition of *element* is trickier. Is granite an element? No. A closer look reveals lots of minerals, including feldspar. So, is feldspar an element? No again. It too contains more than one substance—such as potassium. Is potassium an element? Yes, but you need to look very, very closely to see it. Potassium is made up of only one type of atom—potassium atoms.

life, he is said to have leaped into the volcanic crater at Mount Etna. An anonymous poet wrote:

Great Empedocles, that ardent soul,
Leapt into Etna and was roasted whole.

I don't believe that tale—Empedocles was too rational to jump into a volcano—but I wasn't there. The British poet Matthew Arnold took the story as true and wrote a famous poem called "Empedocles on Etna":

Nature, with equal mind,
Sees all her sons at play;
Sees man control the wind,
The wind sweep man away.

What's important to remember about all this is that the Greeks trusted their brains, and they understood that to

NOT ONE IOTA!

Speaking of small particles, the ninth letter in the Greek alphabet is *iota* (eye-OH-tuh), which means "the tiniest thing possible." So when you say, "I didn't get any ice cream—not one *iota*!" you mean that you didn't get even the littlest bit.

know the large (the universe), they must investigate the small (basic particles and elements). That's exactly what science does today.

Most of Empedocles' writings are lost, but here are a few of the lines that do exist:

This papyrus fragment, which is housed in the Strasbourg Library in France, is a priceless treasure. It's one of several partly preserved passages written by Empedocles. Piecing together the various fragments has enabled scholars to reconstruct Empedocles' theory of the four elements, which interact with what he called the powers Love and Strife to form a never-ending cosmic cycle.

But come, hear my words,
for truly learning causes the mind to grow. For as I said before in declaring the ends of my words . . . at one time there grew to be the one alone out of many, and at another time it separated so that there were many out of the one; fire and water and earth and boundless height of air, and baneful Strife apart from these, balancing each of them, and Love among them, their equal in length and breadth.

Something that is BANEFUL is harmful, destructive, or poisonous. If someone says, "He's the BANE of my existence," that's not a compliment.

If you want to understand that, get out a pencil and write it in your own words. This was the first time anyone said in writing (as far as we know) that matter and its interactions make up all the world and determine how it changes.

Greece: Where the Action Was

A community of thinkers, young and old, is a powerful thing. By living in the same place, they can share ideas, compare results, amass a collection of facts, and challenge each other. It's no surprise that many of the greatest thinkers in ancient Greece knew each other, or knew about each other, or had relatives in common, or had the same teachers. Greece was and is a small country.

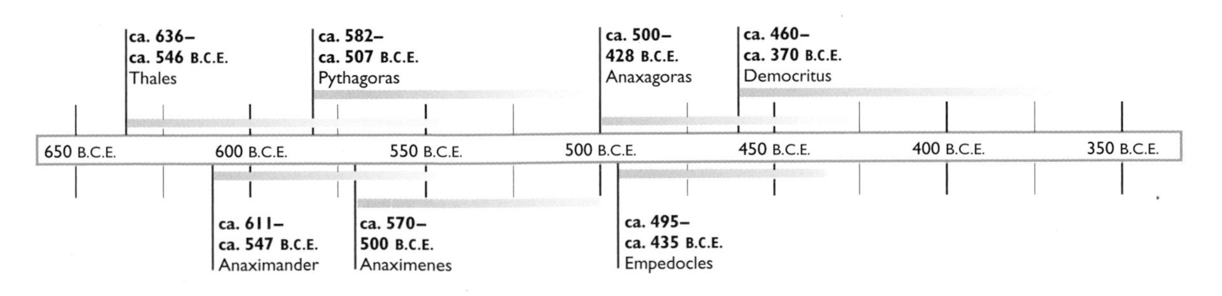

7 Being at Sea

For myself, my duty is to report all that is said, but I am not obliged to believe it all alike—a remark which may be understood to apply to my whole History.
—Herodotus (ca. 484–425 B.C.E.), Greek historian, *The History of Herodotus*

To remain ignorant of things that happened before you were born is to remain a child. What is a human life worth unless it is incorporated into the lives of one's ancestors and set in an historical context?
—Cicero (106–43 B.C.E.), Roman leader whose real name was Marcus Tullius, *The Orator*

History, in every country, is so taught as to magnify that country: children learn to believe that their own country has always been in the right and almost always victorious, that it has produced almost all the great men, and that it is in all respects superior to all other countries.
—Bertrand Russell (1872–1970), English philosopher

Herodotus (hi-ROD-uh-tuhs), who lived in the fifth century B.C.E., is thought to be the world's first real historian; the first to collect information carefully and to organize it into a compelling narrative. (He was a great writer; read him, and you'll be carried away.) You could call Herodotus scientific in his approach to knowledge of the past, which made him wary of big talkers and fanciful tales. Like the Greek natural philosophers (those who dealt with the world of nature), he was looking for verification (proof) and truth. So when he heard that Hanno, a Phoenician sea captain, claimed to have circumnavigated the whole continent of Africa in a voyage that took three years, well, he was skeptical. He knew a tall tale when he heard one. When Hanno reported that, far below the equator, the noonday Sun is in the northern half of the sky, that seemed ridiculous. The Greeks knew the skies well, and no one had ever seen the Sun anywhere but in the south.

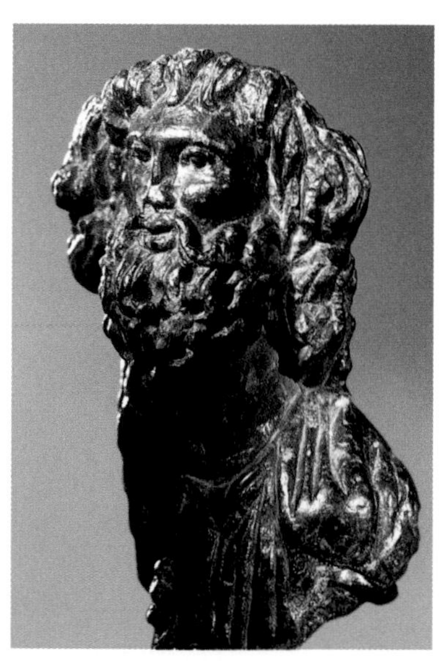

This statue fragment from Alexandria is thought to depict Herodotus, the world's first real historian.

Still, Herodotus reported what he had been told, making it clear that he doubted the whole thing.

Today we know that the Earth's skies are not the same in the Southern Hemisphere as they are north of the equator. Phoenician mariners must have reached southern Africa; the story Hanno told was too wild to make up.

While the Babylonians and Egyptians were mostly farmers and herders, people of land and river, the Phoenicians and Greeks were more independent breeds: they were people of the sea. Many were mariners or merchants or both. In order to do their jobs well, they had to study the heavens. The stars held practical importance for seagoers; they couldn't leave that knowledge to others.

The Phoenicians lived in what is now Lebanon, at the eastern end of the Mediterranean, and also at a satellite city, Carthage, in North Africa. They are said to be the first who went boldly into the open sea (although their greatest contribution to the world may be the alphabet, an idea that no one else had come up with). By 500 B.C.E., when Hanno set out, the Phoenicians

CIRCUMNAVIGATE means "to sail all the way around" something. Since *circum-* is Latin for "around," a **CIRCUMFERENCE** is the distance around the perimeter of a circle. If you draw a square (or any other figure) around the circumference, you are **CIRCUMSCRIBING** the circle. A **MARINER** is a seafarer, from the Latin root *mare*, which means "sea." Galileo, a Renaissance scientist (1564–1642), spotted big dark blotches on the Moon, mistook them for seas, and called them **MARIA** (the plural of *mare*). We now know these "seas" are bone-dry plains, but no one has bothered to change the name.

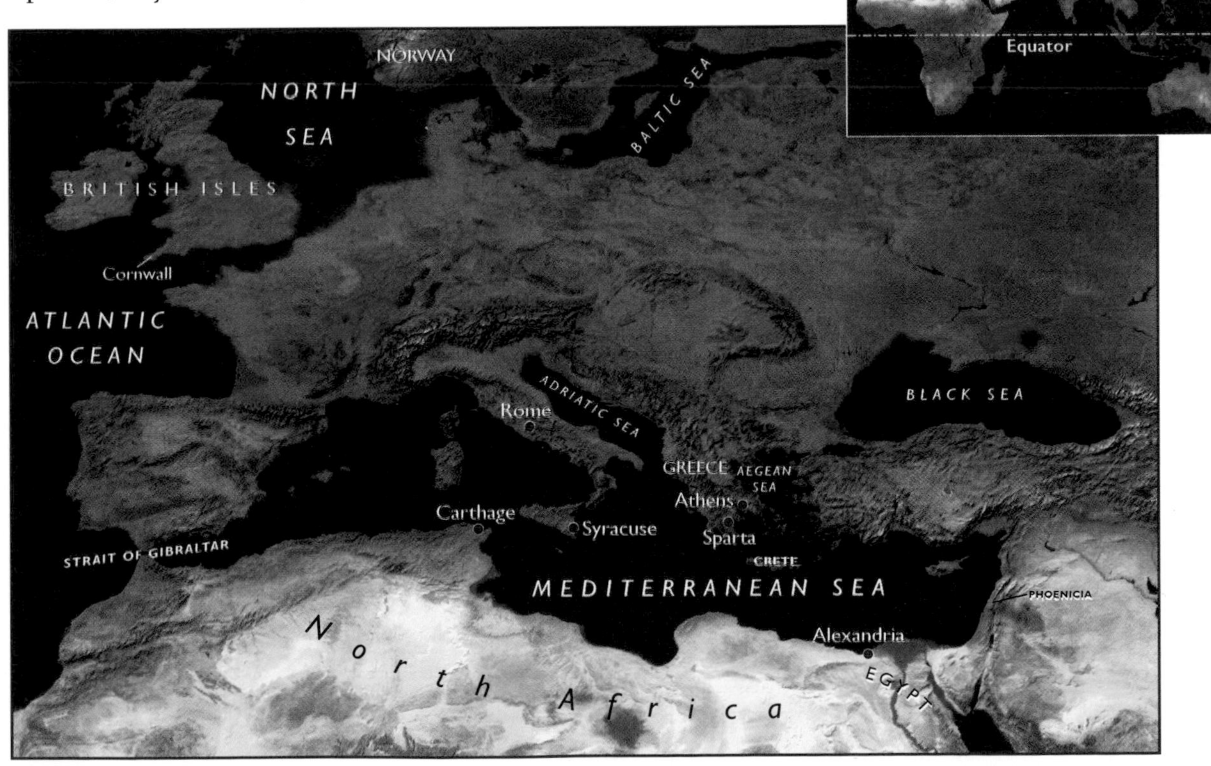

had six centuries of sailing experience to draw on.

Seamen from the island of Crete had been earlier voyagers, but they had mostly hugged the coast in the eastern Mediterranean and Aegean Seas or made short island-to-island runs. The Phoenicians studied the Cretan sailing vessels and the oared Egyptian river craft and built galleys: big ships with both sails and oarsmen. Well before 500 B.C.E., they had gone beyond the Pillars of Hercules at Gibraltar and found a source of tin on an island somewhere that they kept secret. Tin was in short supply, but it was essential in making bronze; and bronze was a hot commodity, used for weapons and statuary and much more. Herodotus had this to say:

> *I can learn nothing about the islands from which our tin comes, and though I have asked everywhere I have met no one who has seen a sea on the west side of Europe. The truth is no one has discovered if Europe is surrounded by water or not.*
>
> *I smile at those who, with no sure knowledge to guide them, describe the ocean flowing around a perfectly circular earth.*

A cross-staff (top) is a surprisingly simple tool for measuring the height of the Sun. The navigator slides the short vertical stick forward and back until the bottom lines up with the horizon and the top with the Sun. The position is marked in degrees on the long horizontal stick. The navigator uses a table of numbers to convert the degrees into latitude.

Below, a Greek vase (ca. 560 B.C.E.) shows a war galley filled with warriors, oarsmen, and helmsmen.

Herodotus had been to India and Egypt and lots of other places and wrote of them with wit and energy. But he hadn't made it to Europe's west coast, and he wasn't going to report hearsay. And those merchants who had sailed into the ocean weren't about to reveal their secrets. Today we think their undisclosed source of tin was probably in England, specifically Cornwall, where the ore was abundant. Wherever it was, it gave the Phoenicians control of the tin market.

How could they find their way from the Mediterranean to England and back again without a compass in a world that was uncharted?

They looked to the heavens. Like all good sailors living in the Northern Hemisphere, the Phoenicians knew that the seven stars of the Big Dipper are visible all night long in every

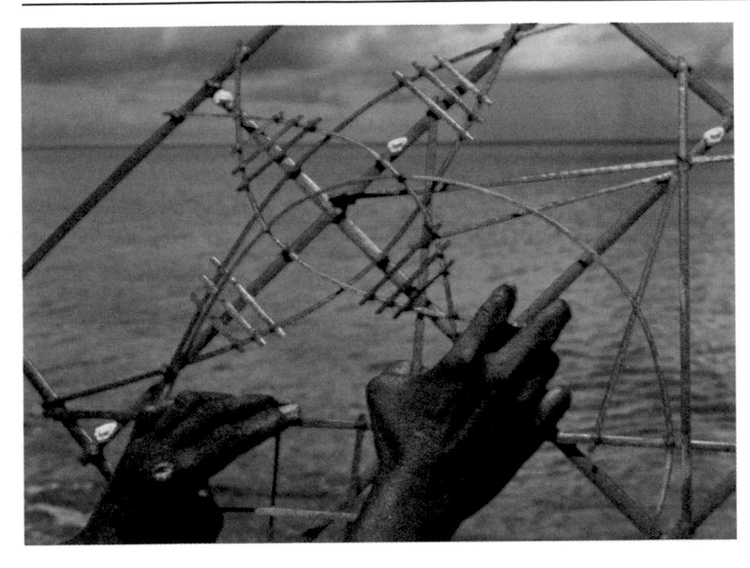

For thousands of years, the skilled sailors of Micronesia, islands in the Pacific Ocean, navigated by reading patterns of ocean currents and swells. To memorize the complex images, they built *mattangs* (left), wooden models of waves striking a canoe, a reef, or a beach from different angles.

That Pytheas connected tides to the Moon is amazing. He probably noticed that sea levels rise and fall dramatically during new and full moons. The cause remained a mystery until Isaac Newton (1642–1727) described gravity as a universal force of attraction. The Moon, tugging on Earth, draws the ocean waters toward it.

season, and that they are always in the north. If you know north, you can find the other directions. (See page 63.)

We have a report from Pytheas (PITH-ee-uhs), a Greek navigator who sailed in about 300 B.C.E. with Phoenicians and said he went to Britain and beyond, to "Thule," which may have been Norway. From there he explored northern Europe and went into the Baltic Sea. Pytheas came home and wrote of roiling seas and tides that came and went in a regular pattern. Astonishingly, he seemed to have figured out that the Moon influences the tides. The Greeks had no experience with the daily rise and fall of water that tides bring—the Mediterranean is a sea without regular tides—so how could they believe that seaman's yarn? Like Hanno (and Marco Polo 15 centuries later), Pytheas too was thought to be a teller of tall tales.

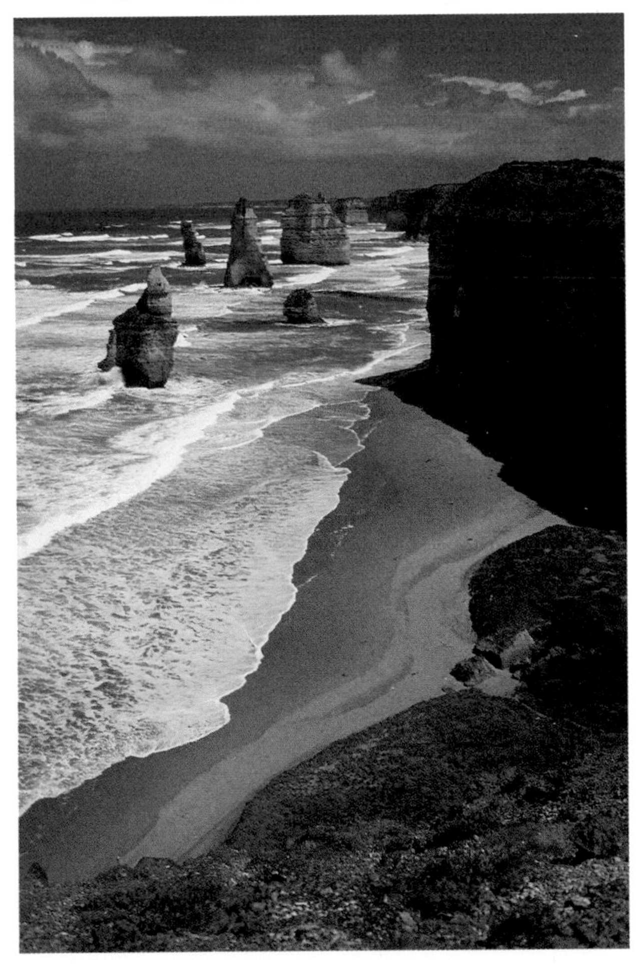

WHY DO POLESTARS TAKE TURNS?

ORION, THE HUNTER

If you turn a globe upside down, the first thing you'll notice is that the Southern Hemisphere is almost a water world. The bulk of Earth's land is in the north, and it was northerners who came up with the names we use for the constellations. Those shapes appear upside down to Australians, South Americans, and southern Africans. Orion, the Hunter, with his raised club and three-star belt, is one of the biggest and brightest examples. He looms over the equator, and so is visible in both hemispheres at least part of the year.

If you were on the opposite hemisphere, your view of Orion would be upside down (below).

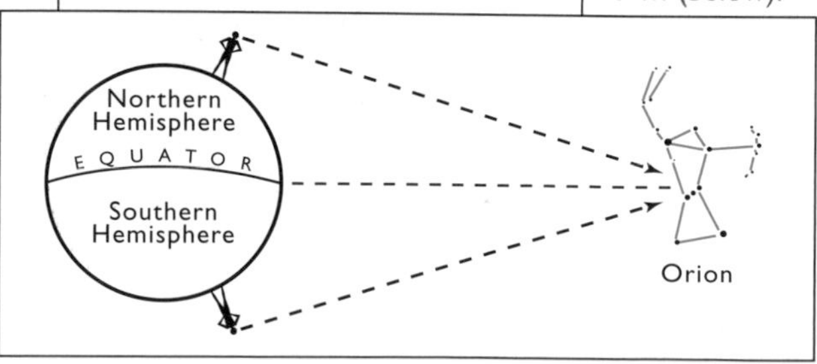

The North Star (called Polaris) isn't the brightest bulb in the Northern sky. But it's the only star that stays in place (more or less—it does move a little) while the others spin around it once every 24 hours. It has center stage because it's perched above the North Pole, which is in line with Earth's axis of rotation. If you're at the North Pole, Polaris is directly overhead. Greece is about halfway between the North Pole and the equator, and so there, the North Star is about halfway up in the Northern sky. At the equator, Polaris appears on the horizon.

Stars shift, but it takes centuries to notice it. The ancient Greeks didn't navigate by Polaris. Their North Star was Kochab, which was the star above the North Pole in 500 B.C.E. (*Kochab* means "the star" in Arabic.) To the ancient Egyptians, the North Star was Thuban. (We can still see both those stars; they're just not over the North Pole anymore.) The title of North Star passes from star to star because Earth wobbles, like an unsteady top, causing the North Pole to move in a big slow loop.

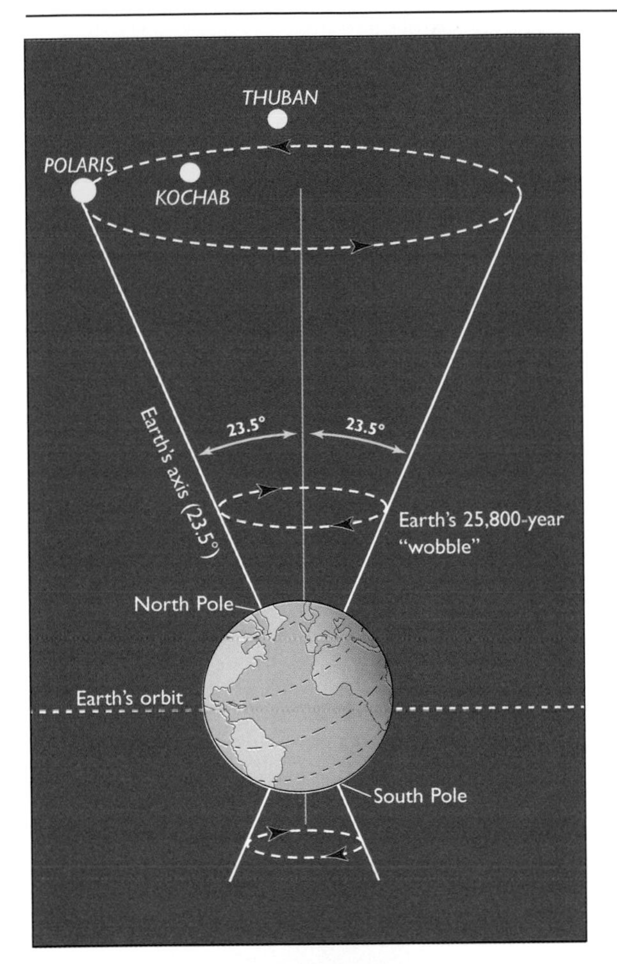

Earth's 25,800-year wobble (left) is called precession. The cause is torque, a twisting force created by the uneven tug of the Sun and Moon on Earth. (Spin a top and watch it wobble. Like the North Pole, the upper tip moves in a circle.)

If you stand on the North Pole, Polaris is directly overhead (see bottom left diagram). All the other stars appear to circle it slowly. The closest stars to Polaris, called circumpolar stars, make the smallest circles, and the farthest ones make the biggest circles. If you move to the equator, your sky view changes. Polaris and the circumpolar stars are hidden below the horizon. Instead of a spinning circle of stars, you see a parade of them, crossing the sky from east to west and then setting below the horizon.

One wobble—and therefore one full loop—takes about 25,800 years.

People in the Southern Hemisphere never see the North Star; their night sky has some different players. As Hanno navigated around Africa's southern tip, he couldn't help but notice an eye-catching constellation: the Southern Cross. Like the Big Dipper, it's famous. The long bar of the Southern Cross points to where a South Pole star ought to be, but isn't. People of the Southern Hemisphere don't have a polestar to show the way.

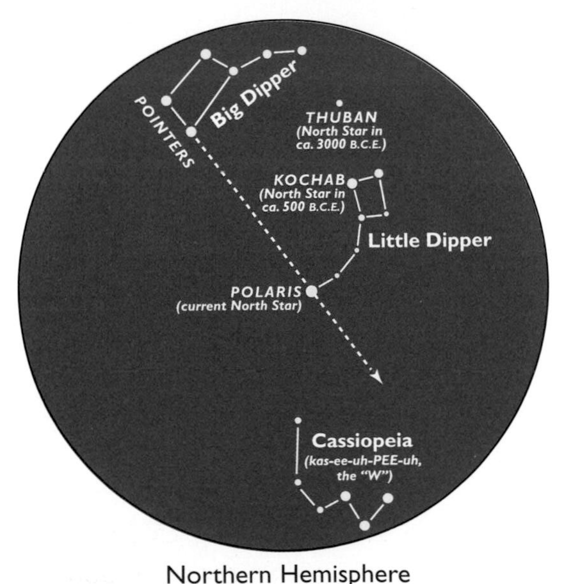

Northern Hemisphere

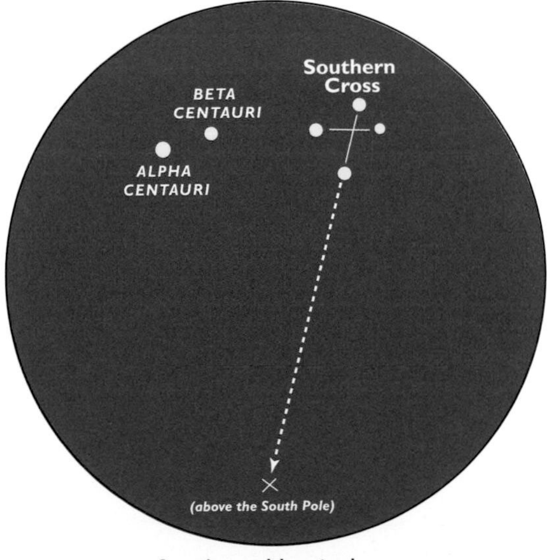

Southern Hemisphere

8 Worshiping Numbers

Philosophy is written in this grand book—I mean the universe—which stands continually open to our gaze, but it cannot be understood unless one first learns to comprehend the language and interpret the characters in which it is written. It is written in the language of mathematics, and its characters are triangles, circles, and other geometrical figures, without which it is humanly impossible to understand a single word of it; without which one wanders vainly in a dark labyrinth.

—Galileo Galilei (1564–1642), Italian scientist, *Il Saggiatore* (*"The Assayer"*)

But the creative principle resides in mathematics. In a certain sense, therefore, I hold it true that pure thought can grasp reality, as the ancients dreamed.

—Albert Einstein (1879–1955), quoted in *The Cosmic Code* by H. R. Pagels

A sense of number—knowing the difference between 2 and 25—seems to be part of what makes us human. It's like communicating with language or making music; even the most primitive peoples do those things. Tablets and murals tell us the very ancients knew more than just how to count and add and subtract: many of them understood ratios and fractions and could do complicated figuring. There were important reasons for it.

The Babylonians used numbers for commercial purposes. Without arithmetic, they would have had a difficult time trading goods or keeping records or dividing up land or work.

The Egyptians found geometry essential for measuring; without it, men and women would have been unable to build pyramids and temples.

The shapes and markings of accounting tokens (ca. 3000 B.C.E.) from Elam (now Susa, Iran) stood for oil, sheep, clothes, and so on.

THE REAL THING?

If you count five pebbles into your hand, you are doing concrete math—using real things that you can touch and manipulate. The early Greeks represented numbers with arrangements of pebbles—like the patterns on dominos. That's a step toward abstract math: those pebbles could stand for soldiers or cattle or sacks of grain.

Numerals like 5 are even more abstract; 5 is one symbol that stands for five things. The algebra equation $x = 5$, where x stands for some unknown quantity, takes a further leap into the abstract.

Pythagoras had it both ways. He did mind-stretching abstract mathematics, but for him, numerals were not just symbols; they were always very real. (By the way, the word *calculate* comes from the Latin *calculus*, meaning "pebble.")

An Italian bust portrays a thoughtful Pythagoras.

But mathematics can be something more than everyday practical. Mathematicians can use numbers as a language to reason and understand the world. The Greeks came to realize that. Why did science succeed so spectacularly in the Western world and not elsewhere? It may be because the Greeks married science and mathematics. They made mathematics the language of science. It is the most important single thing they did to put Western science on a different path from that of the rest of the world.

Those who came to Delphi built temples—called treasuries—and filled them with statues of marble and gold and ivory. (The columns of Apollo's temple are on the left, above.) The bigger the gift, the more attention the giver got. Most visitors stayed for the Pythian Games (named for the awesome Python), which had competitions in music, poetry, and athletics. The Olympic Games, for athletics only, were held at Olympia, not to be confused with Mount Olympus, the home of the gods.

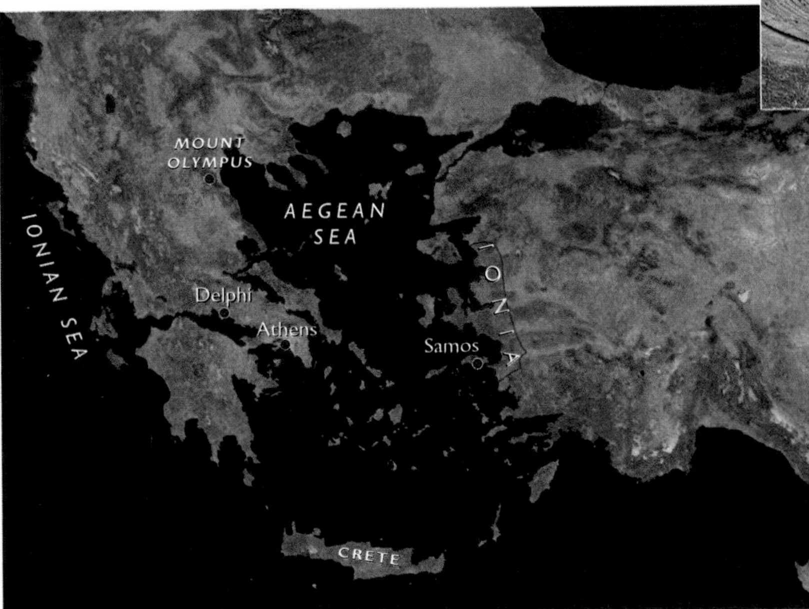

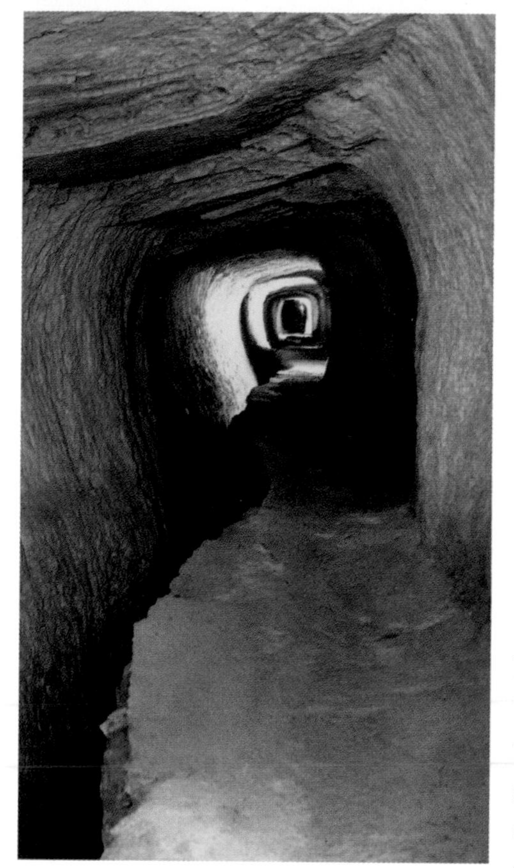

The tunnel of Samos was found in the twentieth century and is now a tourist attraction. It was built with water pipes and a pathway for inspections. If you go to Samos, you can visit it and also Pythagoras's cave. According to the fourth-century C.E. Syrian historian Iamblichus, "Outside the city he [Pythagoras] made a cave the private site of his own philosophical teaching, spending most of the night and daytime there and doing research into the uses of mathematics."

I have written thus at length of the Samians [people from Samos], because they are the makers of the three greatest building and engineering feats in the Greek world: the first is a tunnel nearly a mile long, eight feet wide and eight feet high, driven clean through the base of a hill nine hundred feet in height. The whole length of it carries a second cutting thirty feet deep and three broad, along which water from an abundant source is led through pipes into town.
—Herodotus (ca. 484–425 B.C.E.), Greek historian

Pythagoras (pi-THAG-uh-ruhs), who was born in ca. 582 B.C.E., is often called the world's first great mathematician. He was lucky in the time and place of his growing up. His home was on the island of Samos, just a mile's swim in the Aegean Sea from the coast of Turkey. For thousands of years, the Greek islands have had a special aura; there, the light is clear and vision is exact.

When Pythagoras was a boy, Samos was a big-time prosperous port. Ships carrying new ideas seemed to blow in on almost every breeze. The Greek historian Herodotus tells us that the people of Samos "are responsible for three of the greatest building and engineering feats in the Greek world." If we take ourselves to Samos in the sixth century B.C.E., you can see those engineering marvels: a spectacular tunnel channeling water pipes through a big hill, a man-made harbor, and the largest of all known Greek temples.

It is an age of ferment—and genius. Besides Thales and the "A" team in Ionia, there are Confucius (kuhn-FYOO-shuhs) and Lao-tzu (rhymes with "now-duh") in China, the Pharaoh Necho (NEE-ko) in Egypt, Zoroaster (ZOR-oh-as-ter) in Persia, the Jewish prophets in Israel, and Gautama Buddha in India. Pythagoras has a mind that can hold its own with any of them.

VAPORS are smoke, steam, fog, mist, or gases. To VAPORIZE means to turn something (like a meteorite or a comic-strip villain) into gas.

His story actually begins at Delphi, a small Greek village nested high on a mountainside with spectacular views of valley, sea, and snow-topped mountain peaks. In addition to its wondrous beauty, Delphi sits on a crack in the Earth. Vapors rise from the Earth's interior. Those vapors, it is said,

NUMBER MAGIC

The Greeks weren't the only ones fascinated with numbers. Almost every culture turned to numbers to help explain the mysteries of the universe. *Nine* was a special number to the ancient Chinese. According to an old tale, a magic diagram of the Earth came from heaven and first appeared on the back of a turtle. It was a square divided into nine parts. Each square held a number, and each column, row, and diagonal of the magic square added up to 15.

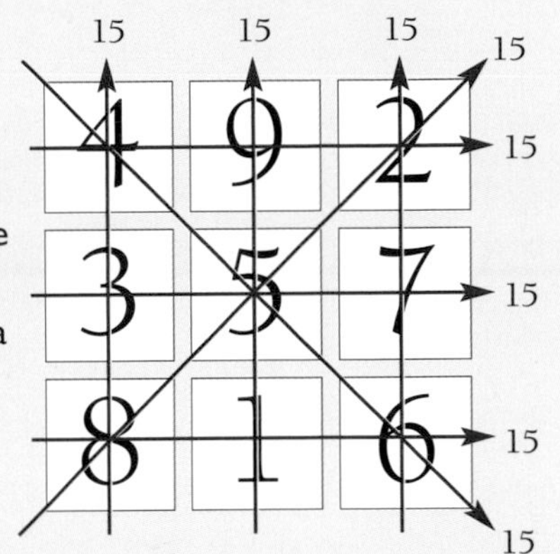

52	61	4	13	20	29	36	45
14	3	62	51	46	35	30	19
53	60	5	12	21	28	37	44
11	6	59	54	43	38	27	22
55	58	7	10	23	26	39	42
9	8	57	56	41	40	25	24
50	63	2	15	18	31	34	47
16	1	64	49	48	33	32	17

Albrecht Dürer (1471–1528), a German artist, placed a 4-by-4 magic square in his engraving *Melancholia* (below). Every row, column, or diagonal adds up to 34. About 200 years later, Benjamin Franklin created an 8-by-8 magic square (left). What's his magic number? (Answer is on this page.)

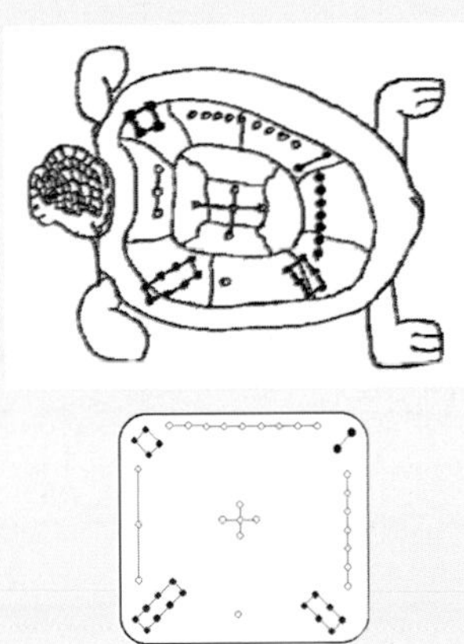

Magic squares have been around for more than 3,000 years. They are descendants of the oldest known number mystery, the Chinese legend of Lo Shu, found in a book titled *Yih King*. The story, adapted from a translation by scholar Philip Lei, goes something like this:

> *In ancient China, there was a huge flood. The people tried to offer sacrifices to the god of the Lo River to calm his anger. However, for each sacrifice, a turtle came from the river and walked around the offering, meaning the river god didn't accept it. One time, a child noticed a curious figure on the turtle's shell. From that figure, a magic square, the people realized the correct number of offerings to make—15.*

Answer to magic square:
Franklin's magic number is 260.

Joseph Mallord William Turner, a nineteenth-century English artist, painted a vulnerable-looking Apollo defeating the monstrous Python.

cause some people to have visions and claim to see the future. And so, from the time before memory or written records, Delphi was a holy place. At first it was Gaea (JEE-uh), the Earth goddess, who was worshiped there. Then, when the god Apollo came from Mount Olympus in search of a site for his temple, he picked Delphi (or so the Greek tales say). In order to take possession of the place, Apollo had to win it from the awful dragon Python, who, depending on the story you read, was Gaea's son or daughter or grandchild. Python (who took a serpent's shape) not only molested humans, he kept the gods from going freely to Delphi.

Using arrows and cleverness, Apollo killed the terrible Python. But he knew murder was wrong. So to cleanse himself of the sin and to set an example for mortals, he served as a slave to King Admetus (ad-MEE-tuhs). Then, crowned with a wreath of laurel leaves, Apollo returned to Delphi, pure and worthy of love.

Thus it was Apollo, the god of prophecy, who was worshiped at Delphi. People came there to receive wisdom

Mouthpiece of the Gods

The oracle at Delphi spoke through a priestess said to be "the mouthpiece of the gods." Known as the Pythia, she answered questions and had the power of prophecy. The answers were often puzzling—a bit like riddles—and could be interpreted in more than one way.

The word *oracle* comes to us by way of Latin; *orare* means "to speak." So in English we have *oral, orate,* and *orator,* as well as *oracle,* which still means "a source of wisdom." William Shakespeare, in the play *The Merchant of Venice,* wrote: "I am sir Oracle, And when I ope my lips, let no dog bark!"

This Athenian cup (ca. 440 B.C.E.) has the only surviving depiction of the Pythia (seated on a tripod).

from the oracle, who was said to speak Apollo's thoughts through a human agent, the Pythia, who was always a woman. Sometimes she also spoke words of the dead serpent, Python.

Mnesarchus and Parthenis, a young couple from Samos who would become Pythagoras's parents, made the sea-and-land journey to consult the oracle. It was worth it. She gave them astounding news: They were to have a son who "would change the world." Naturally they did what she told them to do: they went to Sidon in Phoenicia (now Lebanon) for the lad's birth (then they returned to Samos).

And so, even before he was born, Pythagoras was marked as someone special. As a boy, he was sent to nearby Miletus to study with Thales (who was now an old man) and probably with Anaximander. After that, he traveled widely—as wealthy, educated Greeks did—to Babylon, maybe India, and definitely Egypt, where he spent many years. Stories tell of his psychic powers: he seemed to know things that others did not. Perhaps he was just fortunate; perhaps it was the confidence that the oracle's prediction had given him; certainly, he had rare gifts.

He came back to Greece and soon had disciples who studied and taught and spread his thoughts. They were called Pythagoreans. The oracle was right: their ideas would change the world.

Psychic (SY-kik) has to do with unexplainable matters of the mind or soul. It comes from the Greek for "of the soul." Psychic things are out of the realm of science.

Disciples are followers who are true believers and devoted students. In Latin, the word *discere* means "to learn." **Discipline** comes from the same root. Pythagoras believed the life of the mind is superior to the pleasures of the senses and insisted that his disciples be "self-disciplined" and lead monklike lives.

CHEWING ON PI—OR TASTING ONE OF MATH'S MYSTERIES

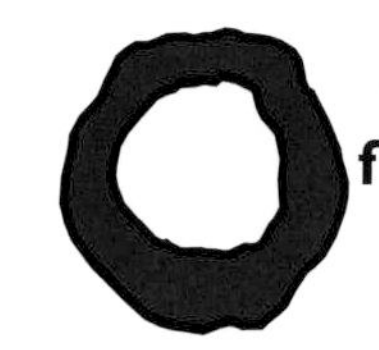

f all the geometric forms, nothing seemed quite as perfect as the circle. The ancients were sure there was a perfect mathematical expression for that perfect form. Finding it became an important goal. Besides, circles were much used. But how do you measure the area of a circle? And how do you measure its circumference (the length of the outside edge)?

Early on, architects and workmen in Egypt, China, and elsewhere found that there is an exact number relationship, or ratio, between the diameter of a circle and its circumference. That ratio of circumference to diameter is constant and unfailing. No matter what size the circle—whether it is dime-size or moon-size—the number you get when you divide circumference by diameter is always the same. Always. Mathematicians call it a constant. It is now known by the sixteenth letter in the Greek alphabet, π, or *pi* (pronounced "pie").

If you know the number for π, you can multiply it by any circle's diameter to get the circle's circumference. That makes it easy to measure huge circles without

This jade pi disk, a simple circle with a hole, is from China's Han dynasty (202 B.C.E.–220 C.E.). People wore pi disks and were buried with them to ensure a smooth journey of the spirit in the heavens. The hole, which represents the Earth in relation to the sky, has to be less than one-third of the disk's diameter.

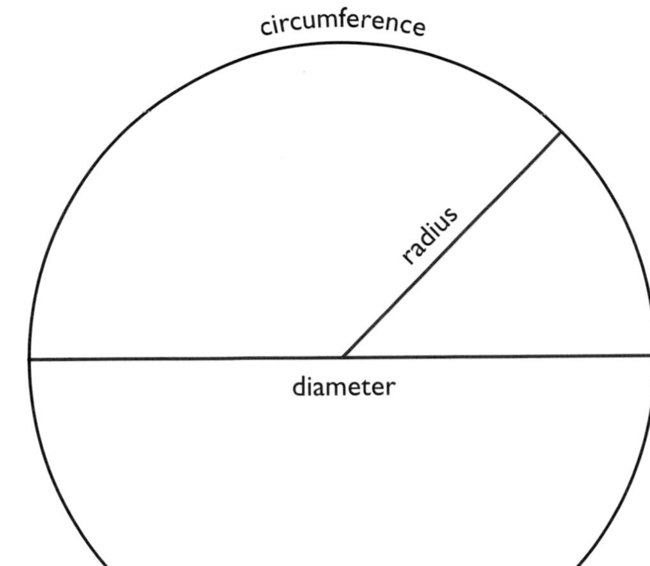

moving from your desk. But in the ancient world, there was an unexpected glitch. No one knew the exact number for π. Everyone, from the Egyptians to Pythagoras and beyond, seemed to work on the problem.

Today, we know if you divide circumference by diameter you will get 3.14159265.... That's π. Notice those dots. Three indicate that the number goes on and on and on, the fourth dot is a period to end the sentence. In some cases, three dots show that a number is just too long to write down easily. But not here. There is no end digit to π; it goes on forever. Besides that, it doesn't repeat digits in a regular pattern. So there is no way to write the value of π exactly. Nor can you express the decimal part of π in a fraction. That makes it an irrational number, but no one understood irrationals before Pythagoras. (More on this in the next chapter.)

Still, using the estimates of π that the ancients had, it became very useful.

With π you can easily measure the area of a circle. The formula is $a = \pi r^2$. It means: the area of a circle equals π multiplied by the square of the circle's *radius* (*r*, which is half its diameter).

But some ancients wanted to be able to figure a circle's area without worrying about π. They called it "squaring the circle" and were convinced that you could construct a square the same size as a circle and thus easily find a circle's area. The historian Plutarch wrote that Anaxagoras had "squared the circle," but Plutarch didn't tell how Anaxagoras did it, and no one else has ever done it.

Other ancients kept trying to find an exact number for π.

Today's computers have calculated more than a million digits after the decimal point in π. Still, no one has discovered its precise size. (No one ever will; mathematicians are now convinced of that.)

Hold on until chapter 17, where you'll find a Greek who came closer to finding exact π than anyone else in the ancient world.

9

Pythagoras Knows It's Round

We have in the end come back to a version of the doctrine of old Pythagoras, from whom mathematics, and mathematical physics, took their rise. He discovered the importance of dealing with abstractions; and in particular directed attention to number.... Truly, Pythagoras in founding European philosophy and European mathematics, endowed them with the luckiest of lucky guesses—or, was it a flash of divine genius, penetrating to the inmost nature of things?

—Alfred North Whitehead (1861–1947), English mathematician and philosopher, *Science and the Modern World*

[The Pythagoreans at Samos] think one should discuss questions about goodness, justice and expediency... [because Pythagoras] made all those subjects his business.

—Iamblichus of Apamea (ca. 250–ca. 330 C.E.) Syrian author, *Life of Pythagoras*

Pythagoras nurtured a cult based upon numbers that was to become a weird mixture of the sublime and the ridiculous.

—John D. Barrow (1952–), English astronomy professor, *Pi in the Sky*

Samos was part of Ionia, but no one thinks of Pythagoras as an Ionian. He stands alone.

How do you make sense of the universe? Do you do it by considering mountains of information—observing this, observing that—adding one block of knowledge to another? Believe that, and you're an Ionian-style scientist.

Or, is it an orderly, perfect creation that can be understood through mathematical formulas and headwork? Believe that, and you're thinking like Pythagoras.

Actually, today's scientific method combines both approaches, pure thinking along with observation, as well as something essential to modern science—experimentation that leads to proofs. But it took a long time to get that

A DODECAHEDRON is a solid with 12 sides that are each polygons.

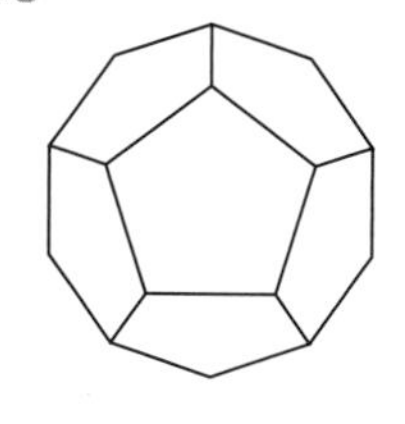

This is a corner from one of the world's great paintings, *The School of Athens* by Raphael (1483–1520). The artist wanted to connect his time (the Renaissance) with the achievements of the Greeks. So Raphael used the sculptor Michelangelo as a model for Heraclitus, the man leaning on his elbow. Leonardo da Vinci was the model for Plato, who is standing in the red robe at the upper right. That's Socrates, in green, in the upper center. The warrior, upper left, may be Alexander, although no one is quite sure of that. But everyone agrees, it is Pythagoras working on a manuscript at the lower left.

method working. The Greeks didn't do much experimenting.

For Pythagoras, the way to understand the universe was by searching for things that are absolutely true, and numbers

DIGGING NUMBERS

"Are there any precious stones in [the mine]?" asked Milo excitedly.

"... I'll say there are. Look here."

[The Mathemagician] reached into one of the carts and pulled out a small object....

"But that's a five," objected Milo, for that was certainly what it was.

"Exactly," agreed the Mathemagician; "as valuable a jewel as you'll find anywhere. Look at some of the others."

He scooped up a great handful of stones and poured them into Milo's arms. They included all the numbers from one to nine, and even an assortment of zeros.

"We dig them and polish them right here," volunteered the Dodecahedron....

"They are exceptional," said Tock, who had a special fondness for numbers.

"So that's where they come from," said Milo, looking in awe at the glittering collection of numbers. He returned them to the Dodecahedron ... but, as he did, one dropped to the floor with a smash and broke in two....

"Oh, don't worry about that," said the Mathemagician as he scooped up the pieces. "We use the broken ones for fractions."

—Norton Juster, *The Phantom Tollbooth*, an imaginative look at numbers

THE NUMBER GODS

We see numbers as descriptions of things or of relationships between things. The Pythagoreans held a mystical belief that numbers themselves have power and significance. They believed that numbers actually are out there holding the universe together.

They worshiped numbers as divine, especially the *tetraktys*, the first four numbers. *One* was reason, *two* was argument (it's easy to guess why), *three* was harmony, and *four*—the sum of equals—was justice. The numbers *one, two, three, four* also represented the four points of the compass and the four basics of geometry (point, line, surface, solid). Plato later identified them with the four elements.

Because one, two, three, four add up to ten, ten was "the mother of all." It represented the universe. They even prayed to ten: "Bless us, divine number, thou who generates gods and men! O holy, holy *tetraktys*, thou ... source of the eternally flowing creation!"

Pythagoras is believed to have said to a new disciple, "See, what you thought to be four was really ten and a complete triangle and our password."

seemed perfect for that quest. "All is number," he said. And he meant it. Everything in the world, he believed, could be explained through mathematics. He went still further: he thought numbers were divine, an expression of God's mind.

We don't have any books or papers or words that Pythagoras actually wrote. All of his work has been lost. But we know enough from the writings of others, who tell of him and his achievements, to realize he was one of the most influential people of all time.

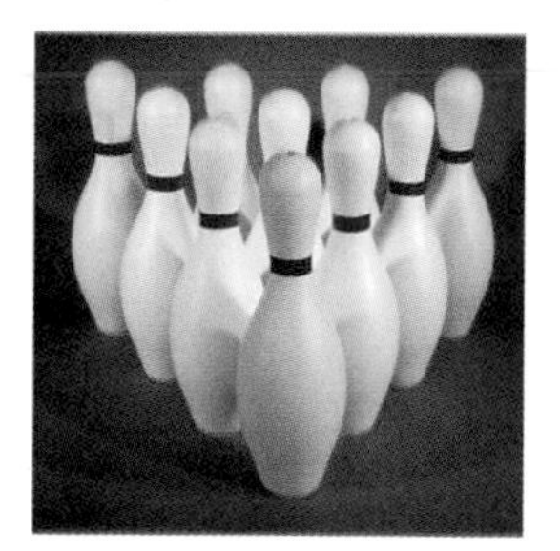

Who knows if Pythagoras would have loved bowling, but he definitely would have approved of the tenpin setup. The pins are arranged in a "holy ten" equilateral (equal-sided) triangle using the *tetraktys* numbers: *one, two, three, four.*

Pythagoras organized his followers into a society of vegetarians who lived an austere communal life and worshiped him as a kind of god. But some people of his time thought he was weird; some thought him dangerous. Powerful thinkers can be frightening.

There was nothing democratic about his community. Ordinary people were supposed to follow the leaders and not ask disturbing questions. The leaders were expected to be trustworthy and take care of their followers. An intellectual elite was schooled for leadership.

Pythagoras thought there were three kinds of people: those who were concerned with making money, those who cared about attention and fame, and those who were interested in the life of the mind. (You can guess whom he favored.) He didn't think most people were smart enough to understand or care about difficult ideas. So, in his society, only easily understood numbers and concepts were taught to the average person. It was the leaders who delved into advanced math and philosophy. Pythagoras was trying to create an ideal culture, and this is the way he saw it.

Pythagoras and his followers formed a sect with secret rituals; today it might be called a cult. They believed in reincarnation. While they didn't quite believe in the equality of women, they did acknowledge that women (and slaves) had brains—and they encouraged their use.

Reincarnation is the belief that at death, the soul is reborn as another human or animal. Sometimes it is called "transmigration of souls."

In a famous story, Pythagoras is supposed to have seen a man whipping a dog and asked him to stop because in the dog's bark he recognized the voice of an old friend.

Pythagoras wore trousers—an Eastern fashion—while the rest of the Greeks were still in loose-fitting robes. He did what he wanted to do and didn't pay attention to conventional rules. But he expected his disciples to follow *his* rules. Pythagoras seemed to see all of nature as intertwined. He thought beans (yes, beans!) held mystical powers; he wouldn't touch them.

How do you think the other people of Samos felt about the Pythagoreans? Well, their secrecy and aloofness made many distrust them. Polycrates (poh-LIK-ruh-teez), a heavy-handed despot who controlled Samos, didn't want them on the island and forced them out. (Later Polycrates, who got rich on piracy, was trapped by Persians and crucified.)

Pythagoras and his band of devoted followers left for Croton, Italy, where there was a Greek colony. Although Pythagoras lived and taught there for many years, his intensity may have scared the people of Croton. Stories say that he was burned to death in a raid that destroyed a follower's home. The follower was the Olympic superstar Milo, a sports hero of his day. No one is sure if it really happened.

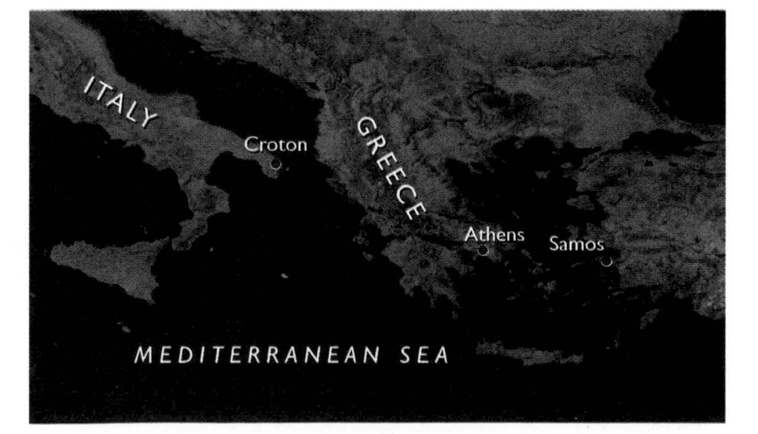

Remember, the Pythagoreans wouldn't share their knowledge with outsiders. They believed devoutly that numbers are the essence of the universe and that they are real and tangible,

DIFFERENT LANGUAGES

Sometimes everyday English and scientific English can seem like different languages. Take that word ***irrational.*** My dictionary defines it as "not endowed with reason," or "loss of usual or normal mental clarity." In other words, if you are acting irrationally, you aren't making sense.

That has little to do with the mathematical meaning of *irrational.* To a mathematician, an **irrational number** is one that *cannot* be turned into a ratio of two integers. (Integers are zero and the positive and negative whole numbers.) A rational number can be expressed as a ratio of integers. (Remember you can't divide by zero.)

Pythagoras is said to have conducted many experiments on the relationship between music and numbers. Here, in a drawing from a music book published in 1496, he has attached different weights to strings to produce notes on the musical scale. The greater the weight, the higher the note is—just as tightening guitar strings raises their pitch.

and so they prayed to them. If someone had an idea that challenged Pythagorean concepts, that idea was suppressed. When a disciple named Hippasus talked openly about the properties of the dodecahedron, he was thrown out of the community.

So imagine what it must have been like for them to find a number that didn't seem to be quite real. A number that can't be exactly written or measured.

Well, that's what happened. And it must have shaken their faith. It's easy to imagine, as the Pythagoreans did, that all numbers are rational numbers. As it turned out, there is a kind of number that isn't rational. It can't be expressed as a ratio of integers. The Pythagoreans tried to keep knowledge of those strange numbers—now called irrationals—a secret.

But put all that aside. Pythagoras had an astonishing mind and he gave us a new way of looking at the universe—through numbers.

By plucking musical strings of different but carefully measured lengths with the same tension, he found that sounds have exact number relationships. This isn't a haphazard thing. A string twice the length of another has a sound that is exactly an octave lower (if both have the same tension). If the length of two strings is a ratio of

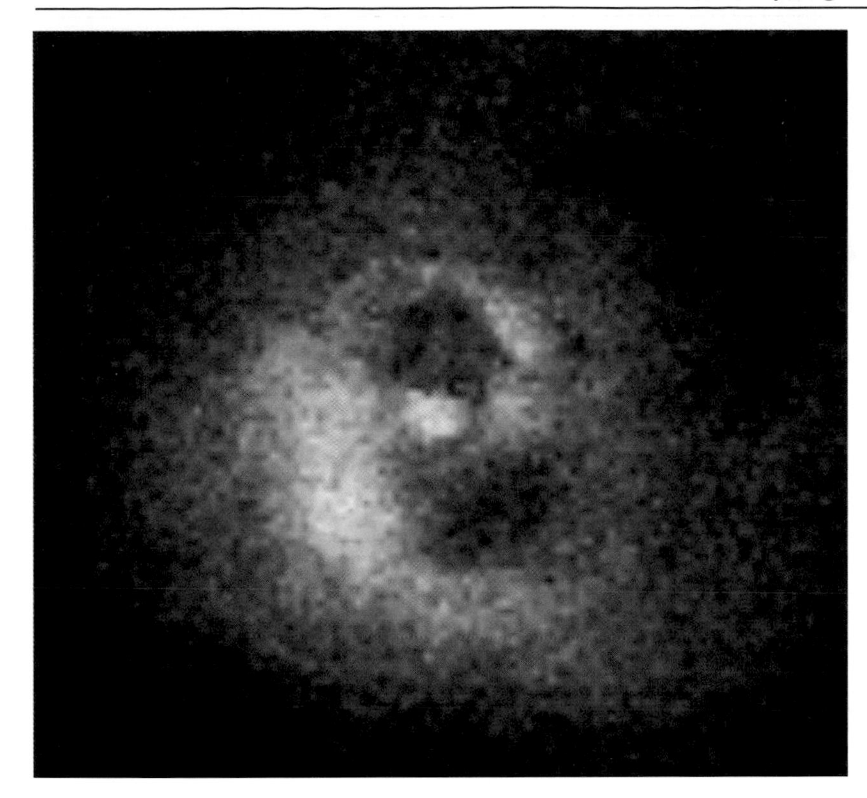

Music of the spheres? Pythagoras may have been right, according to modern astronomers, who say they have heard the sound of a black hole singing. Pressure waves from a black hole in the Perseus A Galaxy seem to be reverberating in B flat—a B flat that is 57 octaves lower than middle C. It's "the lowest note in the universe," said Dr. Andrew Fabian of Cambridge University in England. This image was taken by the Chandra X-ray telescope orbiting Earth. It reveals a high-energy area in the center (in light blue) and two dark spots thought to be bubbles of particles produced by energy released from the black hole.

three to two, the musical interval they produce is called a fifth. If it is four to three, the interval is known as a fourth. Increasing the tension of a string raises its pitch in a predictable way. (The Pythagorean conclusions on sound remain valid today.)

Pythagoras's findings gave order to music that no one before had imagined. If music can be explained mathematically, why can't other things?

Pythagoras thought the cosmos was like an orchestra. He saw the planets as celestial instruments playing in perfect mathematical and musical harmony. This was not a metaphor: he believed the music was out there, but only the gods could hear it.

Pythagoras was the first cosmic guy. It was he (and not Carl Sagan*) who coined the word *kosmos* to refer to everything in our universe, from human beings to the earth to the whirling stars overhead. *Kosmos* is an untranslatable Greek word that denotes the qualities of order and beauty. The universe is a *kosmos*, he said, an ordered whole, and each of us humans is also a *kosmos* (some more than others).

—Leon Lederman (1922–), American physicist and Nobel Prize–winner, *The God Particle*

(*Astronomer Carl Sagan wrote a terrific book called *Cosmos*.)

The Great Pyramids at Giza were built ca. 2550 B.C.E. during Egypt's Old Kingdom. (By the time Cleopatra ruled Egypt, these monuments were 2,500-year-old tourist attractions.) Their dimensions, whether by design or by accident, are mathematically pleasing. The hypotenuse of the right triangle (inset) equals phi, the Golden Ratio (see page 84).

It may sound strange, but it was actually an imaginative unifying idea—all the parts of the universe interacting like instruments in an orchestra to make a greater whole!

Having shown that sound can be understood through numbers, Pythagoras turned to the world of vision. He may have focused on the horizon: If he cut through that horizontal line with a vertical (straight up and down) line, he had a right angle. Pythagoras probably played with right angles in his mind. He is identified with a theorem that seems simple to us now but is one of the great intellectual achievements of all time. **The square of the hypotenuse (the longest side) of a right triangle equals the sum of the squares of the other two sides**, said Pythagoras. It's called the Pythagorean Theorem.

The Egyptians must have used this formula or they couldn't have built their pyramids, but they never expressed it as a useful theory. The Babylonians may have also known the theorem. But it takes a leap of mind, and mathematical proof, to go from an idea that helps you build a structure to

To this day, the theorem of Pythagoras remains the most important single theorem in the whole of mathematics.

—Jacob Bronowski (1908–1974), Polish physicist, biologist, poet, and author, *Science and Human Values*

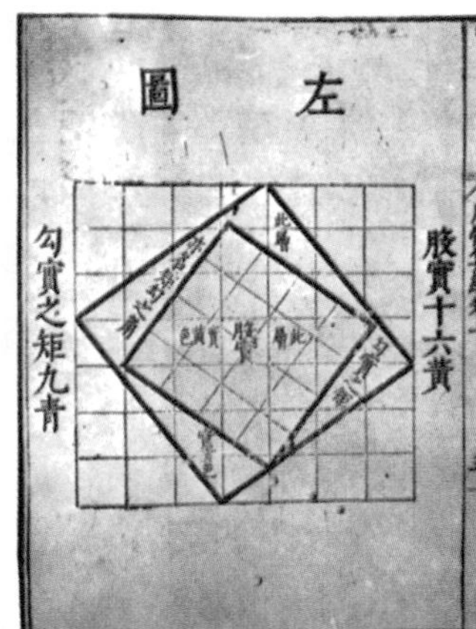

This block print (above) illustrates a Chinese proof of the Gougu Theorem. (It's the same as the Pythagorean Theorem.) It first appeared in China in an ancient work of geometry called the *Zhou Bi Suan Jing*, written in the first century B.C.E.

Paul Klee (1879–1940) had a hard time deciding whether to be a musician or an artist. Art won out. Klee, who was Swiss, was also interested in math and what he called the "science of design." He combined inventiveness with abstraction in work that seems innocent and unsophisticated but is neither. Why do you think he's in this chapter? Do you see the right angles in his painting (left)?

UNIVERSAL MATH

The Pythagorean Theorem for right triangles is $a^2 + b^2 = c^2$. This simple equation has been called a universal language, which means it should make sense on any planet anywhere. If there are other beings living in the cosmos, this formula has been suggested as a way to start communicating.

To perfection-loving Greeks, an elegant example of the theorem is: $3^2 + 4^2 = 5^2$ (which means: $9 + 16 = 25$).

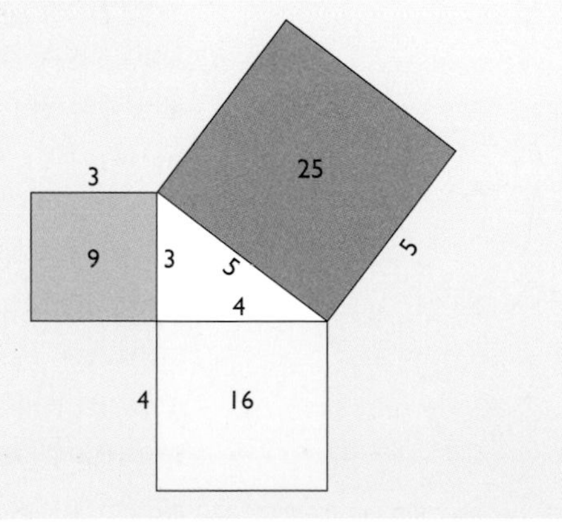

UNTHINKABLE HORRORS

The Pythagoreans discovered that the square root of 2 (a never-ending 1.414213...) can't be written as a ratio of natural numbers. Horrors! How could that be? A number that can't be expressed exactly challenged the Pythagorean idea that numbers rule the universe. This bewildering square root turned out to be an unexpected kind of number—an irrational number. Irrationals created a split between the arithmetic scholars, who discovered these strange numbers, and the geometers, who couldn't find an exact measure for them. To add to the controversy, the Pythagoreans soon discovered that the $\sqrt{2}$ isn't unique. There is an infinity of irrationals. In fact, most numbers are irrational.

an understanding that it is an unchanging formula that works no matter where or how it is applied. Pythagoras took that leap (also working out a proof of the theorem) and introduced it to the Greek-speaking world.

There is an exactness to the world, an orderliness, and it follows rules that can be understood with numbers: that's what Pythagoras told us, and it has been confirmed again and again.

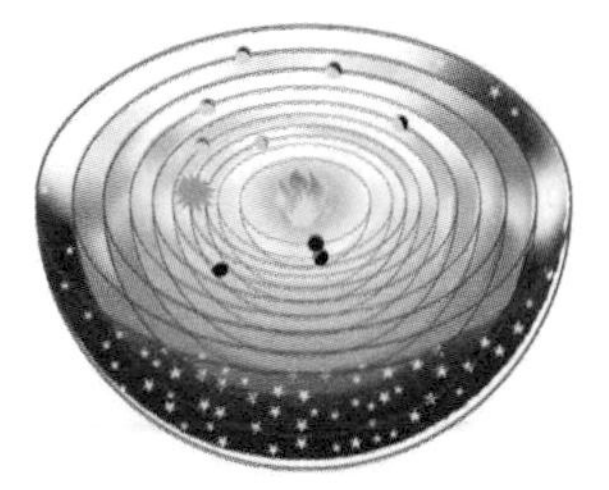

[The Pythagoreans] say... that fire is at the center and that the earth is one of the stars...And they assume yet another earth opposite this which they call the counter-earth."
—Aristotle, *Physics*

If all this isn't enough, Pythagoras is thought to be the first person to teach that the Earth is a sphere. (He took the next step after Anaximander's squatty cylinder.)

What an astonishing feat, to understand that Earth is a big ball (a slightly out-of-shape ball, as we now know). Would you have figured out its spherical shape if you hadn't been told? And that still isn't all: Pythagoras understood that Earth moves!

Anaximander had explained the motion of the Sun and stars by saying that they are all attached to a heavenly sphere that rotates, carrying them in its grip. Pythagoras accepted that idea, but in his cosmology, the Earth, the Sun, and the planets are not attached to the same heavenly sphere as the stars: they follow different paths; some, like Earth, are circling a great celestial fireball. Take note of that fireball—Earth is not in the center in this universe.

Something else: Pythagoras introduced the idea of multiple spheres. For way more than 1,000 years afterward, astronomers would worry about separate spheres for the Sun and planets. (Finally, modern science came up with gravity, smashing those crystal spheres.)

MAKING UP NUMBERS: USE YOUR IMAGINATION

Modern mathematicians call all the numbers on the number line real numbers. But some numbers, called imaginary, don't fit nicely on a number line. So what are imaginary numbers? They are numbers such as $\sqrt{-1}$ (the square root of minus one), or the square root of any negative number.

Mathematicians use imaginaries to solve certain problems. The square roots of negative numbers are used in dealing with alternating current. Imaginary numbers are also used to create fractals. A fractal (from the Latin *fractus,* meaning "broken") is a jagged-edged geometric shape with an infinite number of tinier and tinier repetitions that look the same as the overall shape. The repetitions are generated by plugging an endless succession of numbers—including imaginaries—into an equation. As it happens, computer fractals always seem to be gorgeous works of art, but there's a practical side, too. A jagged coastline, with its zigs and zags, is tough to measure precisely. In fact, as you zoom in closer, you see more zigs and zags, and the measured length grows longer. Fractals, with their infinitely smaller repetitions, offer a way to deal with this problem. (Read up on Gaston Julia and Benoit Mandelbrot if you want to know how.) Fractals help create those fantastic computer-generated effects for films. Moviemakers can grow trees, move mountains, and create whole electronic worlds with software based on fractals.

If you had the idea that math is unimaginative, think again. Modern math is creative, fluid, and exciting.

In a fractal, infinitely smaller iterations (repetitions) each look like the overall shape.

Pythagoras was the first Greek to realize that the morning star (called Phosphorus) and the evening star (called Hesperus) are the same. Today we know it as the planet Venus.

This astonishing thinker and observer understood that the structure and relationships of the universe can be described with mathematical formulas. He made mathematics the language of Western science. No one has done more.

THERE'S GOLD IN THOSE IRRATIONALS

IRRATIONALS ON A NUMBER LINE

The Greeks knew that there is a number for each point on a number line, even the irrational numbers. The distance between points 0 and 1 is 1 unit, which could be 1 inch or 1 centimeter or 1 whatever. Notice that we made a square that is 1 unit on each side. Then we drew a diagonal line from one corner of the square to the other, creating two triangles. Here's an equation for what we did: $1^2 + 1^2 = x^2$, or $x = \sqrt{2}$ (x is the diagonal). Now here's the zinger: $\sqrt{2}$—the square root of 2, which is the length of that diagonal line—is an irrational number. In other words, $\sqrt{2} = 1.414213\ldots$. The decimal digits in $\sqrt{2}$ go on and on with no repeating pattern. We can put $\sqrt{2}$ on our number line and use it for calculations; we just can't write it exactly.

The Pythagoreans believed that numbers can explain everything, but how do you deal with a number, like π, that you can't write exactly as a decimal number? Numbers are supposed to be precise and perfect. The Greeks were horrified at the very idea of irrational numbers.

Those pesky numbers follow a decimal point forever without any repeating pattern. A few fractions, such as $\frac{1}{3}$ or $\frac{1}{7}$, are *repeating* fractions; they go on and on, but in a repeating pattern. Write $\frac{1}{3}$ as a decimal (.3333…), and you'll see. It is not an irrational number. The digits of an irrational follow each other in a *nonrepeating* sequence.

Do we need irrational numbers? Yes, as it turns out, we do. Suppose you know

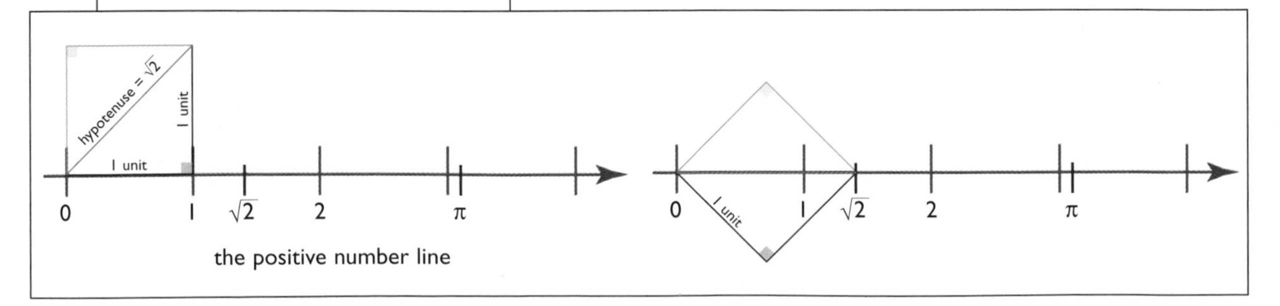

that the area of a square is 1. Now you want to know the length of a line drawn from one corner of the square to the other, which equals $\sqrt{2}$ (see the box on the opposite page). The Pythagoreans could construct a square whose area is 1, but when they measured the length of its diagonal line, they found it was a very strange number. It was close to 1.414, but not exactly. There was no number they could write, no matter how simple or how complex, that when multiplied by itself gave them 2. The only numbers they knew were rational numbers, but there is no rational number whose square is 2. Something didn't make sense.

Nobody ever had a reason to know about such numbers before Pythagoras stated his theorem. Then they had to deal with a whole new kind of number. It took a while before they realized it, but irrationals are useful numbers. They would learn that with irrationals you can do calculations that can't be done otherwise. Modern mathematics, science, and technology would be impossible without irrationals.

AN IRRATIONAL WELCOME

An irrational cousin of π (3.141592...) and $\sqrt{2}$ (1.414213...) is e, which has a decimal value of 2.71828.... Now, e is much loved by engineers, who often use it, rather than 10, as a base for calculations. The Gateway Arch in St. Louis, Missouri, a symbolic welcome to the West, is also an engineer's monument to e. Its curve is an upside-down catenary (from the Latin *catena*, meaning "chain"), a curve that you can form just by holding the two ends of a chain (or jump rope) at the same height.

The Gateway Arch's catenary is elegant: 192 meters (210 yards) tall and 192 meters (210 yards) wide. But a catenary can have other shapes too. Imagine stretching the ends of the arch (or your jump rope) farther apart; as you do so, the curve flattens out. Put the ends side by side, and you can't help but form an extremely steep curve. As with any ratio (see page 43), changing one measurement (the distance between the ends) changes the other (the curve) in a way that can be calculated with an equation. The equation for a catenary curve rests on the irrational shoulders of e.

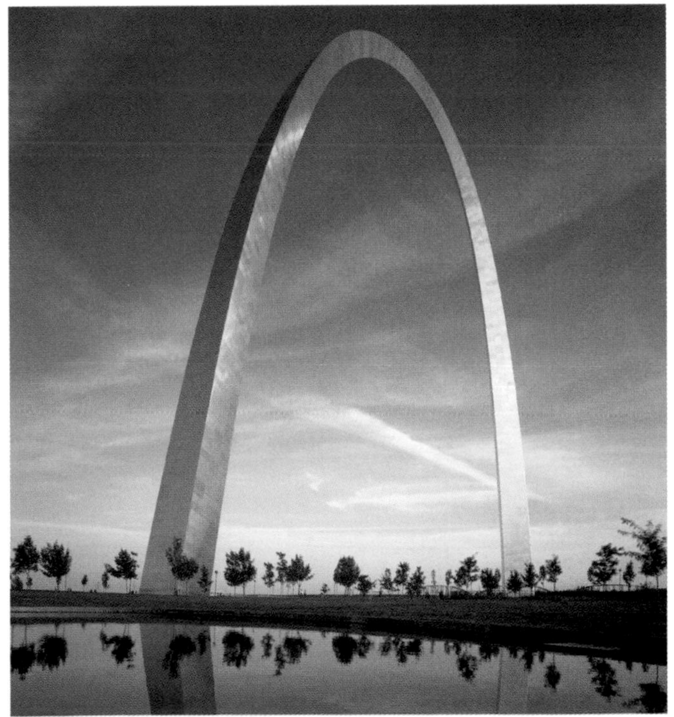

The top of the Gateway Arch offers a spectacular view of land west and east of the Mississippi River.

(more about irrationals on next page)

Ancient Greek and Renaissance artists used phi, the Golden Ratio, by design to create ideal human forms. This statue is by the Greek sculptor Phidias, whose name inspired the symbol for phi. What's golden about it? The total height of the statue divided by its height at the navel is equal to phi. Are real human bodies this golden? Not everyone shares the statue's ideal proportion, but if you measure scores of bodies, men or women, the average ratio turns out to be very close to golden.

An Irrational Beauty

It's easy to confuse the irrational *pi* with an irrational called *phi.* But don't do it. Phi is special in its own right. It is named after the Greek sculptor Phidias and is known as the Golden Ratio. The mathematical symbol for phi is ϕ. Mario Livio, in a book called *The Golden Ratio*, says "It is probably fair to say that the Golden Ratio has inspired thinkers of all disciplines like no other number in the history of mathematics."

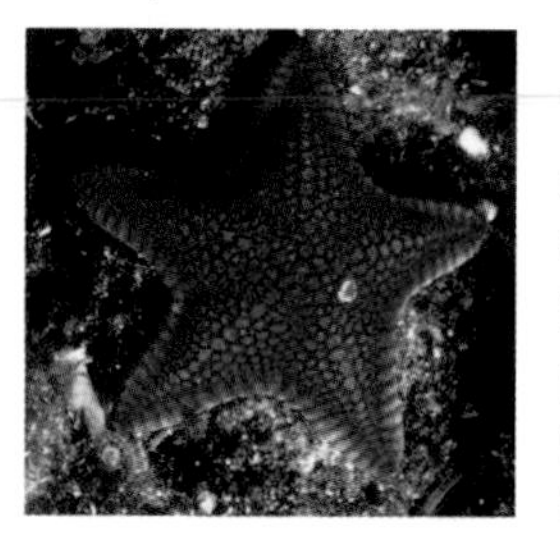

Why the fuss? To begin, that ratio seems to pop up all over the place. You can find its proportions played out in the five-pointed pattern (pentagram) of an apple's seeds or in the five arms of a starfish. The swirls on pineapples and pinecones are golden in their proportions. So are many other ratios in nature. As to Phidias, like many artists, he knew the Golden Ratio and used its proportions in his work. We humans seem to find that the Golden Ratio brings balance, harmony, and beauty where it is found. (For instance, in the ratio of height to length in a building.)

The next time you look at a rose, note the spacing of its petals—they aren't placed one on top of another—the pattern of their placement is golden. (Nature isn't always perfect, so you won't find ϕ in every rose, but take a group of them, and it will dominate.)

length

Golden Ratio = $\frac{\text{length}}{\text{width}}$ = 1.618...

width

Here's some math for you to ponder. The relationship between two sides of a golden rectangle, or two parts of a golden line, is this: The small side is to the large side as the large side is to the sum of the two sides. If the large side is 1, then the small side is 0.61803.... If the small side is 1, the large side is 1.61803.... Note something unusual here. The Golden Ratio is 1.618... and its reciprocal is .618.... Mathematically, it looks like this: $1/\text{phi} = \text{phi} - 1$. Both figures pop up eerily often in nature (check out the sunflower on page 226) and by design in architecture. The front of the Parthenon (above left) in Athens is a golden rectangle, and the spaces between the pillars are in golden proportions.

The Greeks often talked of "mean" and "extreme" when dealing with the Golden Ratio. It's usually illustrated with a simple line divided into a smaller section—the mean—and the whole—the extreme. And it is irrational: you can't write the Golden Ratio as an exact fraction. The equation for phi relies on another irrational—the square root of 5: $\text{phi} = \frac{\sqrt{5} + 1}{2}$

No wonder the perfection-loving Greeks had a hard time dealing with it. How *could* such a messy number be so "beautiful"? (For more on the Golden Ratio—and some concrete numbers that produce it—see chapter 25.)

A pentagon with lines connecting the vertices (the intersections of the angles) creates a pentagram, or five-sided star. The Pythagoreans were fascinated with the geometric properties of the pentagram and used it as a symbol of their sect. They called it "health."

If you look at the line segments in order of decreasing length (A, B, C, D), each is smaller than its predecessor in proportion to the Golden Ratio, ϕ. Line A forms one side of an isosceles triangle (a triangle with two equal sides). There are five identical isosceles triangles in the pentagram. Can you see them?

Also notice that inside the pentagram is another pentagon, in which you could draw another star, which would contain another pentagon, and so on. This process of creating nesting pentagons and pentagrams could go on indefinitely. It's yet another proof that the sides of a pentagon and the lines connecting the vertices are incommensurable, which is another way of saying they are irrational.

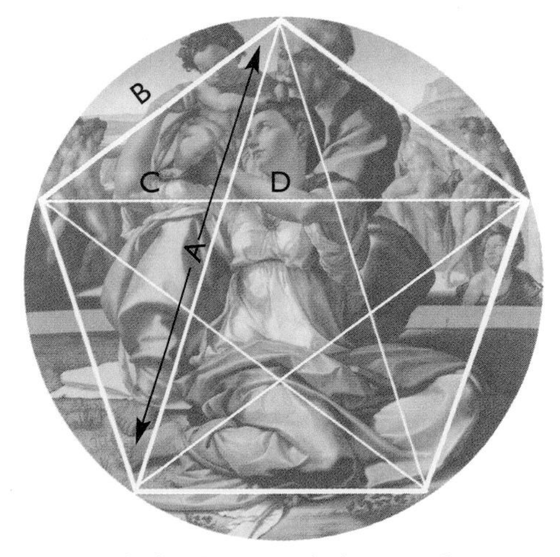

In this tondo (circle painting), the great Renaissance artist Michelangelo (1475–1564) arranged the Holy Family in a pentagram.

10 Getting Atom

Colors, sweetness, bitterness, these exist by convention; in truth there are atoms and the void.

—Democritus (ca. 460–ca. 370 B.C.E.), Greek philosopher

And the atoms move continuously for all time, some of them falling straight down, others swerving, and others recoiling from their collisions. . . . And these motions have no beginning, since the atoms and the void are the cause.

—Epicurus (341–270 B.C.E.), Greek philosopher, "Letter to Herodotus"

"I would rather understand one cause than be King of Persia," said Democritus (di-MAHK-ri-tuhs), who was born about 100 years after Pythagoras (we think in 460 B.C.E.). Now the king of Persia had about as much power as anyone could have, and he was fabulously wealthy too, so only those who understood the power of ideas would get what Democritus was saying.

Some people probably laughed when they heard his comment and noted that Democritus was born in Thrace, which was an unfashionable, out-of-the-way place for a philosopher. "What can you expect from someone born in Thrace?"

"Happiness resides not in possessions, and not in gold; happiness dwells in the soul," said Democritus. French artist Antoine Coypel (1661–1722) portrayed him as a cheerful fellow.

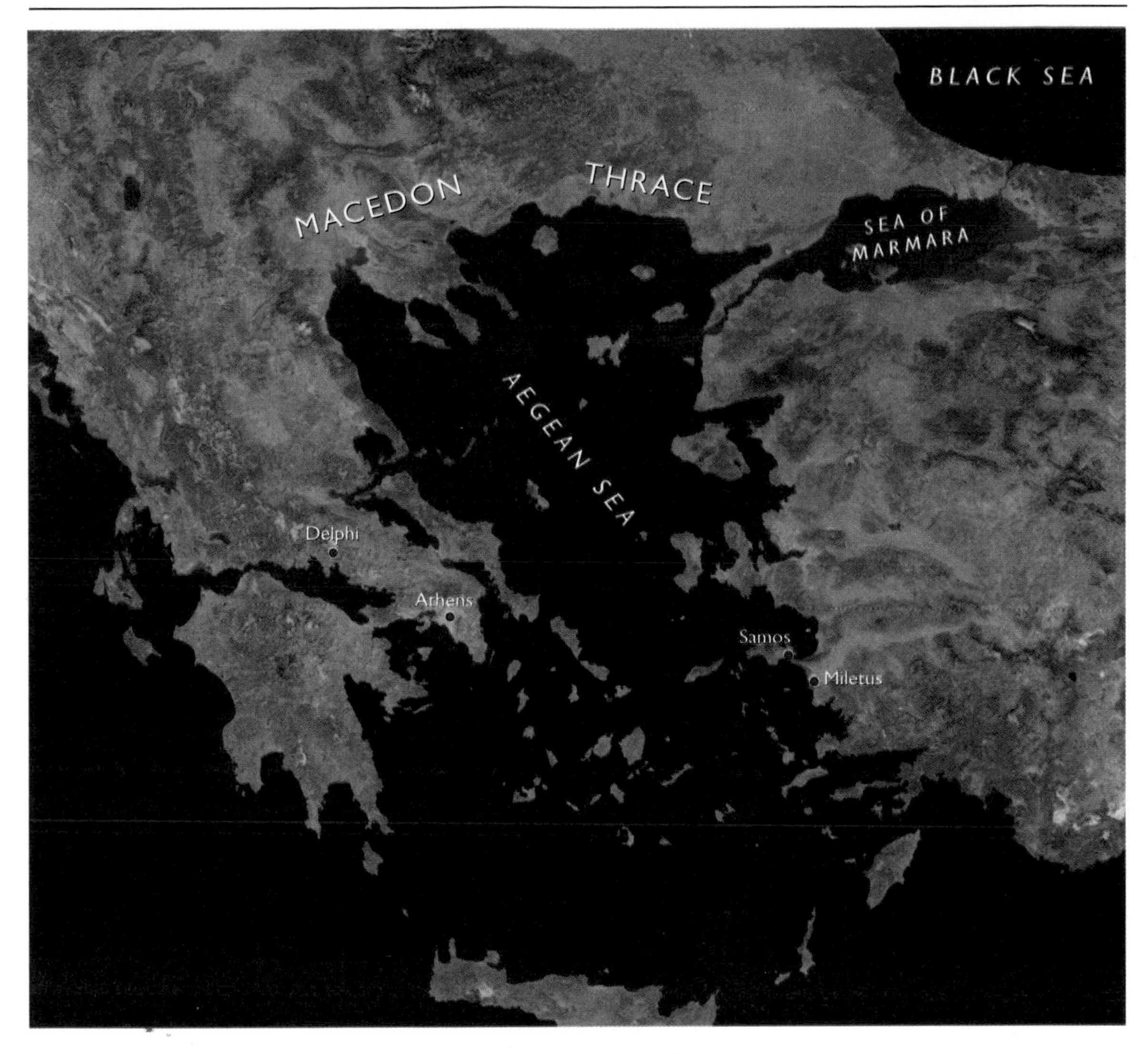

they may have said. Thrace was west of the Black Sea and north of the Aegean (see map, above), and it was not a center of philosophy, as Miletus, Athens, and Samos were. Some people used to chuckle if you said you were from Thrace. But that never stopped Democritus, who was another powerful thinker.

Democritus believed that to understand the universe you need to know what it is made of. The Ionians had come up with those four basic elements: earth, air, fire, and water. Democritus thought there must be something still smaller, something that unified the "elements." Something they all had in common.

Democritus's work on the void was revolutionary. He knew, for instance, that there is no top, bottom, or middle in space. Although the idea was first suggested by Anaximander, it was still quite an accomplishment for a human born on this planet with its geocentric populace.... One of Democritus's further-out beliefs was that there are innumerable worlds of different sizes.
—Leon Lederman (1922–),
American physicist,
The God Particle

ATOMIC SEEDS

Anaxagoras believed the universe originated through the action of an abstract "Mind" upon an infinite number of "seeds." A fellow Ionian named Leucippus (loo-SIP-uhs) is said to be the inventor of the idea of atoms. Can you imagine looking at trees and rocks and people and figuring out that they are all made of atoms?

Leucippus, who was Democritus's teacher, said atoms are solid and indestructible. He also believed that they assume geometric forms (which accounts for the variety of life) and that they are perpetually in motion. (Take note of those ideas; you'll hear them again.) As to motion, he was right! It would be the twentieth century before that idea was understood again.

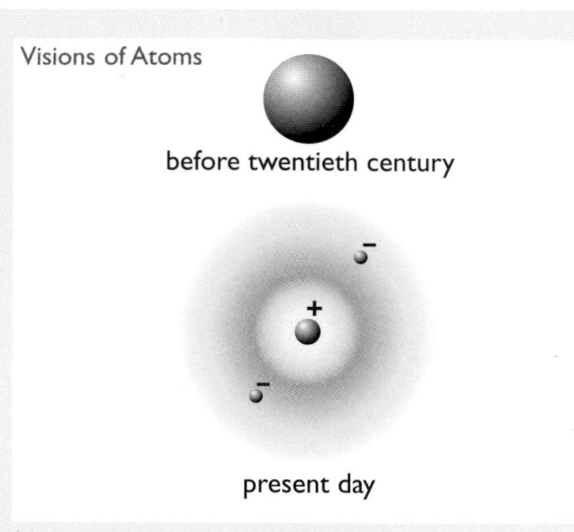

Atoms aren't solid balls (top). They have a nucleus (core) with at least one proton (positive charge). The nucleus is orbited by electrons (negative charge).

A VOID is empty space or nothingness. It's the opposite of matter.

He said there had to be a smallest substance in the universe that can't be cut up or destroyed and is basic to everything else. He called that smallest substance an atom. Nothing exists, said Democritus, but "atoms and the void." The atoms that Democritus had in his mind were solid, hard, and compact. Nothing could penetrate them. And they were too small to be seen.

Much of what we know of Democritus is hearsay; except for a few words, his writings have been lost. (In those days before printing, all books had to be hand copied, so there weren't many copies.) We do know that he traveled widely—to Egypt and to the lands east of Greece. He is said to have lived a long life and to have written 73 books.

Was he right? Is there a basic building block of life? A smallest of the small out of which comes everything? It's a question we're still considering.

The Greek believers in atoms had no way to prove that those tiny particles actually existed. Microscopes and other scientific technology were far in the future. So they relied on their brains, using them in ways that still astound us.

How do you explain change in a substance when you add or subtract energy (heating a liquid, for example)?

CUTTING-EDGE ATOMS

In Greek, *a* means "not" and *tom* means "cut." So to the ancient Greeks, *atom* meant "something that can't be cut." In the nineteenth century, we found the modern atom. We discovered that all the atoms in an element (such as gold) are the same, but they are different from the atoms in every other element. For a while, we thought those elemental atoms were basic and "uncuttable." That turned out to be wrong.

Atoms are little worlds in themselves, each with a nucleus and still smaller particles called protons, neutrons, and electrons. In the 1930s, scientists learned how to split a nucleus—a process called nuclear fission. (That's what's happening in the photo—the reactor core of a nuclear power plant.) The reverse turned out to be doable too: In nuclear fusion, the nuclei of atoms join together.

The investigation into smaller-than-atom particles would open up a whole new field of discovery. Many subatomic particles, such as quarks, leptons, and neutrinos, have been found. Does that mean Democritus was wrong? Or is there something that unites all those subatomic particles? No one is sure, but many physicists are betting on Democritus and his hypothesis. They are searching for the smallest unifying particles within all matter. So far, there are clues but no proof.

The Greeks inferred that atoms existed and that atomic action and reaction helped account for change. Their information was scattered, but promising. And then, for religious and cultural reasons, scientific progress stopped for almost 1,000 years.

Besides, brains and imagination can only take you so far in science, and then you hit a wall. Without the technology to experiment and test, you can't confirm your theories. That was the problem the Greeks faced. There didn't seem to be anyplace to go with science. Democritus was astonishing, but he couldn't prove his ideas; no one knew if they were right or wrong. It was hopeless to look for atoms: if they existed, they were too small to be seen. And studying the stars with your naked eye? You miss most of the universe.

So the next generations headed in a different direction. They began to study human emotions and thought. The

The word NUCLEUS (NOO-klee-us) is Latin for "little nut." Like a nut in a shell, it's a mass at the core of something—a comet, a skin cell, an atom. The plural is NUCLEI. NUCLEAR FISSION involves splitting the nucleus of a heavy atom, such as uranium (92 protons, 92 electrons), into light atoms, such as barium (56 protons, 56 electrons). In NUCLEAR FUSION, the nuclei of light atoms join to make a heavy nucleus. Both fission and fusion create energy—as in an atomic bomb or power for a city.

Even today, you can't see atoms with a light microscope (one that magnifies objects in visible light). It wasn't until the 1980s that the first images were taken with an electron microscope. This amazing machine scans a surface, and when it detects an electron—a subatomic particle—it records a dot on a map. Dot-by-dot, the electron microscope maps out an image of atoms. Artists used the technology to create this electronic work of art, titled *Quantum Corral.* They arranged 48 iron atoms into a ring, trapping electrons inside.

Ionian idea—that the universe is knowable and follows orderly laws men and women can understand—that idea was mostly set aside (to be rediscovered 1,000 years later). Almost all the books written by the Ionians were lost or destroyed. Their ideas have come to us in fragments or in the writings of others.

Socrates (SAHK-ruh-teez), who was called the wisest man in the world by the oracle at Delphi, turned from physical science to a study of the human soul. "Know thyself," he told his followers. Those words, which may have come from the oracle, are good advice, but they don't do much for scientific research.

Democritus is said to have gone to Athens to meet Socrates. (They were only five years apart in age.) But when he got there, he was too shy to introduce himself. Too bad. Socrates never got interested in atoms. Neither did his famous student Plato or Plato's famous student Aristotle. They all thought that the "elements" could be cut up endlessly—and that you'd never get anything else. They didn't believe there was any underlying basic particle. They thought earth, air, fire, and water were as basic as you could get. Democritus's hard, impenetrable atoms didn't have much appeal.

Aristotle, who was born when Democritus was a very old man, became the most renowned of all the Greek scientists. He was an organizer, a classifier, and an original thinker too. But he never believed in atoms.

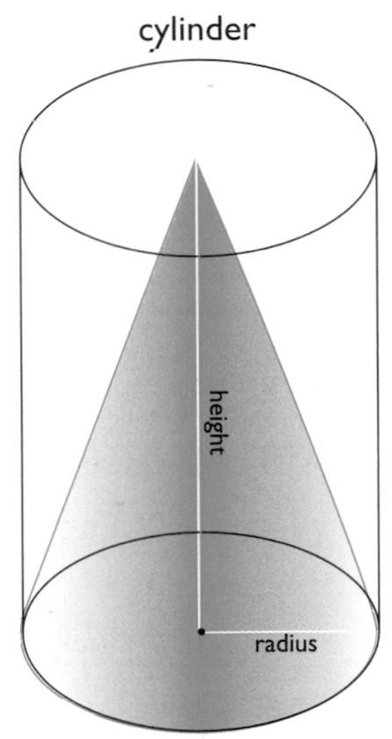

Democritus developed a formula for the volume of a cone by mathematically stacking circles that were smaller and smaller, one on top of another. Other Greeks built on his thinking. Unfortunately his writings were destroyed by religious zealots, so we can't read them. But we do know his formula for a cone: Multiply the area of a circle (πr^2) by its height and then divide by 3. (If you forget to divide by 3, you have the volume of a cylinder.) It is illustrated above.

This Syrian painting appears in an early thirteenth-century book of sayings. It shows Socrates (seated on a rock) with his students. All three wear Arab clothing, even though the subject of the painting is Greek.

ODE TO AN ATOM

Lucretius (loo-KREE-shuhs), a Roman poet who lived in the century before Christ, took Democritus's idea about atoms and used it as a central idea in a remarkable poem. Here is some of it. Note what he has to say about atoms. It will take about 17 centuries for atoms to get their next champion. Today, atomic theory is central to science. To say Democritus and Lucretius were ahead of their time is a big understatement.

On the Nature of the Universe (or **On the Nature of Things**)

So you can see
That actions never exist all by themselves
As matter does, or void, but rather are
By-products, both of matter and of space.

Bodies are partly basic elements
Of things, and partly compounds of the same.
The basic elements no force can shatter
Since, being solid, they resist destruction.
Yet it seems difficult to believe that objects
Are ever found to be completely solid.
A thunderbolt goes through the walls of houses,
As noise and voices do, and iron whitens
In fire, and steam at boiling point splits rocks,
Gold's hardnesses are pliant under heat,
The ice of bronze melts in the flame, and silver
Succumbs to warmth or chill, as our sense tells us
With the cup in our hands, and water, hot or cold,
Poured into wine. No, there is nothing solid
In things, or so it seems; reason, however,
And science are compelling forces—therefore
Stay with me; it will not take many verses
For me to explain that there are things with bodies
Solid and everlasting; these we call
Seeds of things, firstlings, atoms, and in them lies
The sum of all created things.

Here's part of the atom poem by Lucretius in Latin, hand copied and illustrated in the fifteenth century for Pope Sixtus IV. Though Lucretius, an ancient Roman, was considered a pagan, his words spread across Christian Italy.

A COMPOUND is two or more elements joined together into a substance that's the same throughout. A compound that we all know and drink is water—two hydrogen atoms and one oxygen atom (H_2O).

11 Aristotle and His Teacher

The fate of empires depends on the education of youth.
—Aristotle (384–322 B.C.E.), Greek philosopher

Whereas the rattle is a suitable occupation for infant children, education serves as a rattle for young people when older.
—Aristotle, *Politics*

Amicus Plato, sed magis amica veritas. **("Plato is dear to me, but dearer still is truth.")**
—Latin words ascribed to Aristotle, who said them in Greek

By the time Aristotle came along—almost 200 years after Pythagoras—sky gazers had figured out that the round shape on the Moon during a lunar eclipse is our planet's shadow. Since those sky gazers were apt to be the wisest of scholars, it was a heavy nail in the flat-Earth coffin.

An astronomer friend tells me sky gazers *still are* the wisest of scholars!

There were others. The Greeks traveled widely in the Mediterranean lands. They knew that the North Star is lower in the sky in southern Egypt than it is in northern Greece. A round Earth explains that. And they also knew that the first one sees of a ship coming over the horizon is its sail; only later does the hull come into view. That seemed to confirm the round-world theory.

But other puzzles in the sky needed answering. Aristotle rolled up his sleeves—well, that may have been difficult in a flowing toga—whatever he did, he got to work.

Aristotle was another brainy Greek, perhaps the brainiest of them all. When he was born, in 384 B.C.E., Greece was in the midst of a period of creativity that would change the world forever. (It's known as Greece's classic period.)

A nineteenth-century marble sculpture captures a young Aristotle lost in thought.

The Earth's shadow doesn't fall on the Moon any old time. A lunar eclipse happens only during a full moon, when Earth is directly between the Sun and the Moon (see page 27). The ancient Greeks watched without telescopes as this shadow passed over the full moon (it happens in minutes) and saw that the edge was slightly curved. They concluded that Earth, like its shadow, must be curved, too.

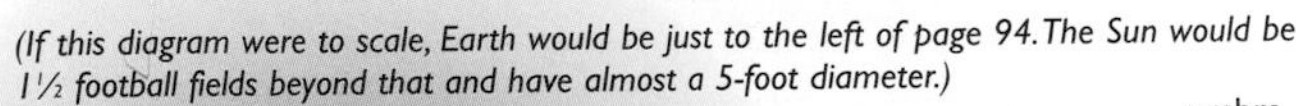

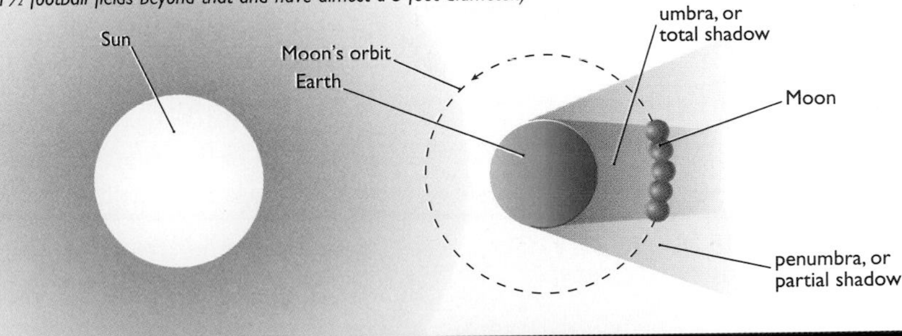

Classical Greek achievements in art, architecture, literature, and philosophy set standards of beauty and excellence that still leave us awed. We, today, are what we are in great part because of the ancient Greeks. (Look at a Greek statue from the classic period and see what you think.)

That classic period was long after Thales' time, and although the early Ionians were not forgotten, they must have seemed a bit old-fashioned. New winds were blowing: power and influence had crossed the Aegean Sea from the Ionian coast of Turkey to mainland Greece.

Aristotle was born in Stagira, a northern city located on a mountainous peninsula with fingers that extend into the Aegean Sea. He was fortunate in being able to go to good schools where he studied the work of both Thales and Pythagoras. Then, at age 18, he went off to Athens, the leading city-state in Greece, to study at the academy of the greatest philosopher of his day—Plato.

Unlike the Ionians, Plato (ca. 427–347 B.C.E.) didn't much trust his eyes and ears and other senses. If he wanted to understand something, he withdrew to the world of pure thought. Plato was looking for beauty, truth, and clarity. He didn't believe they could be found on Earth, so he urged his students to study mathematics and the stars. In both studies, he discovered patterns and order that seemed astonishing.

This series of photos shows the total lunar eclipse of January 21, 2000, which lasted for 1 hour and 18 minutes. The photos were snapped at 20-minute intervals. The Moon is in full phase throughout; the shifting darkness on its surface is the Earth's shadow. The umbra (the darkest part of the shadow) doesn't obscure the Moon completely, because Earth's atmosphere refracts (bends) some sunlight onto the Moon's surface. As the light passes through the atmosphere, the blue wavelengths scatter, so the eclipsed Moon appears orange.

A revered Greek named Academus once owned the land under Plato's Academy. Ever since, the word academy has been used for schools.

Lewis Carroll, who wrote *Alice in Wonderland*, dedicated his book *Symbolic Logic* to the memory of Aristotle, which isn't surprising: Aristotle is called the Father of Logic. Logic is a way of using the mind to solve problems. *Symbolic Logic* was written for young people. I recommend it.

Mathematics had a purity that lifted it above mundane, flawed earthly things. As to the stars, he thought they were made of different stuff from the Earth and were an example of God's perfection.

Plato was influenced by Pythagoras's mathematics, which focused on perfect shapes and perfect harmony. Plato was hooked on perfection. He thought about the perfect table, the perfect cat, the perfect flower—or the perfect whatever. He was searching for "ideal forms." He knew perfection can't be found on this Earth, but he believed if we try to imagine perfection and then strive for it, we will lead the best lives possible. It is his search for the ideal—the most beautiful and harmonious—that makes Plato a favorite of artists, poets, and mathematicians.

Plato's real name was Aristocles, and he was a descendant

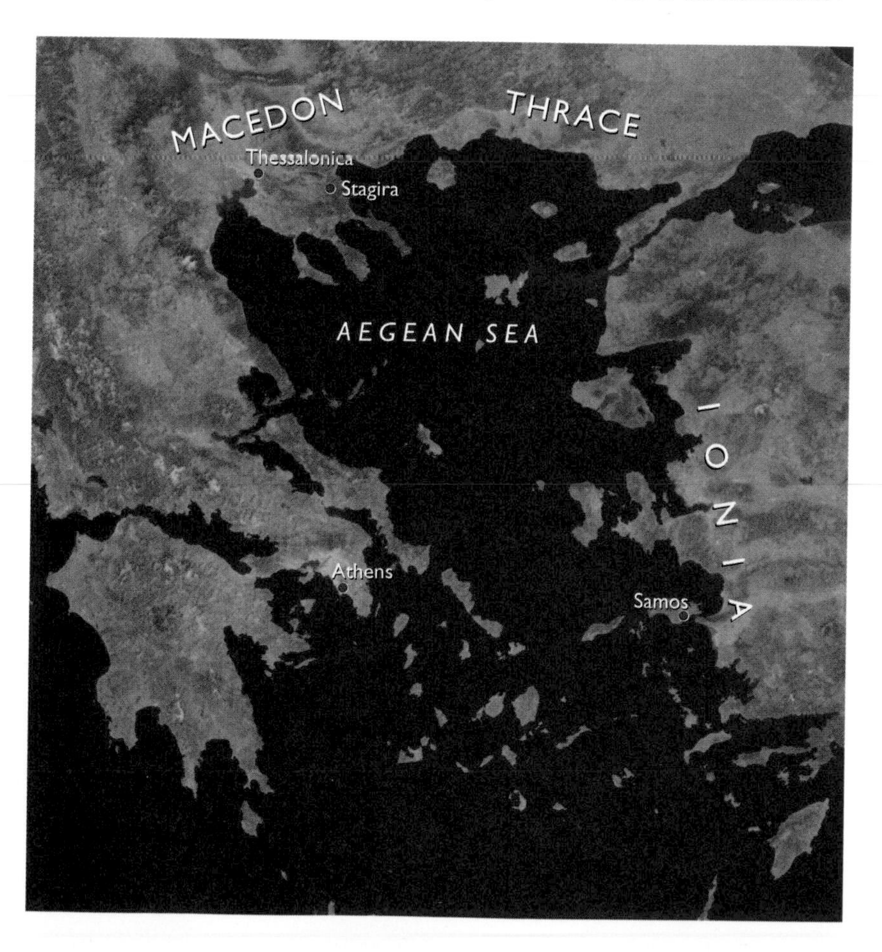

of the early kings of Athens. Because he had broad shoulders, he was nicknamed Platon in school (it means "broad"), and that's the name he carried through the rest of his life. As a young man, Plato traveled and got involved with politics. Later he devoted himself to philosophy, writing, and teaching. His writings are mostly in the form of dialogues—discussions between his teacher Socrates and others.

DUELING IDEAS

The differences between Plato and Aristotle are important; they keep appearing in the history of ideas. Plato was searching for "ideal forms." Aristotle asked questions and then examined existing objects. Plato's philosophy led to deep thinking; Aristotle's, to observation and eventually to experimentation.

Both are necessary. But should one method carry more weight than the other?

Plato (left) and Aristotle talk in Raphael's *The School of Athens* (see also page 73).

The safest general characterization of the European philosophical tradition is that it consists of a series of footnotes to Plato.

—Alfred North Whitehead (1861–1947), English philosopher, *Process and Reality*

GREEK MINDS MEET AGAIN

Plato was a student of Socrates and a big admirer of Pythagoras. Aristotle was a student of Plato. So what's the point? Young people learn from old people, but sometimes it's the other way around. Science needs new, fresh ideas and approaches to keep moving ahead. It needs the wisdom of experience to use those ideas well.

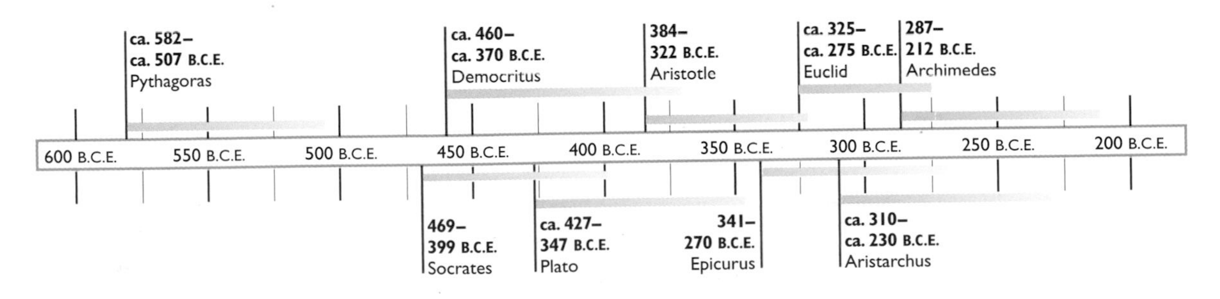

PLATO, MATH, AND PERFECT NUMBERS

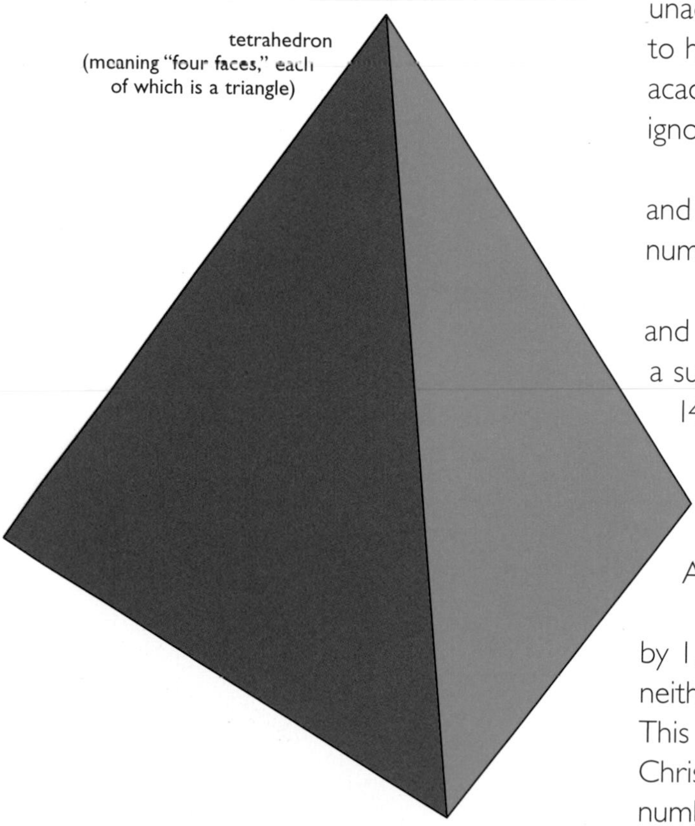

tetrahedron
(meaning "four faces," each of which is a triangle)

Plato looked down his aristocratic nose at most natural philosophy (what we call science). It seemed too practical for his mind. He was searching for pure knowledge. So mathematics, which he considered unadulterated thought, had a special appeal to him. Over the door of his famous academy were the words "Let no one ignorant of mathematics enter here."

Plato was fascinated with perfection, and he identified what he called perfect numbers. What's a perfect number?

Consider 14. It can be divided by 1, 2, and 7. Add up those divisors, and you get a sum that is less than 14. Thus, said Plato, 14 is *not* a perfect number. It is known as excessive.

How about 12? Its divisors (1, 2, 3, 4, 6) add up to more than 12. Another imperfect number—a defective.

Now, try 6. It can be divided evenly by 1, 2, and 3. Add them up and—ahh—neither excess nor deficiency. Perfection. This idea outlasted the ancient Greeks. The Christian Saint Augustine wrote: "Six is a number perfect in itself, and not because

Archimedes (287–212 B.C.E.) discovered a total of 13 solids. One was a soccer ball—except that he called it a truncated icosahedron (eye-KAH-suh-hee-druhn), meaning it's an icosahedron with the corners lopped off. What did Archimedes' other 12 solids look like? Think about it along the way to chapter 17.

God created all things in six days; rather the converse is true. God created the world in six days because this number is perfect."

The next perfect number after 6 is 28. Then comes 496. After that is 8,128. And after that, 33,550,336. All of them are even numbers, and every perfect that we know ends in 6 or 8.

Now if you're hooked on perfection—and some numbers appear that not only seem imperfect, but are totally weird and unexplainable—well, that could be disturbing. And that's just what happened when the Greeks ran into those irrational numbers (see page 82). You can understand why they found them so upsetting.

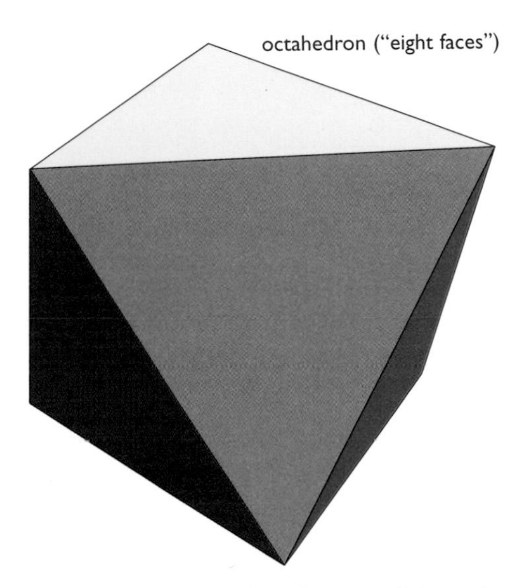

octahedron ("eight faces")

It was actually Pythagoras who discovered these shapes, but Plato made them famous in a dialogue called *Timaeus*, and they are now called Platonic solids. Plato said these are the five (and only five) regular solids with equal lines and angles and faces (all triangles, for example). He saw these solids as perfect shapes, and he was sure the heavenly bodies are exact geometric forms like them. He believed the stars and planets exist in perfect geometric harmony. (Not true, but geometry helps a lot in understanding them.)

hexahedron
("six faces"—Go ahead and call it a cube.)

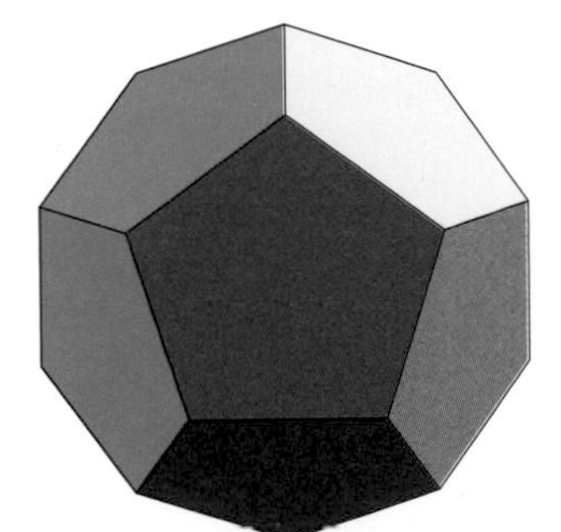

dodecahedron
("twelve faces")

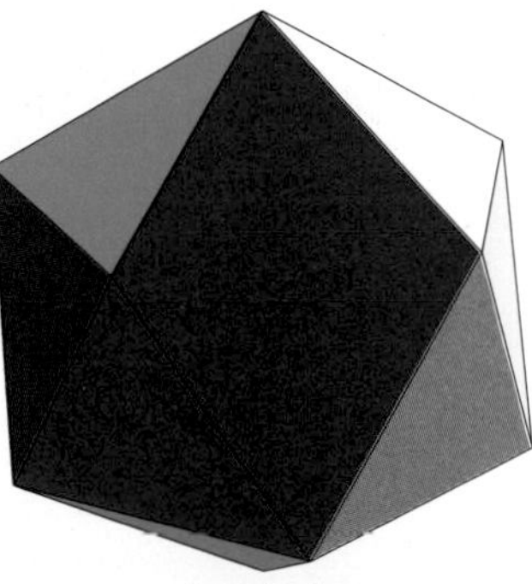

icosahedron
("twenty faces")

An ancient Roman floor mosaic in Pompeii, dating from the first century B.C.E., depicts Plato (in the center, pointing a stick) with his students at the academy.

A PREMISE is a statement or idea backed up by arguments. The Latin verb *mittere* means "to send," and so a premise is something sent before.

A marble bust of Alexander the Great (356–323 B.C.E.), made two centuries after he lived, shows him as a youth. Aristotle tutored the soon-to-be "King of Kings" in science and philosophy.

They discussed government, education, ethics and other topics. Plato is probably the most widely read philosopher of all time. As to the academy he founded, it is a model that educators still turn to, agreeing or disagreeing with its premises, but unable to ignore them.

Aristotle was the most renowned of all the academy's students. Plato is said to have called him "the intelligence of the school."

But while Plato was obsessed with mathematics and perfection, scientists toil with the real world, and Aristotle was a scientist. So although he was pulled by Plato's ideas of purity and abstraction, Aristotle had a practical, feet-on-the-ground mind. Aristotle was fascinated with the everyday world around him.

Maybe it was his father's legacy that made Aristotle think as he did. His father had been court physician to the king of Macedon, but he died when his son was young. Doctors consider problems one at a time. Aristotle would become famous for his well-organized, practical mind. He developed the principles of logic—a system of reasoning that was an orderly way to study astronomy, biology, chemistry, and most of the other sciences. But before he did that, he stayed for many years studying with Plato.

Then, after Plato died, Aristotle was summoned north by King Phillip II of Macedon. (Athens and Macedon were rivals, so this must have complicated the move.) King Phillip wanted the now-famous son of his father's physician to teach his heir, Prince Alexander. That royal tutoring job lasted for eight years and is said to have profoundly influenced the future ruler. When Phillip died, Alexander marched off to become a famous conqueror who would be known as Alexander the Great, a "King of Kings."

Meanwhile, Aristotle had gone back to Athens and started his own academy. He was said to be miffed that Plato had not appointed him his successor. The ideas of those two men—Plato and Aristotle—would eventually lead to different schools of thought. Sometimes there would be harmony between them; sometimes, conflict.

Platonists focus on the life of the mind and those ideal forms. Aristotelians observe the world around them and use

These ruins in ancient Mieza, which was near Thessalonica in Macedon (see map on page 96), are believed to be the remnants of a school for boys from ages 10 to 15. Aristotle taught at the Mieza boys' school from 343 to 340 B.C.E. before joining the faculty at the Lyceum in Athens.

Aristotle sparked in Alexander an omnivorous interest in the world around him. An illustration from a fifteenth-century French manuscript dramatizes Alexander's mythical descent in a glass barrel (a crude diving bell) into the Indian Ocean to observe undersea life.

that as a guide for their thoughts. Keep in mind, in ancient Greece, no one seemed to understand the importance of experiments as a way to prove ideas. For the Greeks, including Aristotle, all profound truths could be found in the mind. (Scientists today observe nature, use their minds to form hypotheses, and do experiments to prove or disprove their ideas. It's that last step that the Greeks never quite got. Maybe they couldn't: they didn't have the technology to go far with experimenting.)

Aristotle didn't limit himself to one subject; he considered everything possible—poetry, art, music, math, warfare, ethics, religion, and science. He had a mind like an encyclopedia. Aristotle has been called a great synthesizer. (That word means more than an electronic keyboard.) He made lists of *everything* he could find in nature, and then (this is important) he organized, analyzed, and connected that knowledge. It was a monumental achievement.

Among other things, you can think of Aristotle as the

world's first great biologist. He dissected hundreds of specimens and then wrote about what he saw. He looked at a three-day-old chicken egg, watched the embryo's tiny heart beating, and saw it pumping blood to vessels extending into the yolk. Many of his contemporaries thought he was wasting his time. They believed that only humans were worth studying. Aristotle said, "If any person thinks the examination of the rest of the animal kingdom an unworthy task, he must hold in like disesteem the study of man."

DISESTEEM is low regard or low opinion. **SELF-ESTEEM** is about how you regard yourself. Someone or something you **ESTEEM** is respected or prized.

Aristotle's student, the great Alexander, appreciated what his teacher was doing. So when he became king, Alexander sent men around the Greek world and beyond collecting animals. Aristotle put them in what is thought to be the world's first zoo. That gave Aristotle access to a variety of animals that no one had seen together before.

"We must not have a childish disgust for the examination of less admirable animals. For in all natural things there is something wondrous," he wrote.

But knowing the parts of a living thing was not enough for Aristotle: he also wanted to understand how the whole is put together. "For we should not be content with saying that the couch was made of bronze or wood, . . . but should try to

WHERE ARISTOTLE HAD IT WRONG

Aristotle thought the Earth was at the center of the universe. He said gravity was something that made all objects seek the center of the Earth. He believed the natural state of all objects is to be at rest unless moved by force. He thought heavy objects fall faster than light objects. (This moon experiment, at right, proved otherwise.)

None of that is true. But because Aristotle was so brilliant and was right on many things, his ideas led thinkers astray for many generations. Aristotle came up with some wrong answers, but the important thing to remember is that he asked questions and organized data. A good process for thinking and observing helps us eventually find the right answers—and Aristotle gave us that.

In 1971, astronaut David Scott refuted Aristotle by dropping a hammer and a feather, which both hit the Moon's surface at the same time.

describe its design or mode of composition.... For a couch is such and such a form embodied in this or that matter." Aristotle realized that life is more than a collection of parts.

Like the other Greeks, Aristotle thought about light and vision. The Pythagoreans had believed a "visual ray" came out of the eye and hit an object, and that caused sight. (It's actually the other way around: light reflects off objects and enters the eyes.) Empedocles said the eye was like a lantern with an internal fire that went out to illuminate the world. But Aristotle questioned that. (He was a great question asker, as are all good scientists.) He asked: If the eye is a lantern, why don't we see at night? Although he didn't get it right, Aristotle thought hard about light and vision. (Light, so vital to life and yet so difficult to capture and analyze, would become central to twentieth-century physics.)

When it came to astronomy, Aristotle's ideas were mostly wrong, perhaps because he accepted the idea of heavenly spheres, which those who came before him had theorized. Aristotle even added to the total of spheres. He thought there were 54 of them out there rotating and holding the stars and planets. He agreed with the Pythagoreans and Plato that Earth and heaven are different realms following different laws of nature. (Today we know that's not so.) Those wrong thoughts of Aristotle's would be accepted as truth for centuries and centuries to come. Often they would hold back scientific progress. Despite that, we owe him a great debt for the breadth of his vision and the depth of his work.

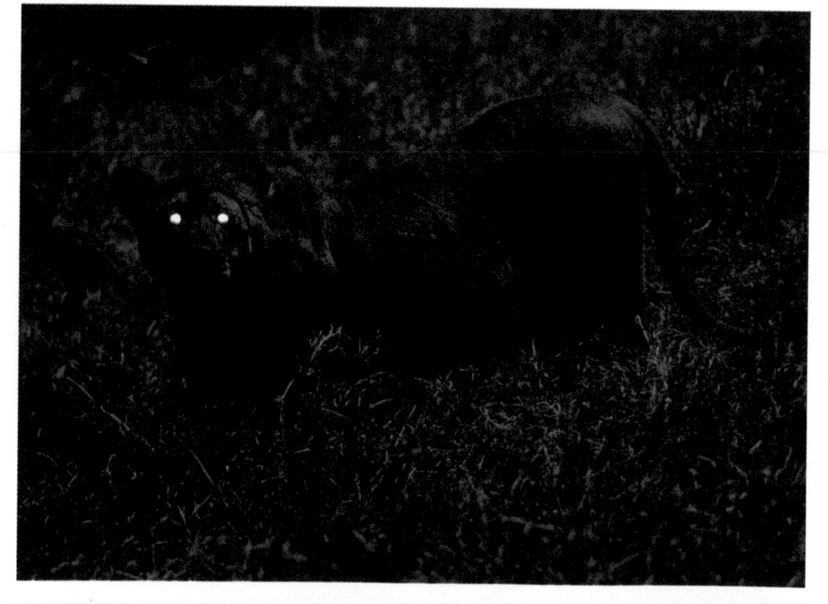

When animal eyes are shining, it's easy to believe there are "lanterns" or "visual rays" inside. But put this lioness in a pitch-dark room (no light whatsoever), and you won't see eyes—or anything. Neither will the lioness. Light has to be present and entering our eyes in order for us to see.

Aristotle took the ideas of those who had come before him and melded them into a grand theory that attempted to explain

Getting to Know Socrates

He wasn't a scientist, but Socrates was certainly someone to know. We're told that he was ugly—short and stout with a big nose—but his personality, wit, intelligence, and good humor won him wide admiration. He was a talker, and his conversation was famously fascinating. He didn't care about clothes or wealth or worldly goods. So maybe his wife, Xanthippe, had reason to complain. She has come down to us as a shrew.

French painter François-Louis-Joseph Watteau (1758–1823) shows Socrates calmly reaching for a cup of poison while his grief-stricken disciples helplessly look on.

We think Socrates was a student of Anaxagoras. But he was far more interested in questions of ethics (ways to behave) than in learning about the universe. We know he was uncommonly brave. He didn't believe in the ancient Greek gods, who were revered, and he didn't think much of democracy as a system of government. Remember, he was a great talker and he influenced people. How would you feel about him if you were an official in Athens's democracy? Socrates was accused of atheism and treason and corrupting the young; he was brought to trial.

Perhaps Socrates could have saved himself, but he refused to do so. He didn't speak in his defense, and a jury of 500 men sentenced him to death. Then he refused to escape when it might have been easy and calmly drank a cup of poison hemlock. His pupil Plato would spend much of his life trying to understand and explain the meaning of Socrates' life and death. We're still considering issues he raised.

and classify everything known. It was an enormous accomplishment, and it set a base for science that is still with us today. Even though many of his ideas turned out to be wrong, the important thing is that he gave thinkers a starting point—something to work with and examine and agree or disagree with. Which is exactly what good thinkers do.

12 Does It Change? No Way, Says A

The movement of that which is divine must be eternal. But such is the heaven, viz. a divine body, and for that reason to it is given the circular body whose nature it is to move always in a circle. Why, then, is not the whole body of the heaven of the same character as that part? Because there must be something at rest at the center of the revolving body.... Earth then has to exist; for it is earth which is at rest at the center.

—Aristotle (384–322 B.C.E.), Greek philosopher, *On the Heavens*

It is not once nor twice but times without number that the same ideas make their appearance in the world.

—Aristotle, *On the Heavens*

Aristotle thought that the earth was stationary and that the sun, the moon, the planets, and the stars moved in circular orbits about the earth. He believed this because he felt, for mystical reasons, that the earth was the center of the universe, and that circular motion was the most perfect.

—Stephen Hawking (1942–), British physicist, *A Brief History of Time*

Funny thing about Aristotle: he may be the most influential scientist of all time, but today he wouldn't win a popularity contest with most cosmologists. They say he held back progress in their field for almost 2,000 years. See what you think.

Aristotle believed that the universe is fixed, motionless, and everlasting. "There is necessarily some change in the whole world, but not in the way of coming into existence or perishing (for the universe is permanent)," he wrote.

And, "In the whole range of time past, so far as our inherited records reach, no change appears to have taken

Cosmologists aren't experts in makeup. (Makeup artists are cosmetologists—as in "cosmetics.") Cosmologists are experts on the cosmos—the universe and the way it works.

In 1660, a teacher and mathematician named Andreas Cellarius published a stunning color atlas of the universe called *Harmonia Macrocosmica* (*"Harmony of the Universe"*) in Amsterdam, Holland. His hand-engraved maps include both Earth-centered (Ptolemaic) and Sun-centered (Copernican) systems. Here, orbiting the Earth (closest to farthest), are the Moon, Mercury, Venus, the Sun, Mars, Jupiter, Saturn, and the zodiac constellations.

place either in the whole scheme of the outermost heaven or in any of its proper parts."

The universe is the way it is, and it has always been that way, and that is that, said Aristotle.

The challenge was to understand it. He led the way when he created a model of the Earth and the stars. Aristotle's plan of the universe seemed to make sense, especially after he explained it in a book called *On the Heavens*, which he wrote in 350 B.C.E.

A heavy body, such as Earth, can't possibly move, said Aristotle, who rejected Pythagoras's idea of the Earth as a

Jesus Christ, as the Prime Mover, sits on his private sphere at the edge of the universe and manipulates heavenly bodies at will. This twelfth-century Byzantine mosaic decorates the Cathedral of Monreale in Sicily, Italy.

planet in motion. In Aristotle's cosmology, Earth stands dead still in the center of the universe with the Sun and stars and planets all circling it while attached to perfect—and perfectly clear—hard crystal spheres. Those encircling cosmic spheres nest one inside the other—54 of them—to create an elegant, shimmering universe. He said the farthest sphere belongs to the Prime Mover: God.

There was one problem in the heavens that no one had been able to explain; it had bothered Plato a lot: the planets follow strange, awkward paths. Usually, they appear a little farther east in the sky from night to night, compared to the fixed backdrop of stars. But sometimes, the planets seem to go backward, stepping from east to west for a while—a process called retrograde motion—and then they turn back around again (see pages 112–113). Plato, who was much influenced by Pythagoras, saw it as a mathematical challenge. Someone had to make sense of those wiggly orbits. Aristotle tried to solve the problem of the planets by attaching each of them to its own special planetary sphere that followed a different route from the starry spheres. It was a good try—for a long time the best there was—and it seemed to work pretty well (although it was very complicated). Some other Greeks put the planets on jointed rods that each had its own cycle.

RETROGRADE comes from the Latin root for "step back." Something that's "So retro!" is a blast from the past, a step back in time.

Today we know all that is wrong, but that's because we've learned that the reason the planets sometimes seem to be moving backward is because they are being viewed from a moving planet (Earth). Since the ancients didn't realize we are moving, they never could understand that retrograde motion. No one could feel the motion of the Earth, so no one

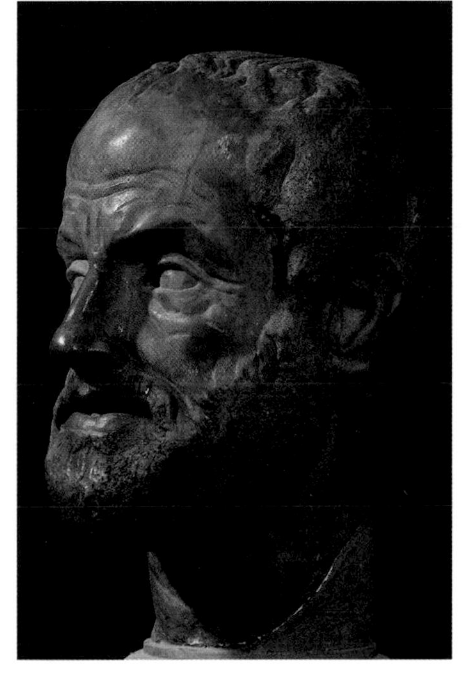

The science of Aristotle was particularly important in biology and taxonomy (classifying things). He and another important Greek, Hippocrates, helped create those sciences. Aristotle (shown here in a second-century B.C.E. bust) observed nature accurately and came up with logical theories. Today's science is more empirical—centered on experimentation.

A LITTLE LEVITY

Aristotle believed the Earth had four elements and two forces:

- **gravity**—downward motion—which he said makes the elements earth and water sink, and
- **levity**—upward motion—which makes the elements air and fire rise.

We still divide the universe into elements and forces. Aristotle's gravity had nothing to do with stars and planets. Today we believe it does, but we're still trying to find out just how gravity works. It's a hot topic in the world of science. And levity? Today that's a joke.

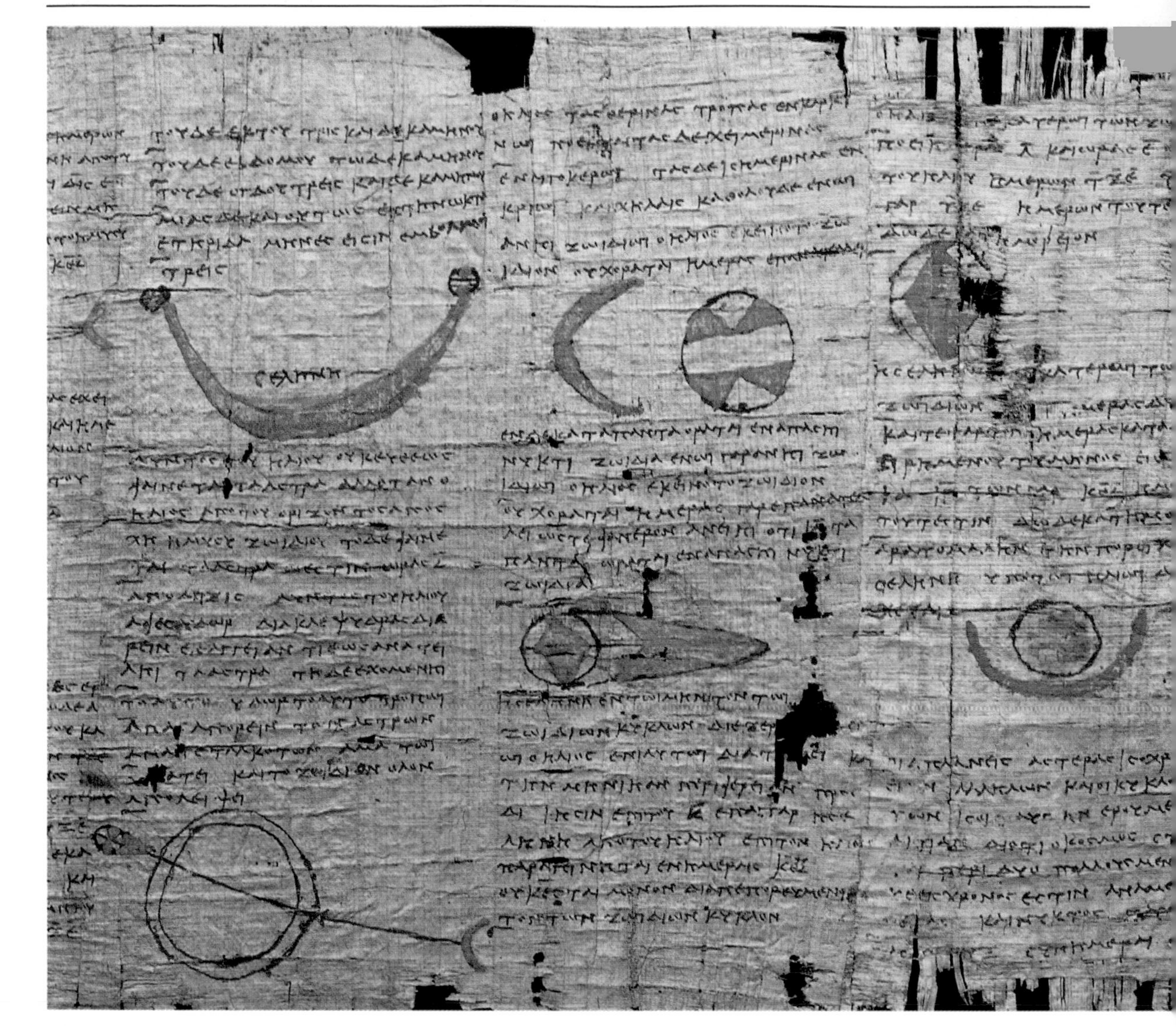

No words written by the hand of Eudoxus survive today. But this second-century B.C.E. papyrus, made 200 years after he lived, describes his geometric approach to astronomy. Eudoxus built one of the first known observatories in his hometown of Cnidus (now in Turkey).

figured out that it was happening. As to the spheres, if you didn't know about gravity, how would you explain the way the stars stay in the sky? The Greek idea of hard crystal spheres was a reasoned guess. It just didn't happen to be right.

Aristotle made his model of the universe a model of perfection. After all, he was a student of Plato and, although Aristotle didn't agree with him on many things, Plato was a big influence on his thinking. The heavenly bodies are uniform and perfect, Aristotle said. And since the Greeks thought that circular motion is perfect, Aristotle believed

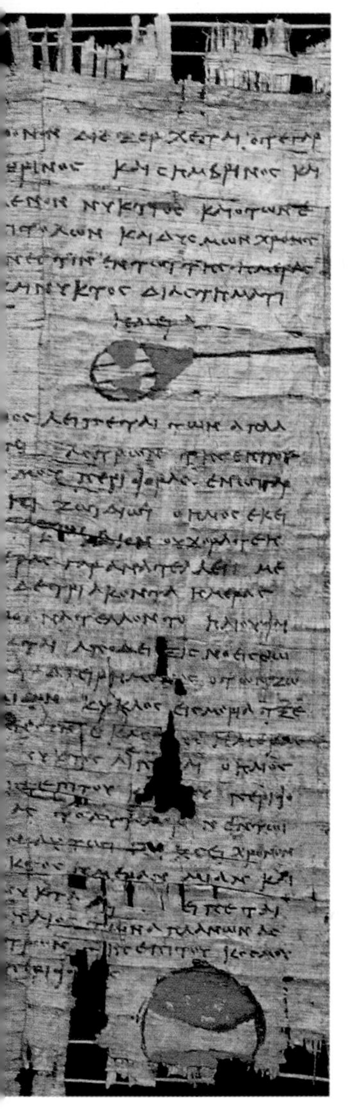

the stars and planets must follow perfectly circular paths.

So in his world, the heavens were perfect and the stars and planets were made of a perfect element, different from anything on Earth—a divine, everlasting fifth element called "aether." Everything about the heavens was eternal, perfect, and static, he said.

AETHER (also spelled ETHER) comes from the Greek root for "burn or glow." It was the perfect element—pure and unchanging. To modern chemists, it's a colorless compound of alcohol and sulfuric acid, once used to knock out patients before surgery.

One piece of Aristotle's picture was not ideal: Earth. The earthly elements—earth, air, fire, and water—are less than perfect. A look around the Earth shows that it is not static; it is a place of change: of birth, growth, decline, death, and decay. According to Aristotle, the celestial world and the earthly world are completely separate and different.

Astronomy compels the soul to look upwards and leads us from this world to another.
—Plato (ca. 427–347 B.C.E.), *The Republic*

All this would turn out to be wrong, but it's important to know about it because for a long, long time everyone believed it. (Besides, it helps you be aware that everything that science says, even today, is not necessarily right.) Aristotle's wrong ideas would lead to other wrong ideas and to some scientific dead ends. Aristotle and those of his time didn't have any technological help—not even a telescope. Imagine yourself in their sandals. All they had to work with were their brains, their eyes, and mathematics. They did astonishing things with what they had.

WHY MARS IS A LITTLE LOOPY

I**f you happen to spot Mars** one night, keep watching. The bright, orange dot looks like one of thousands of cities peppered on a giant map. If you watch long enough—more than a few minutes—you'll notice that this whole sky map, Mars included, appears to slowly move east to west across the night sky.

So, if both stars and planets travel in unison, why did the Greeks call planets "wanderers"? Stay tuned. Look for Mars again a couple weeks later. You'll see that the orange dot isn't like a mapped city after all. It's more like a very slow-moving bug that crawls around the fixed star map in a slow, quirky path.

Tunç Tezel, a Turkish engineer, caught the planet-watching bug as a kid. Now he patiently records their wanderings night after night on film. He photographs the same spot in the sky over several months and then combines the photos into one image. Notice, in his photo series of Mars in 2001 (right), that the stars never change position in relation to each other. On February 18, 2001, Mars looked fixed in place, too, next

EYE TRICKERY AND MOVING OBJECTS

You're in the backseat of a car. When you look out the window, you see trees, buildings, and telephone poles moving by. Of course, you know your eyes are playing tricks on you. The trees can't be moving. It's you and the car that are in motion, and your motion gives a false illusion of speed to things outside the window.

That's exactly what happens when you look at the night sky. All the stars and planets seem to be moving at the same speed. There can only be two reasons for that: Either they really are all moving at a uniform speed, which means they must be attached to a crystal sphere, or it's we on Earth doing the moving (and our eyes are tricking us). A few ancient thinkers understood this. But not many. So for much too long, that crystal-sphere notion won out.

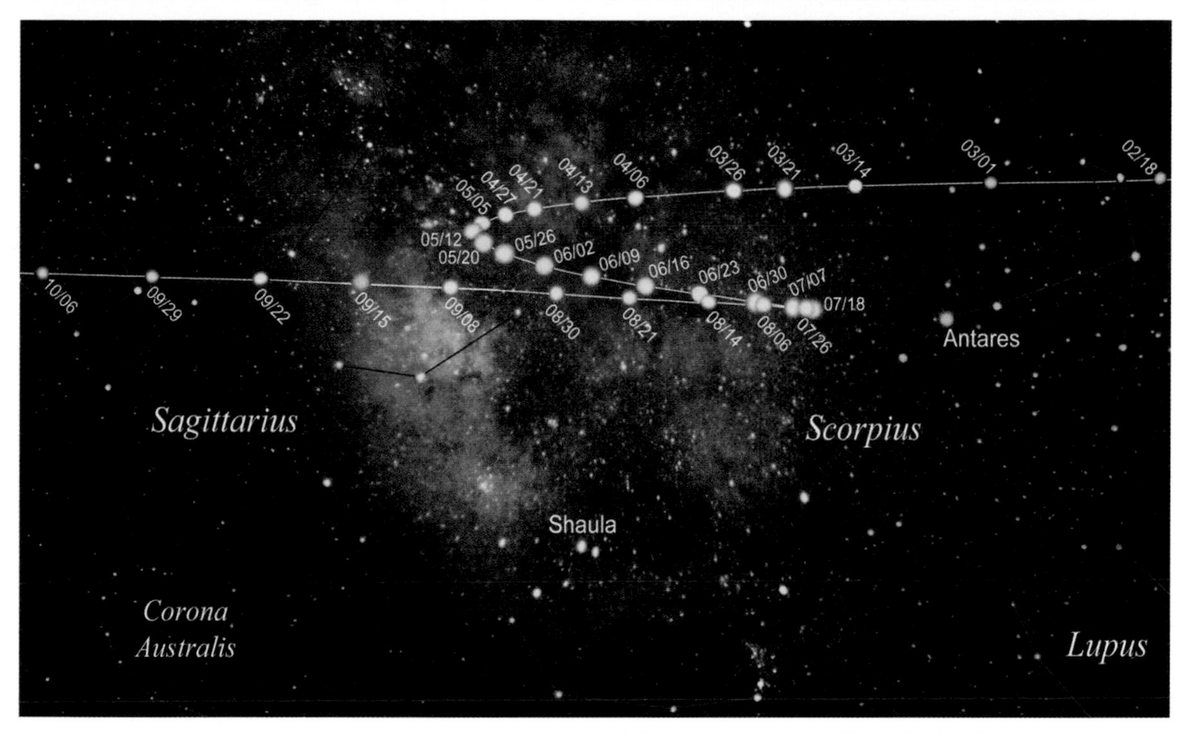

A big fan of astronomy, Tunç Tezel of Turkey patiently took these photos of Mars from February to October of 2001. He's a civil engineer by day and a photographer of the stars by night.

to the tail of the zodiac constellation Scorpius (the Latin name for Scorpio). The Red Planet stayed there all night long. But, by March 1, 2001, Mars had wandered to the opposite side of Scorpio's tail.

Over the next couple months—through March, April, and into May—the planet crept closer to the next zodiac constellation, Sagittarius. Then, in mid-May, something weird happened. The planet began to loop in the opposite direction—away from Sagittarius and back toward Scorpio—in a retrograde (backward) motion. On July 18, Mars took another sharp turn and headed back east toward Sagittarius.

Now, *that's* a wander. Mars and the other planets don't *really* zigzag through space, so why do they appear to do so in our night sky? Many ancient astronomers struggled to explain the loopy motions with complex schemes that never quite worked perfectly. The most lasting of these clever but imperfect solutions was Ptolemy's epicycle plan (see page 178). He envisioned little, curlicue orbits within a big orbit around an unmoving Earth.

For planetary paths to make perfect sense, astronomers first had to accept that the Earth is moving, too—and at a different speed than the other planets. Retrograde motion happens when Earth passes a slower planet (like Mars) or when a faster planet (like Venus) passes us. Our view of the planet shifts dramatically in relation to the fixed stars behind it, something like passing a slow jogger while riding a fast merry-go-round.

13 Aristarchus Got It Right—Well, Almost!

The time will come when diligent research over long periods will bring to light things which now lie hidden. A single lifetime, even though entirely devoted to the sky, would not be enough for the investigation of so vast a subject....And so this knowledge will only be unfolded through long successive ages....Nature does not reveal her mysteries once and for all.

—Lucius Annaeus Seneca (ca. 3 B.C.E.–65 C.E.), also known as Seneca the Younger, Roman philosopher born in Spain, *Natural Questions*

Anyone who thinks it reasonable for the whole universe to move in order to let the earth remain fixed is as irrational as one who climbs to the top of a cupola just to view the city and its environs, and then demands that the whole countryside revolve around him so that he does not have to take the trouble to turn his head.

—Galileo Galilei (1564–1642), Italian scientist, *Dialogue Concerning the Two Chief World Systems*

Aristotle's universe—with Earth in its center—was orderly and reasonable. When we look at the world around us, it seems that Earth is standing still and that the Sun and stars are moving. It's what our senses tell us. It's what Aristotle said. So that's the picture of the universe that most people in the ancient world kept in their minds.

But not Aristarchus, who lived in the century after Aristotle. We think he was born about 310 B.C.E., and he may have died in 230 B.C.E.; we know he was in his prime about 270 B.C.E. Aristarchus was a brilliant mathematician and a meticulous observer of the stars—right out of the Ionian tradition, which isn't surprising, because he was born on Samos.

METICULOUS means "careful about details." But originally, when it was the Latin word *meticulosus*, it meant "fearful and dangerous." Can you see why not getting the details right might make you fearful?

Compare Andreas Cellarius's Copernican map of the universe (above) with the other one he drew (page 107). The Earth and Sun have switched places. Mercury and Venus are now inside Earth's orbit, which helps explain why these two planets behave differently than the others. The Moon no longer has its own orbital ring; it has become a tiny companion to the Earth.

Aristarchus figured out that the Earth revolves around the Sun. In his model, a moving Earth orbits a stationary Sun. Now that was a revolutionary idea! And clearly astonishing to most people. Or was it just silly, they asked?

People spoke of Aristarchus with respect. He had a powerful mind. And yet he seemed to have lost it with that Sun-centered notion. The Earth goes around the Sun? You could look in the sky and see the Sun rise in the east and travel west. Besides that, the great Aristotle had said Earth stands still. It was hard arguing with Aristotle (especially now that he was dead).

Aristarchus wasn't awed by a dead authority. Besides that idea

CHANGING SEASONS

Understanding seasons boils down to a number—23.5° (or 23.45° if you want to get picky). You've seen it before. If you study astronomy or geography, you'll see it again and again. It's the angle that the Earth tilts. Have you noticed that this number is also the latitude for the tropic of Cancer (23.5° N) and the tropic of Capricorn (23.5° S)? That's no coincidence. Sandwiched between these latitudes is Earth's tropical region, a belt around the planet's fattest part, which is the only place where the Sun can be directly overhead. Plenty of direct sunlight makes the Tropics toasty all year round.

The rest of the globe receives weaker, indirect rays part of the year. For instance, Florida is subtropical—just north of the Tropics—and so every once in a while, the orange groves get a winter frost.

If you subtract 23.5° from 90° (the latitude of the poles), you get another number worth remembering—66.5°. That's the latitude of the Arctic Circle (66.5° N) and the Antarctic Circle (66.5° S), the dividing lines between the Temperate Zones and polar regions. What makes the polar regions polar is that the Sun doesn't set (in summer) or doesn't rise (in winter) part of the year. At 66.5° latitude (N or S), that happens on only two days—the solstices. At the poles, that happens all winter or all summer.

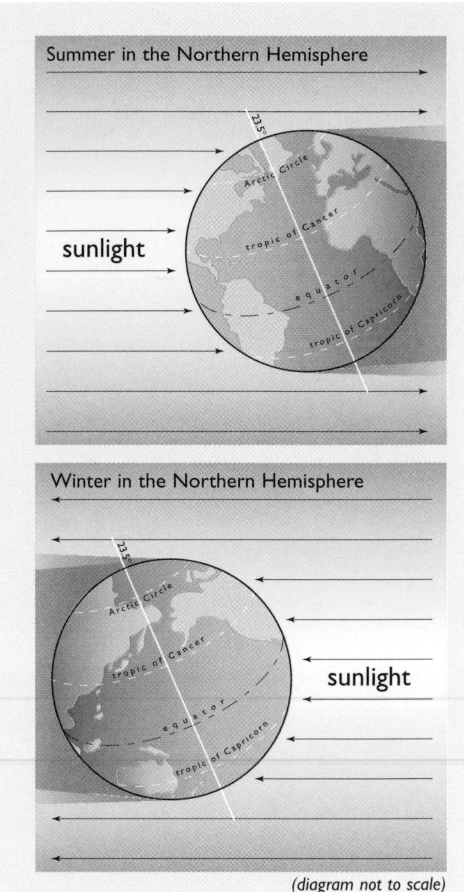

(diagram not to scale)

of his that Earth goes around the Sun, he figured out the size of the Moon and its distance from Earth. How could he possibly do that? By measuring the shadow the Earth throws on the Moon during a lunar eclipse and then doing some mathematics. According to his calculations, the Moon was one-third the size of Earth, which was an overestimation, but not by much.

Then Aristarchus said that the Sun is larger than the Earth! And that Earth rotates on its axis, causing day and night! That wasn't all. He said that Earth's axis is inclined, which means that the Sun's rays reach it at differing angles at different points in its orbit, and that causes the seasons. (He was right.) But all that was too much for most people.

When Aristarchus said that the Earth revolves around the Sun, hardly anyone took that idea seriously. Are there voices today we should be hearing but are not?

Everyone could see that the Sun is tiny—why, you can block it out with one hand. Cleanthes (klee-AN-theez)—

Nine hours recorded on one piece of film is time enough for the stars to slowly rise in the east over the Sierra Nevada, a mountain range in California.

A New Spin

The first person (of whom we know) who suggested that the Earth might be moving around the Sun was Heracleides (ca. 388–315 B.C.E.) He was one of Plato's students, so he must have known Aristotle. Where did Heracleides get this amazing idea? When he looked at the sky, he saw what we see: night after night the same, familiar stars appear to move from east to west. Aristotle, and just about everyone else, thought that motion was real. They thought the stars were revolving in a continuous loop around the Earth.

Heracleides used his imagination and realized that the sky would look the same way if the Earth were spinning on its axis from west to east. That would also explain why Polaris, the star directly above the North Pole, doesn't appear to move along with the others (see page 62). Aristarchus, who was born about the time that Heracleides died, must have learned about this wild conjecture. Most people laughed, but he took it seriously.

another Greek of the time—said Aristarchus should be tried for spreading false and impious ideas. (Remember what happened to Anaxagoras?) Being ahead of your time can be dangerous. Aristarchus's ideas about the Sun and the Earth were laughed at and rejected by most people; he was lucky he wasn't banished or jailed.

Most of his writings got tossed aside, and his ideas were buried in history books. But the next time you're impatient about something, think of Aristarchus. It took 1,700 years, but finally a Polish church canon named Nicolaus Copernicus (1473–1543) learned what Aristarchus had written and paid attention. (What Copernicus did with those almost forgotten ideas of Aristarchus helped create modern science.)

The Greeks were wrestling with some important questions. How big is the Earth? And the Moon? And the stars? And how far away are they? These were tough questions to answer. But they had created a society in which—for the most part—they could study and talk about their big ideas.

Aristarchus pointed out, about 260 B.C.E., that the motions of the heavenly bodies could easily be interpreted if it were assumed that all the planets, including the Earth, revolved about the sun. Since the stars seemed motionless, except for the diurnal motion due to the rotating Earth, they must be infinitely far away.

—Isaac Asimov (1920–1992), science writer, *Asimov's Biographical Encyclopedia of Science & Technology*

Diurnal means "daily"—happening in a 24-hour period of time.

HOW FAR THE MOON? IT'S ABOUT TIME

How do you use time to measure distance? In the Space Age, we calculate the Moon's distance by timing how long it takes a laser beam to zip to a mirror on the Moon and back again. The more split seconds it takes, the farther the Moon.

Using time to measure distance is nothing new. The ancients did it, too, but they had to use their heads instead of lasers. We think the ancient astronomers may have waited patiently for a lunar eclipse. Then, they timed how long it took for the Moon to pass through Earth's shadow. The result led to a surprisingly close estimate of the Moon's distance. We can imagine how it might have been done.

Suppose Aristarchus pictured the Moon's orbit as a giant circle (it is, almost). Geometry being a strong suit, he saw that the distance from Earth to the Moon equals the radius of that large circle. So he labeled the distance with a capital R (see diagram at right).

It never fails; the circumference of every circle—including the Moon's orbit—is always $2\pi r$. Since π (pi) is about 3.14...,

How do we know about these dead Greeks? Especially those whose writings are mostly lost? Often we have to rely on the words of others. Archimedes (coming up soon) tells us this about Aristarchus: "His hypotheses are that the fixed stars and the Sun remain motionless, that the Earth revolves about the Sun in the circumference of a circle, the Sun lying in the middle of the orbit." Plutarch, in his book *Of the Face in the Disc of the Moon* wrote that Cleanthes "thought it was the duty of Greeks to indict Aristarchus of Samos on the charge of impiety for putting in motion the Hearth of the Universe."

Aristarchus multiplied by 2 to get **6.28R**. The time it takes the Moon to cover that circumference (one orbit around Earth) is **T = 656 hours** (about 27 days).

To Aristarchus, it was obvious that the diameter of the Earth equals the diameter of its shadow. (It's actually smaller, but close enough.) During an eclipse, the Moon passes through that shadow. So he set the shadowed part of the Moon's orbit equal to **2*r*** (or 1 Earth diameter). Finally, a total lunar eclipse happened, and the Moon passed through Earth's shadow in about **t = 3 hours**.

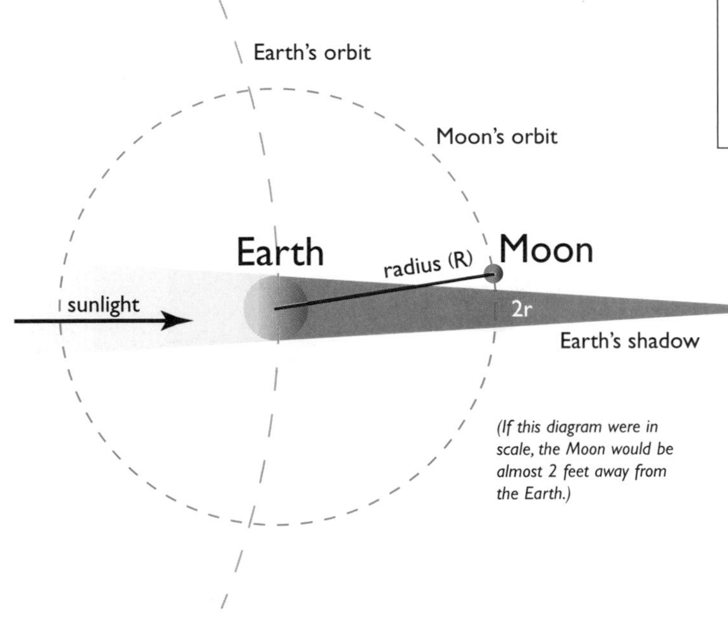

(If this diagram were in scale, the Moon would be almost 2 feet away from the Earth.)

Aristarchus now had all the numbers he needed. He knew that the ratio of a whole orbit (6.28R) to the shadow length (2*r*) is the same as the ratio of the whole orbit time (T) to the shadow time (t). Or:

$$\frac{6.28R}{2r} = \frac{T}{t}$$

Aristarchus entered the two times (T and t):

$$\frac{6.28R}{2r} = \frac{656 \text{ hours}}{3 \text{ hours}}$$

The rest was arithmetic:

$$\frac{6.28R}{2r} = 219$$

$$6.28R = 437r$$

$$R = 70r$$

This simple equation meant that the distance to the moon (R) is about 70*r*—or 35 Earths in a row. We know now that it's closer to 60 radii, but that's not the point. Aristarchus did something no one else had done before, and he did it long before lasers came around.

14

Alexander's City

He [Alexander] was naturally a great lover of all kinds of learning and reading; and...he constantly laid Homer's Iliad...with his dagger under his pillow, declaring that he esteemed it a perfect portable treasure of all military virtue and knowledge....For a while he loved and cherished Aristotle no less, as he was wont to say himself, than if he had been his father, giving this reason for it, that as he had received life from the one, so the other had taught him to live well.

—Plutarch (ca. 46–ca. 120 C.E.), Greek historian, *Alexander*

[In Alexandria] commerce was carried on by free people who were not segregated socially from the scholars....[The scholars] became aware of and involved in the problems facing the people at large....Alexandria produced a kind of mathematics almost opposite in character to that produced by the classical Greek age. The new mathematics was practical, the earlier almost entirely unrelated to application.

—Morris Kline (1908–1992), American math professor and author, *Mathematics in Western Culture*

The Mediterranean world was in turmoil. Aristotle's pupil Alexander of Macedon had defeated a formidable Persian army and then had gone on a conquering spree. One victory topped another until he had put together an enormous empire that stretched from southern Russia to Greece to India to Egypt to Libya. His soldiers called him Alexander the Great, and soon everyone else did too, for he was an astonishing combination of cruel warrior, far-sighted ruler, and brilliant educated man. He founded more than 70 cities—including one in Egypt that he intended as his capital and named (with no modesty) Alexandria.

Read Plutarch's quotation (above). This seventeenth-century painting depicts the scene—Alexander reading Homer's epic.

Alexander wanted to change things. He wanted to break down barriers between peoples, perhaps to rule them more easily. So he placed Alexandria on the Mediterranean Sea near where Asia, Africa, and Europe meet. It was a terrific spot for a city intended as a world center of trade and

A partly preserved floor mosaic (above) from Pompeii glorifies Alexander the Great's victory over Persian King Darius III at Issus in 333 B.C.E. On the left, a strong and determined Alexander rides his favorite horse, Bucephalus, into battle, while a distressed Darius stands in his chariot, on the right. The entire mosaic is made out of more than a million tiny tiles, making it one of the largest and most significant pieces of art from ancient times.

A seventeenth-century painting (left), *The Triumph of Alexander*, by Charles Le Brun, captures Alexander's glorious entrance into the city of Babylon, which he conquered in 331 B.C.E. The victory made him ruler of western Asia and ensured that Greek culture would influence the world for centuries to follow.

culture, which is what it became. Alexander was determined to spread Greek ideas about art and science and philosophy to the lands he conquered. He set out to do it.

But great as he was, he couldn't conquer his fate. While feasting at a banquet in Babylon, he became ill, some say from drinking too much liquor. Eleven days later, Alexander was dead. It was 323 B.C.E. He was 33.

No one else was strong enough to hold the empire together. So it was divided among his officers. A Macedonian general named Ptolemy (TOL-uh-mee) got Egypt. Ptolemy's

MACEDON is the ancient kingdom, but its modern name is MACEDONIA. It is north of Greece.

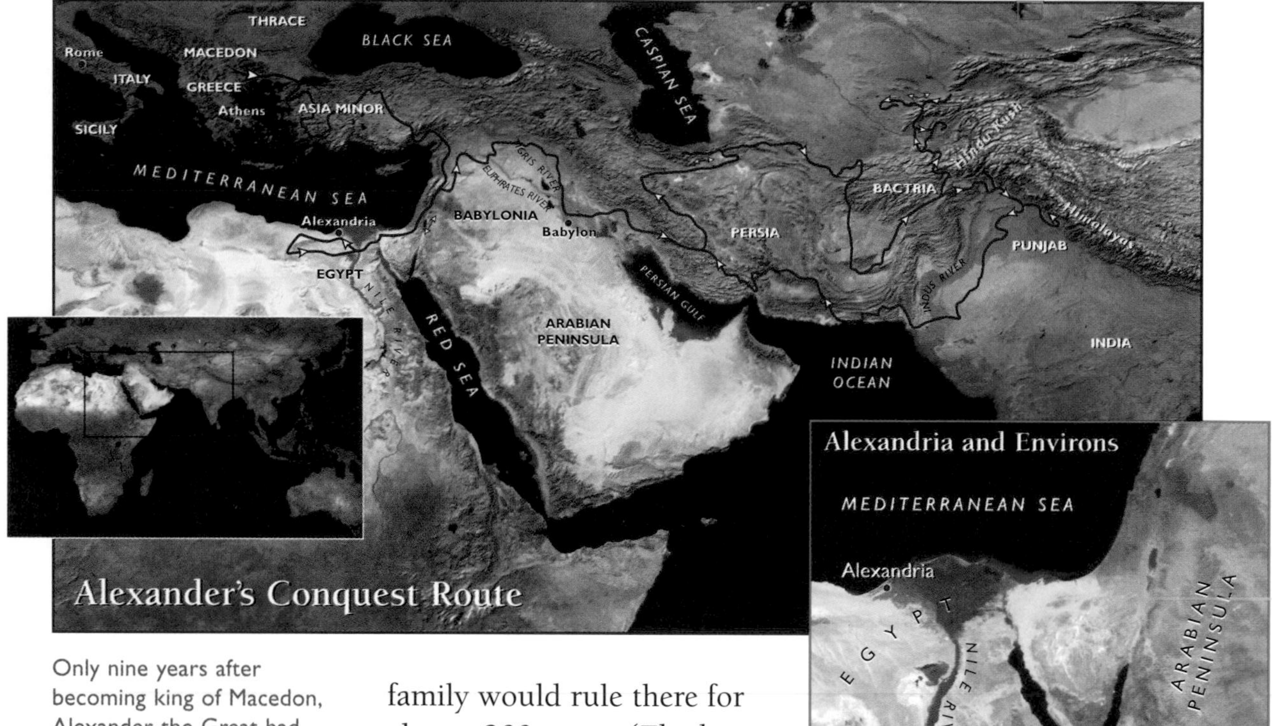

Only nine years after becoming king of Macedon, Alexander the Great had conquered the entire Persian Empire and spread the influence of Greek civilization as far as India. His empire, however, did not survive his death.

family would rule there for almost 300 years. (The last Ptolemaic "king" was a woman: Cleopatra.)

Ptolemy I, known as Ptolemy Soter, brought Greek art, learning, and business ideas to Alexandria. But it wasn't a one-way street. He was much influenced by the rich culture he found in Egypt. He made himself famous for all time by building a spectacular public institution at Alexandria. It was called the Mouseion, after the Muses, but it was actually a fabulous university that included a museum, a library, lecture halls, classrooms, parks, a zoo, and a place to live and eat.

Ptolemy II, known as Ptolemy Philadelphus, went even further than his father. He paid special attention to the library, bringing books together as had never been done before. He also searched for the best thinkers he could find, brought them to Alexandria, and paid them to teach and do research. (Aristarchus was one of them.)

Scouts were sent out across the known world to track down every book and codex that could be found. Scribes got to work copying them. Ships entering Alexandria's harbor

We think of a CODEX as an ancient book whose pages are stitched together; two or more are CODICES. A codex is closer to a modern book than a tablet or scroll. The word comes from Latin "wood split into slabs." Terence (190–159 B.C.E.), a Roman comedy writer, used codex to mean a "blockhead."

A METROPOLIS is a large city. The word is the same in Greek, from *mētr-* ("mother") and *polis* ("city"), so think of it as a "mother city."

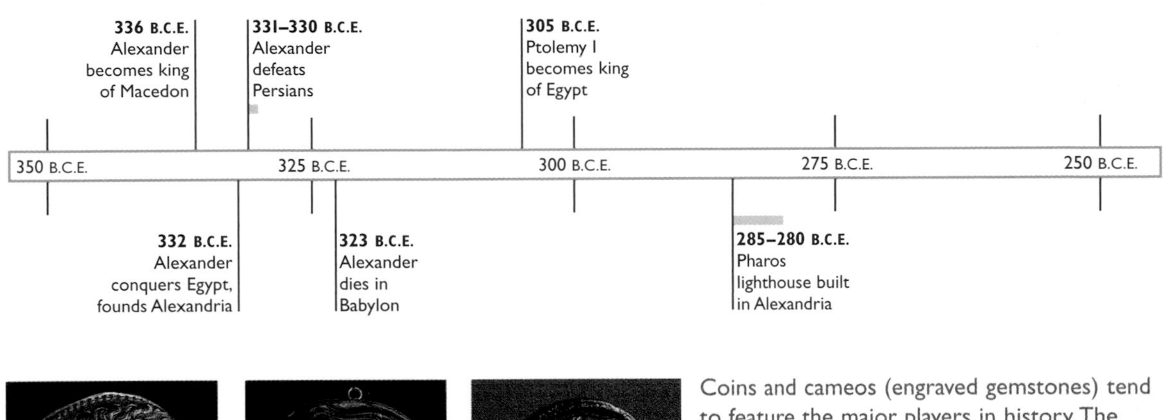

Coins and cameos (engraved gemstones) tend to feature the major players in history. The coin on the far left honors Ptolemy Soter. The middle cameo shows off his children, Ptolemy Philadelphus and Arsinoë. The third coin depicts Cleopatra, ancient Egypt's last ruler before the Roman takeover.

were searched for books. Those books were copied, or sometimes they were just confiscated. Ptolemy II wanted all the world's books to be in the great library at Alexandria. Eventually there would be at least 700,000 volumes. They were all hand copied on scrolls or bound in booklike codices.

Athens had been the intellectual center of the Western world; now it was Alexandria. Here's what astronomer Carl Sagan says of its Mouseion: "This place was once the brain and glory of the greatest city on the planet, the first true research institute in the history of the world. . . . It was in Alexandria, during the 600 years beginning around 300 B.C.E. that human beings, in an important sense, began the intellectual adventure that has led us to the shores of space."

Marvelous as Athens was, it was a small, tight community of Greeks. Alexandria was a metropolis of 600,000 people and a new kind of city. (Europe would not have an urban center on this scale until the eighteenth century, when Paris and London grew large.)

The Alexandrians took ideas from the practical Egyptians and mixed them with the intellectualism of the Greeks, the searching ideas of the Jews, and the energy of the Persians. All that helped create the freest society in the ancient world.

The lighthouse at Pharos, shown on a Roman coin above, was one of the Seven Wonders of the Ancient World (see pages 126–127). On the left is the Egyptian goddess Isis holding a sail and sistrum (a sacred rattle).

The result was a culture called "Hellenistic," after Hellen, a king in Greek mythology who was said to be the grandson of the god Prometheus and the ancestor of all Greeks. Perhaps no city anywhere has ever had a more successful mixture of peoples (maybe not even New York City today). Africans, Persians, Romans, Greeks, Egyptians, Jews, Arabs, Indians, and Phoenicians were all welcome. In Alexandria, it was talent that was admired. And talent could be found anywhere. Plumbers and professors rubbed shoulders in classes and shops. Qualified women were part of the mix. The huge library meant that people could read and learn, using books and instruments provided at public expense. That free atmosphere energized and inspired everyone. It led to increased specialization and to a class of professionals.

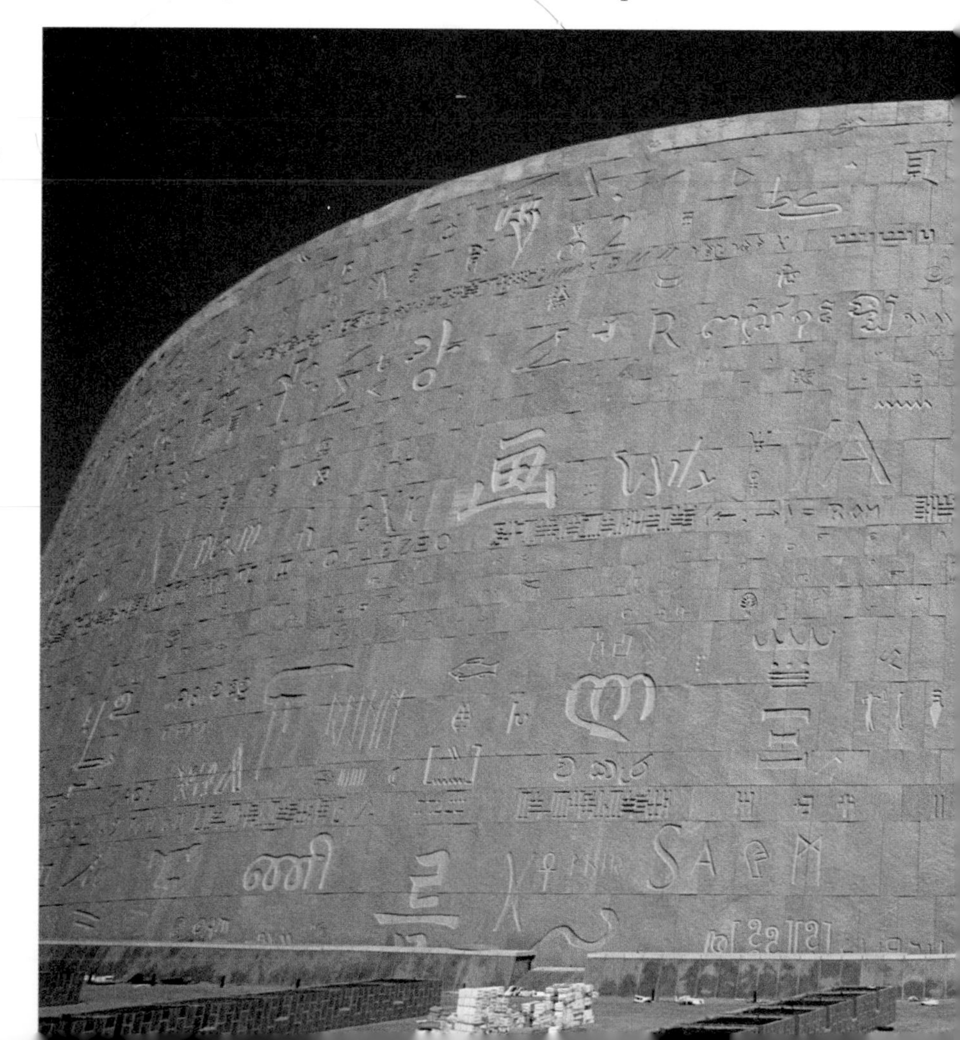

The ancient library of Alexandria burned to the ground under mysterious circumstances. Historians disagree about who did it, but most blame Julius Caesar. During an attack of the city in ca. 48 B.C.E., he ordered his fire ships to set enemy ships aflame. The fire jumped to shore and spread to the city's royal quarter. The new library of Alexandria, called the Bibliotheca Alexandrina, near the same site, was opened in 2002 and has room for 8 million books. The granite exterior (right) is etched with letters from 120 of the known alphabets in the world—ancient and modern.

Alexandria was the place to be if you had an inquiring mind and were looking for ideas and action.

But it wasn't perfect. There were layers to society. If you were Greek, you had big advantages. People on the bottom did much of the work that made life easy for those in charge. And there were slaves. The Ptolemies began to be worshiped as gods, and they acted as if they were divine. Still, the ferment of trade and the encouragement given to the arts and sciences made Alexandria open and exciting in ways that would be special in any place and time.

If you could time-warp yourself anywhere in world history, Alexandria would be a good place to land. In the next chapters, you'll meet some thinkers who were schooled in that astonishing city.

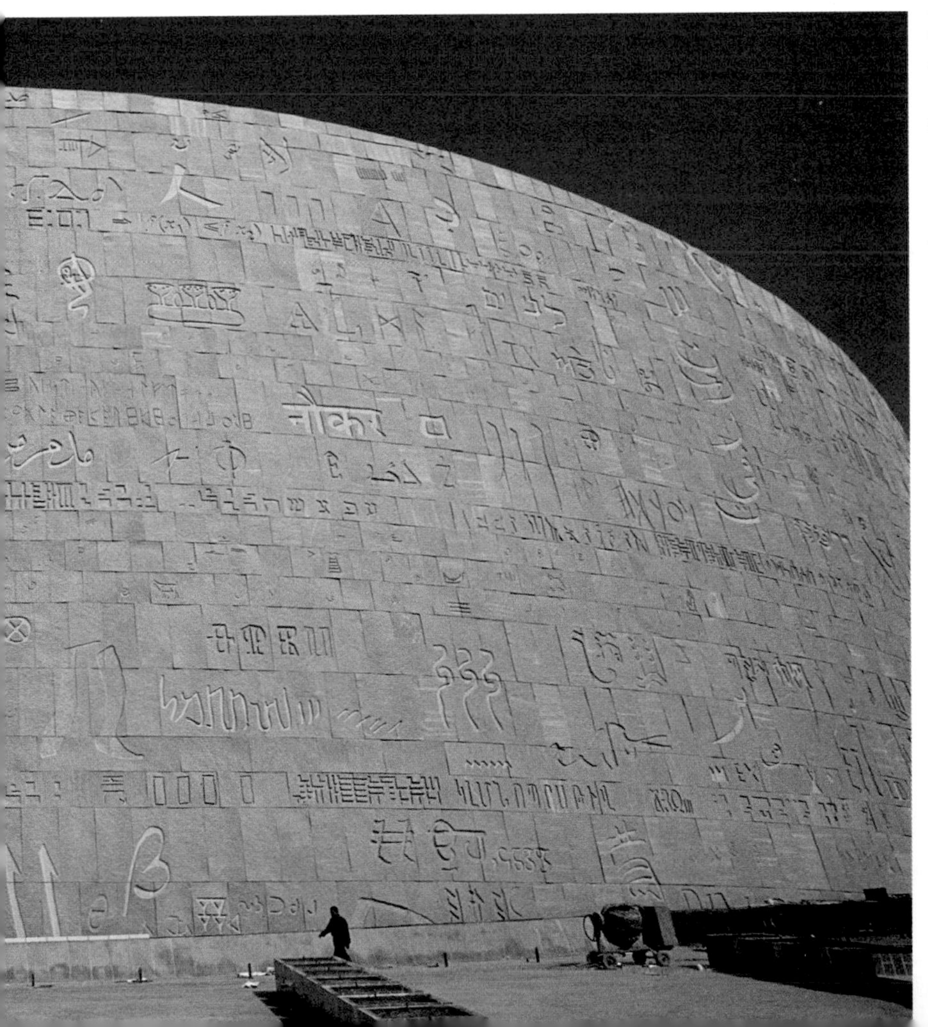

In the world of scholarship, the ancient Library of Alexandria is the ultimate romantic myth, a kind of paradise lost.

—Alexander Stille (1957–), American writer, *The New Yorker* magazine

SMOKE AND MIRRORS

**gigantic lighthouse** on the island of Pharos in Alexandria's harbor was a marvel of engineering and one of the Seven Wonders of the Ancient World. It was a three-layered tower with rectangular windows on all its sides. The bottom floor was a rectangle, the next an octagon, and the third a cylinder. At the very top stood a statue, but its identity is a mystery. Perhaps

The story goes that it was a passage in Homer's *Odyssey* that inspired Alexander to have the Pharos lighthouse built. However, he didn't live long enough to witness the Seventh Wonder of the Ancient World. Construction began under Ptolemy Soter and took more than five years. You could see the beacon from about 50 kilometers (31 miles) away. During the day, the mirrors reflected the sunlight. At night, the fire was a warning signal for sailors.

A fanciful depiction of the lighthouse (right) by sixteenth-century Dutch painter Maerten van Heemskerck gives the landscape and architecture a northern European touch.

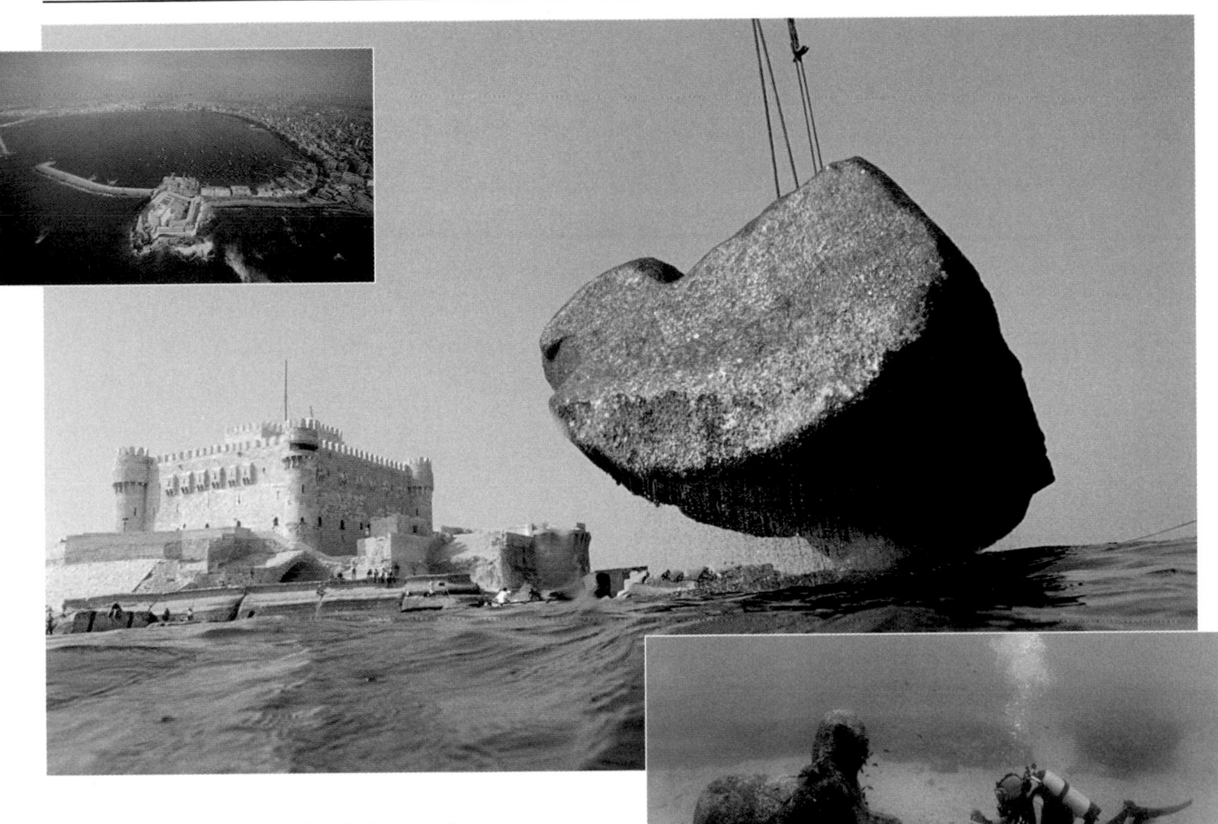

it was Poseidon (god of the sea) or Alexander or a Ptolemy. No one knows.

The Pharos lighthouse was some 117 meters (128 yards) high—about the same as a 36-story skyscraper or the tallest pyramids. It had 300 rooms. Huge statues of royal couples stood at the base of the lighthouse so that everyone who entered the harbor could be impressed with their grandeur.

The lighthouse took more than five years to build and was completed about 280 B.C.E. Earthquakes, not human folly, seem to have destroyed this great architectural marvel. We know of at least 22 quakes starting in 320 C.E. In 796, an earthquake brought down the top floor. By 956, some walls were cracked and crumbling, and the tower was some 18 meters (almost 20 yards) shorter as a result. Earthquakes in 1261 and 1303 caused more stone to fall, but the death blow came in 1326, when violent shaking sent most of the tower into the Mediterranean Sea.

In the mid-1990s, archaeologists began excavating the underwater ruins of Alexandria and the island of Pharos, which is now home to Qaitbay Fort (built in ca. 1477). Among the recovered artifacts, divers found plenty of sphinxes (directly above). Remnants of the lighthouse include huge stone blocks; the one in the center photo weighs as much as 10 elephants. It is thought to have been part of the entrance. Modern Alexandria (upper left) is a busy harbor city.

15 What's a Hero?

By the union of air, earth, fire and water, and the concurrence of three, or four, elementary principles, various combinations are effected, some of which supply the most pressing wants of human life, while others produce amazement and alarm.

—Hero of Alexandria (lived ca. 62 C.E.), *The Pneumatics*

Our inventions are wont to be pretty toys, which distract our attention from serious things. They are but improved means to an unimproved end.

—Henry David Thoreau (1817–1862), American philosopher and author, *Walden*

Man is by nature a pragmatic materialist, a mechanic, a lover of gadgets and gadgetry.

—Elizabeth Gould Davis (1910–1974), American feminist and author, *The First Sex*

For centuries, Alexandria reigned as the Mediterranean's queen city. It was hard to be an educated person without studying in Alexandria. It was hard to be a sophisticated person without at least visiting the city. And when you visited, you couldn't help but be impressed by the public buildings—all made of marble with imposing columned entries. Statues, fountains, and giant monuments added to the aura of majesty and splendor and also provided a place for souvenir sellers, beggars, and tour guides to hawk their wares. Restaurants were varied and numerous, as were shops with valuable items to sell, such as linens, spices, glass, papyrus, books, maps, and slaves. Tourists flocked in from East and West.

PAPYRUS (puh-PY-rus) is a reed that grows abundantly along the Nile River. It was cheap in Alexandria, where it was made into a kind of paper, and so the book-copying business took off.

A sphinx is a mythological creature with a human head and the body of a lion. In ancient Egypt, these statues were everywhere. Here is one made of green stone, dating from the first century C.E.

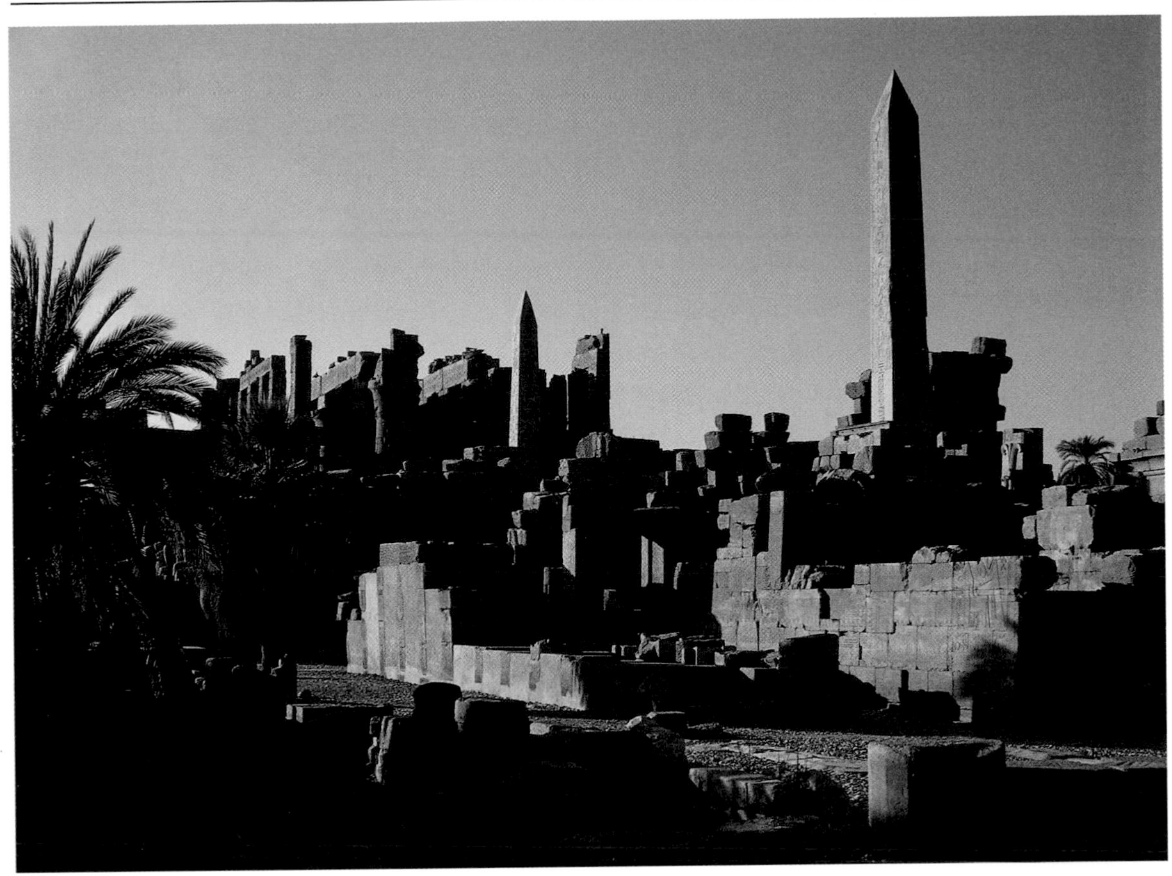

Obelisks are tall, four-sided shafts of stone, usually erected by rulers and inscribed with dedications to gods. Very few ancient obelisks are left in Egypt. Two of the most impressive are in the Temple of Amun at Karnak (Luxor, Egypt). Pharaoh Thutmose I had the one on the right built for him. His daughter, Pharaoh Hatshepsut—the only female Pharaoh—had the one to the left constructed.

Alexander had laid out the city's streets in grids with grand, defining boulevards, so it had an orderly, stately feel. Sphinxes and obelisks towered over temples, royal palaces, and carefully placed open spaces. A broad, central avenue, the Canopic Way, led from the Gate of the Sun at the eastern end of the city to the Gate of the Moon at the western. A first-century B.C.E. historian, Diodorus Siculus, wrote, "By selecting the right angle of the streets, Alexander made the city breathe with . . . winds so that as these blow across a great expanse of sea, they cool the air of the town, and so he provided its inhabitants with a moderate climate and good health."

Many of the splendid statues had been hauled from Heliopolis, the ancient city of the sun god, which was about 10 kilometers (6.2 miles) south of Cairo. Heliopolis had been Egypt's capital during the glory days of the New Kingdom,

CANOPIC comes from Canopus, the second brightest star in the sky (after Sirius). A town east of Alexandria named Canopus was known as a pleasure city. *Canopic* is also an adjective used to describe a fancy jar to hold mummified organs.

way back between 1554 and 1085 B.C.E. The statues linked Alexandria to Egypt's inspired past. Alexander the Great's tomb—a tourist attraction—tied the city to its fourth-century B.C.E. founder.

By the first century C.E., Alexandria was a place of technological wonders. Steam-powered vehicles moved along city streets in religious parades. In the temples, mechanical doves flew to the ceilings, and statues of the gods raised their arms, shed tears, and poured libations (more steam power). If you put a five-drachma coin in an automatic machine, it sprinkled holy water. Organs played music when coins were inserted. And temple doors opened automatically. Hydraulic power activated statues in the parks. In court, water clocks and sundials set time limits on lawyers' speeches.

Many of those inventions were the work of Hero of Alexandria (often known as Heron), who was an inventor, a mathematician, and a writer. He built a steam-powered engine that, in concept, was not that different from a jet engine. Hero was a distinguished scholar, but he is most remembered for his sense of playfulness. He developed siphons and invented a pump, but he used the pump to activate a toy trumpet. (Today siphons and pumps are basic to almost all modern plumbing and sanitation.) Hero used the power of steam to make a toy rattlesnake hiss. (It was almost 18 centuries later, in England, when James Watt built a steam engine, which led to steam-powered trains and ships.) For a long time, scientists and historians didn't take Hero very seriously. They thought he was just a clever tinkerer who had come up with fun gadgets.

But Hero was more than that. He helped develop a formula, still used today, to calculate the area of any triangle (see page 132). In one of his many books, Hero described the six devices that were the basis for all machines until the age of electronics. They are the (1) lever, (2) pulley, (3) wedge, (4) wheel and axle, (5) screw, and (6) siphon.

His books on mathematics, astronomy, land surveying, and mechanics were used as textbooks for many centuries. He

A LIBATION is usually wine poured in honor of a god or in a religious festival. Today it sometimes means just a "drink."

A HYDRAULIC machine is powered by water pressure. *Hudro* is "water" in Greek and *aulos* is "pipe." Originally, it was the name of a musical instrument played by water power.

A SIPHON is a tube for drawing liquids from one place to another.

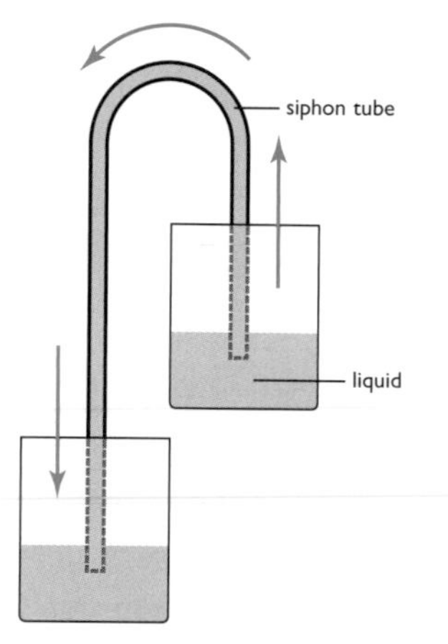

A PUMP is a machine for raising, squeezing, pushing, or drawing liquids. The heart is a pump that squeezes blood.

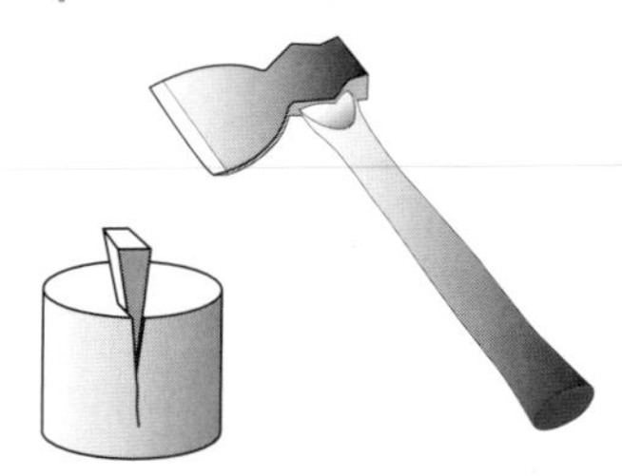

When Hero said **WEDGE**, he meant a simple tool like an ax or an inclined plane (a ramp). Today, a wedge includes any object with two sides that meet at a sharp angle. An ax still qualifies, but so does the prow of a ship.

Diodorus Siculus was born in Sicily, traveled in Asia and Europe, and lived in Rome. An avid historian in the first century B.C.E., he wrote a 40-volume history of the world from the creation of the universe to the wars of Gaius Julius Caesar. Sixteen hundred years later, his books were still of fundamental importance. In this painting, translated editions are being presented to Francis I, king of France from 1515 to 1547.

also wrote how-to books. If you are interested in building catapults, puppet theaters, or war machines, you might want to check Hero.

The name *Hero* was very common in Greece, so for a long time no one was sure which ancient Hero was which. Today we're quite confident that Hero of Alexandria was alive on March 13 in the year 62 C.E., because in one of his books he refers to a lunar eclipse seen at Alexandria on that night. No one knew how far Alexandria is from Rome. The distance couldn't be measured then, because the Mediterranean Sea lies between the two cities. Hero, knowing the local times at which the eclipse of the Moon was observed in each city, was able to figure the distance.

Hero's works were translated into Arabic and, later, into European languages. He was studied until well into the sixteenth century. In 1896, an old manuscript turned up in

All our joys
Are but toys...

—Thomas Campion (1567–1620), British poet, "What if a Day..." (In the spelling of the time, Campion wrote "joyes" and "toyes.")

HERO'S FORMULA

How do you measure the area of a triangle? If you've been reading these math sidebars, you may think, "That's easy. You just do that Pythagorean thing—$a^2 + b^2 = c^2$—to find the lengths of the short sides, a and b. Then just multiply them and divide by 2." And you'd be right—in part. That easy formula for the area of a triangle works every time for triangles that have a 90° angle. That's because right triangles are half of a rectangle or square, so you're basically measuring the area of a rectangle or square ($a \times b$) and cutting it in half.

What about all those triangles without a neat, square corner (like the triangle below right)? They're a little trickier, but Hero the geometer was up to the task. Here's how Hero's Formula works:

1. Add the three sides ($a + b + c = 20$) and divide by 2 ($20 \div 2 = 10$). This number—let's call it s—is the key. (You can call it anything, but s stands for "semiperimeter," half the length of the triangle's border. In this case, s is 10.)
2. Subtract each side (a, b, c) from s ($10 - 5 = 5$, $10 - 7 = 3$, $10 - 8 = 2$).
3. Multiply the answers by each other ($5 \times 3 \times 2 = 30$) and by s ($30 \times 10 = 300$).
4. Grab a calculator, find the square root ($\sqrt{300} = 17.3$), and you'll know the area of the triangle. The sample triangle is 17.3 square units. The units could be anything—inches, kilometers, light years.

Here's what Hero's Formula looks like to algebra enthusiasts:

$s = \frac{1}{2}(a + b + c)$

$\text{Area} = \sqrt{s(s - a)(s - b)(s - c)}$

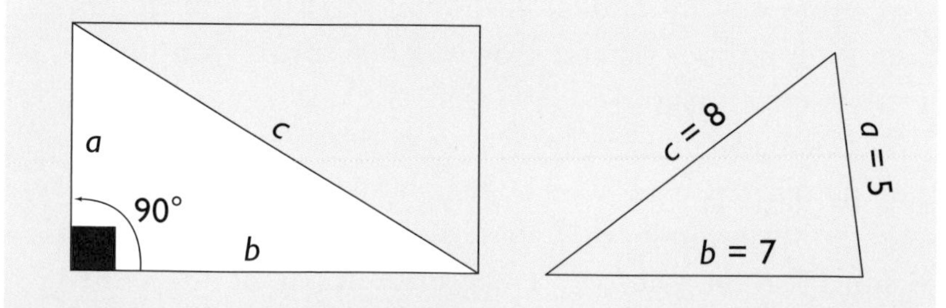

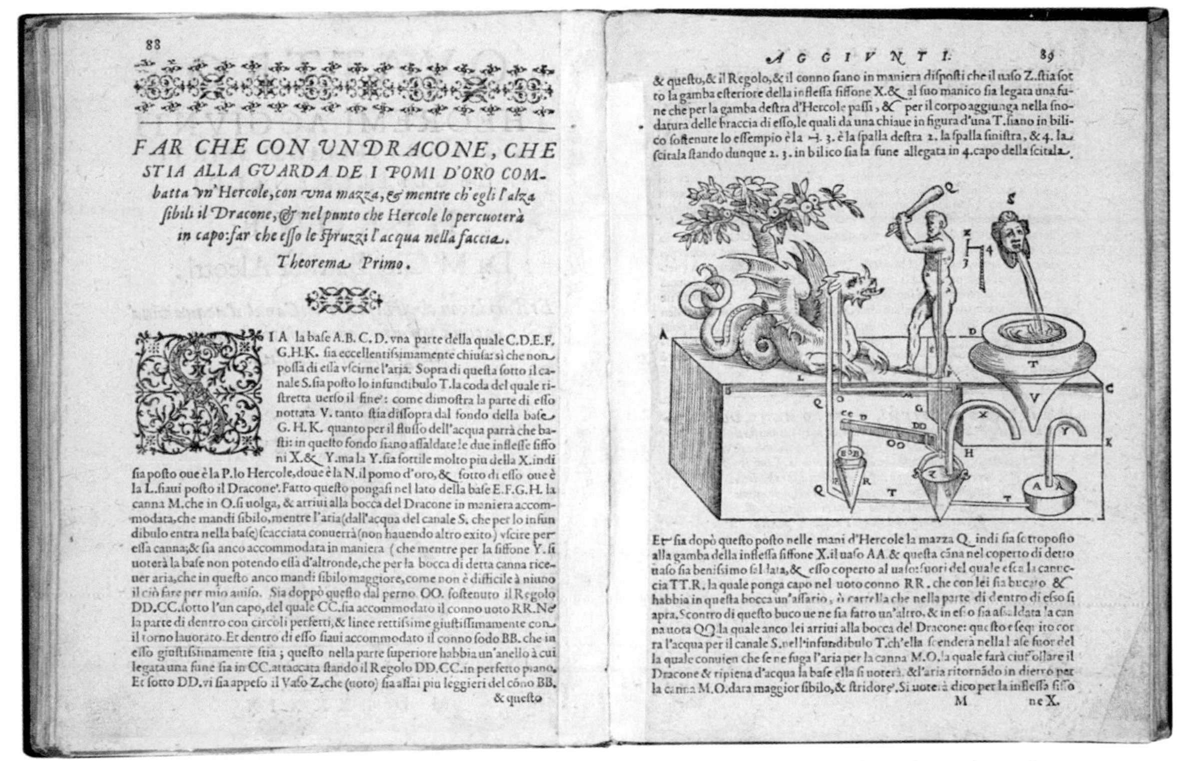

88

FAR CHE CON UN DRACONE, CHE STIA ALLA GUARDA DE I POMI D'ORO COMbatta vn' Hercole, con vna mazza, & mentre ch'egli l'alza ſibili il Dracone, & nel punto che Hercole lo percuoterà in capo: far che eſſo le ſpruzzi l'acqua nella faccia.

Theorema Primo.

SIA la baſe A.B.C.D. vna parte della quale C.D.E.F. G.H.K. ſia eccellentiſsimamente chiuſa: ſi che non poſſa di eſſa vſcirne l'aria. Sopra di queſta ſotto il canale S. ſia poſto lo infundibulo T. la coda del quale riſtretta uerſo il fine: come dimoſtra la parte di eſſo nottata V. tanto ſtia diſſopra dal fondo della baſe G. H. K. quanto per il fluſſo dell'acqua parrà che baſti: in queſto fondo ſiano aſſaldate le due infleſſe ſiffoni X. & Y. ma la Y. ſia ſottile molto piu della X. indi ſia poſto oue è la P. lo Hercole, doue è la N. il pomo d'oro, & ſotto di eſſo oue è la L. ſiaui poſto il Dracone. Fatto queſto pongaſi nel lato della baſe E.F.G.H. la canna M. che in O. ſi uolga, & arriui alla bocca del Dracone in maniera accommodata, che mandi ſibilo, mentre l'aria (dall'acqua del canale S. che per lo infundibulo entra nella baſe) ſcacciata conuerrà (non hauendo altro exito) vſcire per eſſa canna; & ſia anco accommodata in maniera (che mentre per la ſiffone Y. ſi uoterà la baſe non potendo eſſa d'altronde, che per la bocca di detta canna riceuer aria, che in queſto anco mandi ſibilo maggiore, come non è difficile à niuno il ciò fare per mio auiſo. Sia doppò queſto dal perno OO. ſoſtenuto il Regolo DD.CC. ſotto l'un capo, del quale CC. ſia accommodato il conno uoto RR. Ne la parte di dentro con circoli perfetti, & linee rettiſsime giuſtiſſimamente con il torno lauorato. Et dentro di eſſo ſiaui accommodato il conno ſodo BB. che in eſſo giuſtiſsimamente ſtia; queſto nella parte ſuperiore habbia un'anello à cui legata una fune ſia in CC. attaccata ſtando il Regolo DD.CC. in perfetto piano. Et ſotto DD. vi ſia appeſo il Vaſo Z. che (uoto) ſia aſſai piu leggieri del cõno BB.

& queſto

AGGIVNTI. 89

& queſto, & il Regolo, & il conno ſiano in maniera diſpoſti che il uaſo Z. ſtia ſotto la gamba eſteriore della infleſſa ſiffone X. & al ſuo manico ſia legata una fune che per la gamba deſtra d'Hercole paſſi, & per il corpo aggiunga nella ſnodatura delle braccia di eſſo, le quali da una chiaue in figura d'una T. ſiano in bilico ſoſtenute lo eſſempio è la ⊣. 3. è la ſpalla deſtra 2. la ſpalla ſiniſtra, & 4. la ſcitala ſtando dunque 2. 3. in bilico ſia la fune allegata in 4. capo della ſcitala.

Et ſia dopò queſto poſto nelle mani d'Hercole la mazza Q. indi ſia ſottopoſto alla gamba della infleſſa ſiffone X. il uaſo AA. & queſta cãna nel coperto di detto uaſo ſia beniſsimo ſaldata, & eſſo coperto al uaſo: fuori del quale eſca la cannuccia TT.R. la quale ponga capo nel uoto conno RR. che con lei ſia bucato & habbia in queſta bocca un'aſſario, ò cartella che nella parte di dentro di eſſo ſi apra. Scontro di queſto buco ue ne ſia fatto un'altro, & in eſſo ſia aſſaldata la canna uota QQ. la quale anco lei arriui alla bocca del Dracone: queſto eſſequito correrà l'acqua per il canale S. nell'infundibulo T. ch'ella ſcenderà nella baſe fuor della quale conuien che ſe ne fuga l'aria per la canna M.O. la quale farà ciuffollare il Dracone & ripiena d'acqua la baſe ella ſi uoterà. & l'aria ritornãdo in dietro per la canna M.O. dara maggior ſibilo, & ſtridore. Si uoterà dico per la infleſſa ſiffo

M ne X.

Hero's mechanical-how-to book, *The Pneumatics*, was translated into Italian in the late sixteenth century.

Constantinople (now Istanbul). It was one of Hero's books. We now believe he taught at Alexandria, because some of his books seem to be lecture notes. But Hero was no ordinary professor. Of the 78 inventions described in his book *The Pneumatics*, nearly half of them are vessels for pouring wine and water, including a hat that doubles as a drinking cup. Here are his comments on a vessel that holds two liquids: "We may also . . . pour out wine for some, wine and water for others, and mere water for those whom we wish to jest with."

Hero, an Airhead?

Number 50: The Steam Engine
A fire in the pot heats water into steam, which rises through two pipes to a ball. The ball is on pivots (turning pins), so it spins when the steam hits it. The spinning motion can power other machine parts. After building this simple model, Hero improved and elaborated on it. One of his steam machines works like a jet engine.

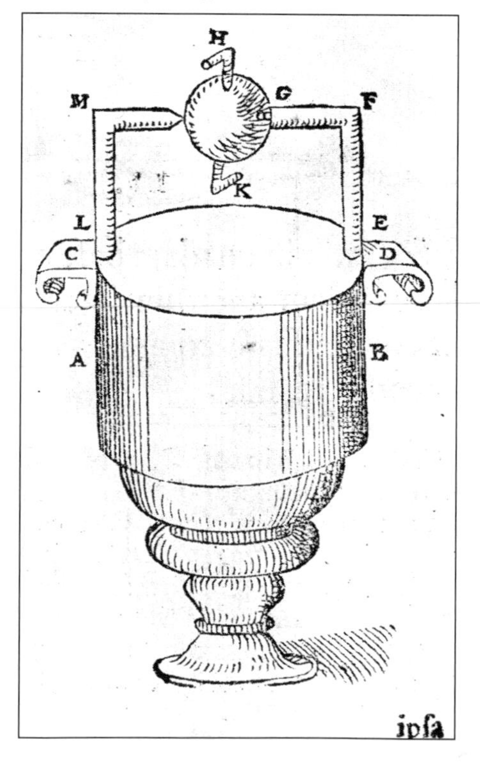

The title of a book by Hero, *The Pneumatics* (noo-MAT-iks), comes from *pneuma*, the Greek word for "air." A fair number of Hero's gadgets rely on compressed (squeezed) air for power. He explains, "Vessels which seem to most men empty are not empty, as they suppose, but full of air. Now the air... is composed of particles minute [my-NOOT, meaning "tiny"] and light, and for the most part invisible." (It would be 16 centuries before the idea that air is composed of minute particles was seriously considered again.) Hero then describes a demonstration that's still popular in science classrooms. He asks readers to invert a vessel (cup), and submerge it in water, "being carefully kept upright;... the air, being matter, and having itself filled all the space in the vessel, does not allow the water to enter...." Raise the vessel and feel the inside, he continues; "we shall find the inner surface of the vessel entirely free from moisture." Hero also made creative use of wind power ("for wind is nothing else but air in motion"), vacuums (an absence of air), fire (number 37), and hydraulics (water pressure). Here are a handful of Hero's gadgets, most of them frivolous and a few of them serious.

- *the first known vending machine*
- *a windmill-driven organ*
- *a hydraulic (water) organ*
- *a pneumatic (air) gun*
- *a mechanical puppet theater*
- *a fire engine*
- *a hose for spraying liquid fire*
- *a syringe*
- *a water clock*
- *a solar-powered water fountain*
- *temple figures activated by steam power*
- *automatic doors*
- *mechanical "singing" birds*
- *an oil lamp with self-trimming wick*

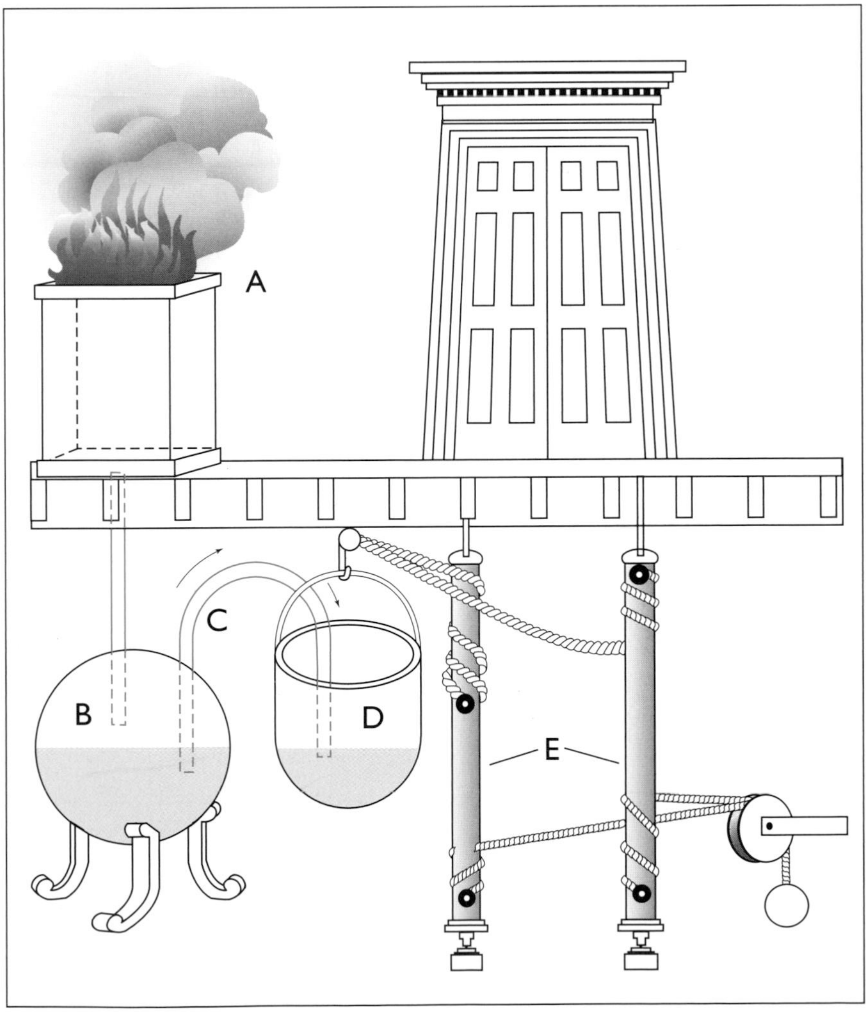

Number 37:
Temple Doors Opened by Fire
Heat from a fire (A) expands air in a globe (B), driving liquid through a siphon (C) into a vessel hanging from a pulley (D). The vessel drops with the added weight of the water, pulling chains that turn cylinders (E) that open the doors. When the fire is extinguished, the doors close again!

16 Euclid in His Elements

Euclid alone has looked on Beauty bare.

—Edna St. Vincent Millay (1892–1950), American poet, *The Harp-Weaver and Other Poems* (Was she kidding? Is there beauty in math?)

Parallel straight lines are straight lines which, being in the same plane and being produced indefinitely in both directions, do not meet one another in either direction.

—Euclid (ca. 325–ca. 270 B.C.E.), *The Elements*

Ivan speaks: **"I tell you that I accept God simply. But you must note this: if God exists and if He really did create the world, then, as we all know, He created it according to the geometry of Euclid."**

—Fyodor Dostoyevsky (1821–1881), Russian novelist, *The Brothers Karamazov*

A math professor named Euclid (YOO-klid) proved that there are infinitely many prime numbers. The Greeks were fascinated with primes (see pages 143–145). Primes fit their atom-centered picture of a universe where all things can be broken down into essential components. And primes *are* essential components; every other whole number is built from primes. The Greeks thought of primes as atoms; all other whole numbers were what we would call "molecules." To those with this world view, Euclid's proof that there are infinitely many primes must have been a real shock.

PYTHAGOREAN THEOREM REFRESHER:

$a^2 + b^2 = c^2$. In any right triangle, the sum of the squares of sides a and b equals the square of side c, the hypotenuse (see page 79).

That's Pythagoras and Euclid in this fifteenth-century marble tile from Florence, Italy.

In this painting by Jacopo de' Barbari (ca. 1440–1516), math professor Luca Pacioli is copying a drawing from Euclid's book. Notice the dodecahedron (bottom right) on top of Pacioli's popular math text, which was based on Euclid and had illustrations by Leonardo da Vinci. A glass polyhedron with 26 faces (a rhombicuboctahedron, to be precise) hangs in the air (left). It's half-filled with water, its transparency symbolizing the clarity of mathematics. The student on the right might be Albrecht Dürer, a German artist who studied math with Pacioli.

Euclid also came up with a solid proof of the Pythagorean Theorem. Pythagoras had made a keen observation about right triangles. He even wrote a simple equation for others to use. But it was Euclid who proved that the equation is always true for every right triangle.

Optics is the science of light, color, lenses, and vision; just ask your optician.

There's more: Euclid made optics a part of geometry by dealing with light rays as if they were straight lines. That opened surprising avenues of discovery. And he wrote what became the most widely used math textbook ever. Who was this guy?

We know that Euclid was alive in 300 B.C.E. We think he attended Plato's Academy in Athens. We're quite sure he studied mathematics in Alexandria and later wrote and taught there. He dedicated his life's work to Ptolemy I, which is a clue that he probably had some contact with the head of state. And that's about all we know of Euclid as a man.

What we do know about is a book he wrote called *The Elements*, which is among the most influential books of all time. (*The Elements* is actually a series of 13 books.)

Johannes Kepler, a famous seventeenth-century scientist, wrote, "Geometry has two great treasures: one is the theorem of Pythagoras; the other, the division of a line in extreme and mean ratio. The first we may compare to a measure of gold; the second we may name a precious jewel."

Extreme and mean? That's a description of a line divided into two parts that equal the Golden Ratio (see page 85).

SIGHT LINES AND AIRLINES

Plato's definition of a line is like that old street-crossing rule: "Look both ways." This road is straight as far as the eye can see. If you look the opposite way and see another straight path, said Plato, you've got yourself a line.

Plato said that a line was "that of which the middle covers the ends." It sounds like a riddle, maybe something the oracle of Delphi would say, but Plato was just talking about a line of sight. He meant that if you stand anywhere on a line, and if you can look straight down it to either end, then you know it's a line.

Euclid's line axiom—*a straight line is the shortest distance between two points*—came from this same idea of two ends and a middle. It's so simple that it sounds almost like a challenge: Is it possible to draw the shortest distance between two points as anything *but* a straight line? If you stick with pencil and paper, you'll be all right—you'll get a straight line every time. Euclidean geometry works well in two dimensions, meaning on a flat plane with length and width.

But we're all standing on Earth, which is a three-dimensional sphere; it has depth too. Suppose you want to fly between two points on its curved surface—from Los Angeles, California, to Tel Aviv, Israel. You *can't* go in a straight line. You'd be flying through solid Earth. You don't even want to head east, the general direction of India. No, the shortest route is north, over the Arctic, and then south, along a curve called a great-circle arc. (To see why, find a globe and connect any two cities with a piece of string. The string, held taut so that it doesn't sag, shows the shortest path between the two points.)

A great circle is any round-the-world circle that has the same center point as the sphere. The equator (0° latitude) is the most famous great circle. Every other degree of latitude is a small circle—off-center in relation to the sphere.

A sphere's surface is one example where non-Euclidean geometry applies. Sophisticated stuff. But not as hard as some people think.

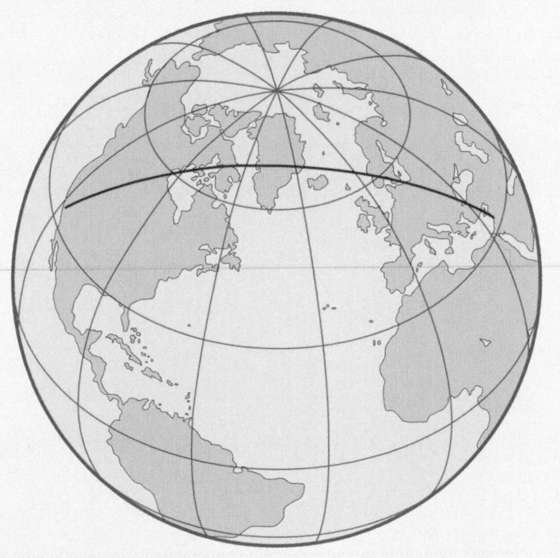

The red line is a great-circle arc, the shortest distance between two points on the surface of a sphere. To travel the whole great circle, just keep following it all the way around the globe to where you started.

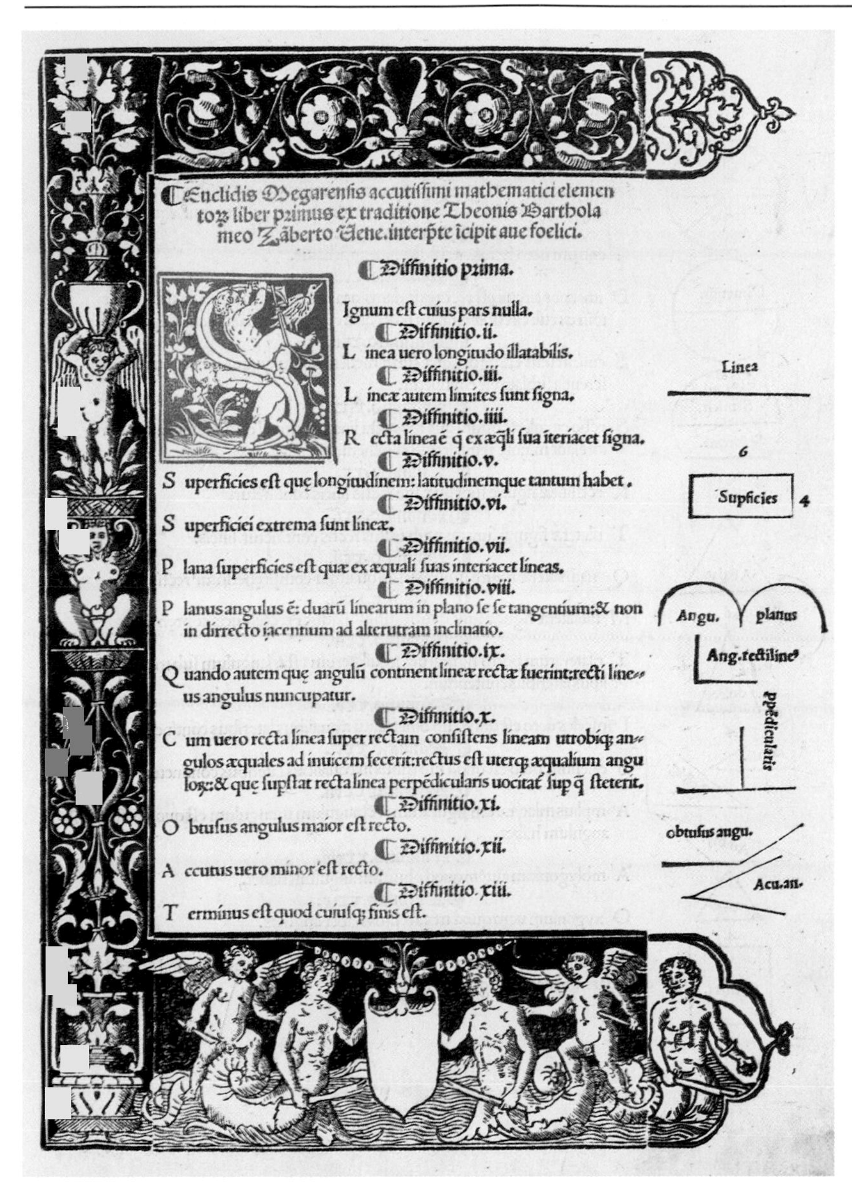
Euclidis Megarenſis accutiſſimi mathematici elementoꝝ liber primus ex traditione Theonis Bartholameo Zāberto Uene. interpte īcipit aue foelici.

Diffinitio prima.

S Ignum eſt cuius pars nulla.

Diffinitio. ii.

L inea uero longitudo illatabilis.

Diffinitio. iii.

L ineæ autem limites ſunt ſigna.

Diffinitio. iiii.

R ecta linea ē q̄ ex æq̄li ſua īteriacet ſigna.

Diffinitio. v.

S uperficies eſt quę longitudinem: latitudinemque tantum habet.

Diffinitio. vi.

S uperficiei extrema ſunt lineæ.

Diffinitio. vii.

P lana ſuperficies eſt quæ ex æquali ſuas interiacet lineas.

Diffinitio. viii.

P lanus angulus ē: duarū linearum in plano ſe ſe tangentium: & non in dirrecto iacentium ad alterutram inclinatio.

Diffinitio. ix.

Q uando autem quę angulū continent lineæ rectæ fuerint: recti lineus angulus nuncupatur.

Diffinitio. x.

C um uero recta linea ſuper rectam conſiſtens lineam utrobiq; angulos æquales ad inuicem fecerit: rectus eſt uterq; æqualium angulorum: & quę ſupſtat recta linea perpēdicularis uocitat̄ ſup q̄ ſteterit.

Diffinitio. xi.

O btuſus angulus maior eſt recto.

Diffinitio. xii.

A ccutus uero minor eſt recto.

Diffinitio. xiii.

T erminus eſt quod cuiuſq; finis eſt.

Euclid's *The Elements* is a spartan list of geometry definitions, axioms (which he called "common notions"), postulates, and other vital math facts. This page is from a decorative edition probably printed in the late fifteenth century—nearly 18 centuries after Euclid's time. How many of today's math books will be around in the year 3800, 18 centuries from now?

Euclid came up with a method of division that allows for a remainder. Thus, 17 divided by 8 leaves a remainder of 1. Is that familiar? You can thank Euclid and Euclidean division.

Euclid was a genius—no doubt about that. His genius was as a scholar and compiler. He seems to have taken all the mathematics since the days of Thales, digested it, and presented it in clearly written form. So *The Elements* is a kind of encyclopedia of world mathematical knowledge. It set standards of precision and mathematical eloquence.

Early math focused on two things: business dealings (essentially arithmetic) and measurement (essentially

The artist Max Ernst (1891–1976) painted a surreal (dreamlike) picture of Euclid as a thinking triangle. How would you paint him?

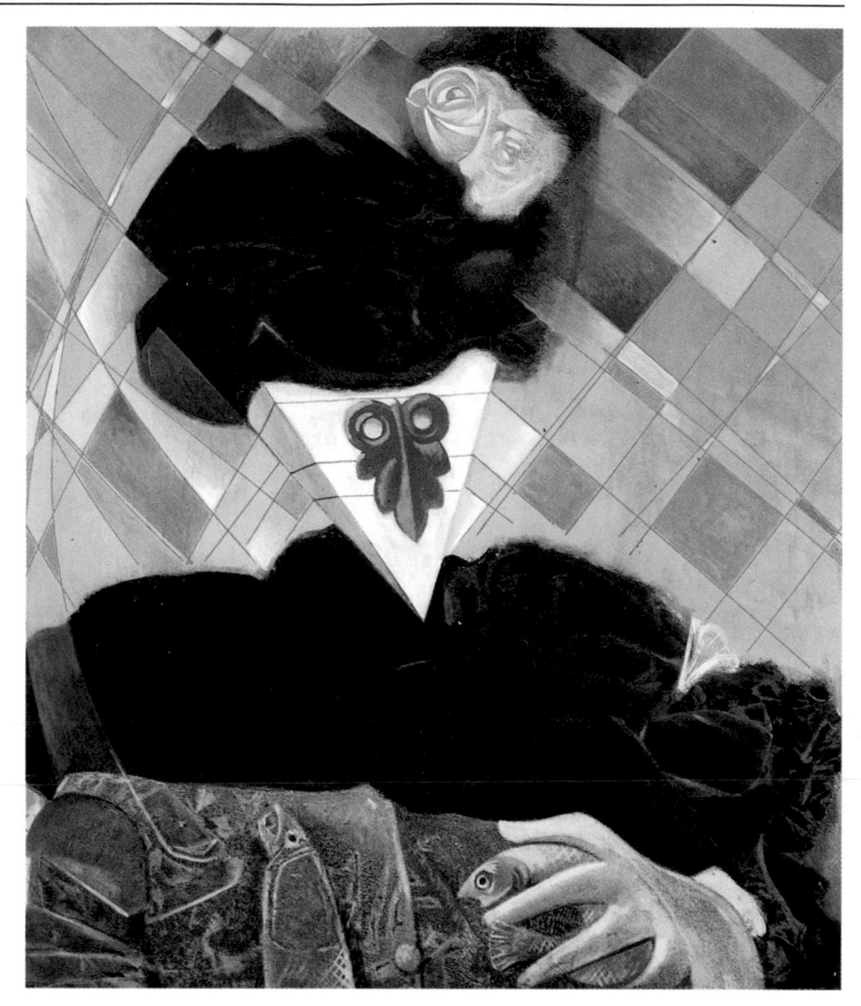

GEOMETRY is about points, lines, angles, circles, spheres, and other shapes too. It's the study of solids and surfaces. Its name comes from two Greek roots: geo for "earth" and metron for "measure." So geometry is about measurement—on Earth, in nature, and in space.

geometry). The Egyptians and the Babylonians had used geometry to solve specific problems. Euclid went further: He thought in terms of general rules and overarching ideas that could provide mathematical guidelines. He began with a series of axioms.

Axioms are statements that mathematicians accept to be true or self-evident: *The whole is greater than the part.* Euclid said we can't be sure of any statements, including axioms, without proof. So he set out to prove them. Pythagoras had observed that $\sqrt{2}$, the square root of 2, is irrational; Euclid came up with a series of logical statements to prove it. A proven statement is a theorem—as in the Pythagorean Theorem.

THINK ICE-CREAM CONE

It wasn't Euclid who was known as the Great Geometer to the ancients. It was someone else. Euclid didn't cover all of geometry's shapes. He missed some special curves–ellipses, parabolas, and hyperbolas–which can't be produced with rulers or compasses, but can be made by cutting through cones.

Cutting cones? Yes, mathematicians find that a fascinating thing to do. (It's kind of interesting to me too.) But before you begin chopping, here's a pause for clarity. As it happens, everyday language and mathematical language can be different. When a mathematician uses the word *cone*, he or she means a double structure. To a mathematician, an ice-cream cone is half a cone. Think two of them balanced on their points, and you have a mathematical cone.

A Greek mathematician named Apollonius of Perga investigated cones and was known in antiquity as the Great Geometer. So if you want to remember Apollonius, just think of him as a conehead. He's famous for slicing through cones and coming up with what we call "conic sections." Apollonius lived in the third century B.C.E.; we believe he was educated in Alexandria, maybe under Archimedes, and that he was a younger contemporary of Eratosthenes. We know he wrote eight books that focused on conic sections. After Euclid's texts, they were math's best-sellers in the ancient world.

Now for some slicing. Take half a cone, cut a flat plane parallel to the cone's base, and you have a circle. Slice at an angle to the base, and you'll get an ellipse (an elongated circle). Cut a slice parallel to a side of the cone, and you'll have an open-ended curve—a parabola; its extensions will go on to infinity. Cut a slice that goes through both halves of the cone, and you get an open double curve known as a hyperbola.

For a long time, conic sections seemed mostly mathematical play (or pure math). Then, 18 centuries after Apollonius, a scientist named Galileo discovered that cannonballs and bullets follow trajectories that are parabolic. The arc of water in a water fountain is a parabola. Johannes Kepler, who was Galileo's contemporary, discovered that most planetary orbits are elliptical. All this was astonishing news to scientists. They began paying attention to conic sections.

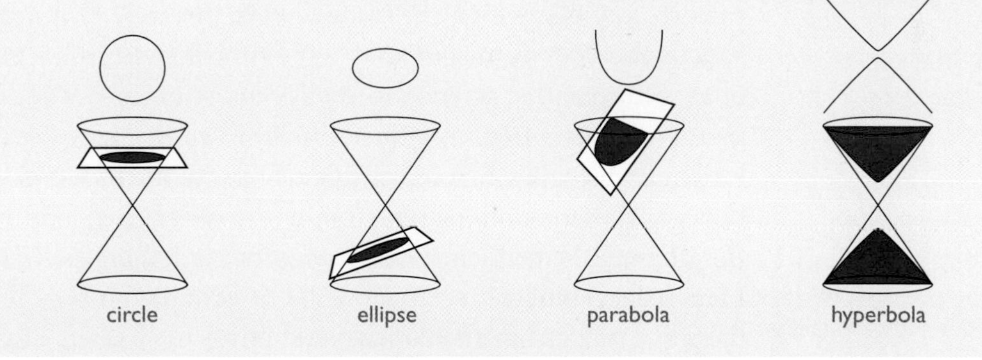

Euclid initiated a whole new way of thinking about mathematics; he set high standards of proof. But there is more to him than that. His axioms and proofs have long been admired for their brevity and clarity. Think of Euclid as someone who drew blueprints. The landscape he sketched was mathematics. It is hard to imagine anyone improving on the presentation.

APOCRYPHAL means "questionable." An apochryphal work was one without a known author. But now the word means "a story that is doubtful or made up." Some Jewish and Christian writings called Apocrypha are controversial and are not part of the established Bible but are taken seriously by many scholars.

THE ONE AND THE MANY

Euclid clearly defined the idea of *one* and *many.* One is the foundation of all numbers and, according to the Greeks, of existence itself. "A unit is that by virtue of which each of the things that exist is called one," he wrote. Out of one comes many. "A number is a multitude composed of ones," said Euclid. When many has a limit, it is called *arithmos*, or number. Its study is arithmetic. When many has no limit, Euclid calls it *plethos*, or multiplicity. In English, *plethora* means "too many to count."

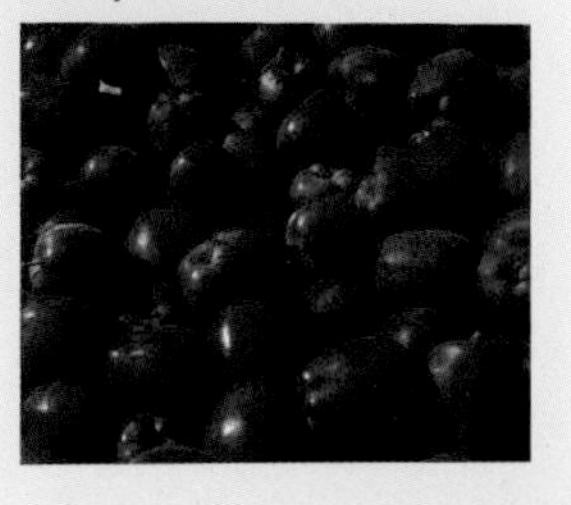

The only personal story that has come to us is of Euclid attempting to instruct King Ptolemy I in geometry. The king asked if Euclid couldn't make things a bit easier for him. "There is no royal road to geometry," he is said to have told the king. But this may be apocryphal (uh-PAHK-ruh-fuhl); the same anecdote is told about Archimedes. (He's in the next chapter.)

After the arrival of mechanical printing (in the late fifteenth century), *The Elements* went through more than 1,000 reprintings, which makes it the most successful textbook of all time. For almost 2,000 years (until the twentieth century), much of geometry was dominated by Euclid and his elegant definitions and proofs. Euclid's work was seen as great truth, beyond questioning. Here are five of his famous axioms, which he called "postulates":

1. You can draw a straight line between any two points.
2. You can extend the line indefinitely.
3. You can draw a circle using any line segment as the radius and one end point as the center.
4. All right angles are equal.
5. Given a line and a point, you can draw only one line through the point that is parallel to the first line.

In the nineteenth century, mathematicians began questioning Euclid's axioms, which they knew were just agreed-upon statements. (Modern scientists have trouble with the idea of absolute truths; science is always questioning.) The mathematicians tried to reduce Euclid's already-spare list even further by using the first four axioms to prove the fifth one. As it turns out, that axiom doesn't follow the others logically, but in the attempt, the mathematicians invented something called non-Euclidean geometry (see page 138). It provided a base for the ideas of a modern mathematician and physicist—Albert Einstein.

But Einstein's ideas were way in the future. When they came, they would help us understand space and the greater universe. Euclid held keys to the down-on-the-Earth, everyday world. Since we don't know much about him, I'm guessing he was a shy, quiet fellow. But his mathematics led to idea explosions and technological wonders. Read the next chapter, and you'll see what I mean.

NUMBERS: IN THEIR PRIME

For fun, mathematician Rich Schwartz drew wacky cartoons to represent numbers. Here's math meeting imagination:

45 = 5 × 3 × 3. A starfish creature represents 5; it overlaps two triangle creatures (3 and 3), slicing through each other.

46 = 2 × 23. A little blue bee, representing a 2, lies flat in the Sun (upper right) and melts down onto a block-headed, one-eyed guy who, by reason of imagination, represents the prime number 23.

55 = 5 × 11. A starfish (5) peels away in spots to reveal a guy with an 11-sided head.

56 = 2 × 2 × 2 × 7. Three bees (look for their antennae), each representing a 2, swarm around a heptagon girl with 7 eyes, who tries to scare them off by biting them.

**o you like puzzles?** Well, prime numbers are puzzlers. They have obsessed a lot of thinkers.

What is a prime number? A prime is a number greater than one that can only be evenly divided by one and by itself. The numbers 2, 3, 5, and 7 are primes. So are 11, 13, 17, 19, and 1,299,007. Two is the only even prime. Primes are sometimes called indivisibles, because no other number can divide them. (Numbers other than primes are called composites or divisibles.) Common sense would have you expect that if a number is big enough, some number can be evenly divided into it. Not true with primes, although when you get to big numbers, it is hard to find them. But there is no end to primes; Euclid proved that there are infinitely many prime numbers.

Natural numbers come one after another in a perfect odd-even sequence, like alternating beads on a necklace. Primes appear among the natural numbers without any known regular sequence—although some thinkers see hints of rhythms in their appearance. Searching for patterns in primes is said to be like searching for an underlying

Math is often about solving puzzles. Look closely at this poster puzzle created by mathematician Rich Schwartz. How do his colorful cartoons represent numbers? Can you find the primes?

musical theme in the number system. Others say primes are just like weeds: they pop up at random. Mathematician Rich Schwartz says, "Some people really do think that primes have a pattern to them—it's just that the pattern is on a very large scale. Looking at the numbers in sequence and deciding if they are prime is something like looking at a painting of a house molecule by molecule; you would never see that it was a house this way because you're too close to it."

Twenty-five percent of the numbers between 1 and 100 are primes. But fewer than 4 percent of those between 1 and 1 trillion are prime. Check your birth year or your home address, and see if they are primes (22,973 is a prime; so is 98,713). How do you find out if a number is prime? That's the catch. There are several approaches; one is called Rabin's test. This fast test (for a computer) takes numbers that are less than the suspected prime and plugs them, one at a time, into an algebraic formula. If just one number fails this test, the suspect is definitely not prime. The more numbers that pass the test, the more sure you can be that the number *is* prime. In 2003, Manindra Agrawal came up with a new test that tells, beyond all doubt, if a number is prime. (For details on how these tests work, ask a friendly math professor.)

One simple method of finding primes was devised by a Greek, Eratosthenes (see chapter 18), and is still in use. It is called Eratosthenes' Sieve. You write down the numbers 2, 3, 4, 5, 6.... You circle 2, and then cross off all multiples of 2. Next, you circle 3 and cross off all multiples of 3. Then you circle the first number that hasn't been crossed off, and cross off all multiples of that number. It isn't very difficult to create a sieve to 100 (see page 270). When you get to big numbers, there are sieves that extend as far as 10,000,000. (If you want to learn more, and you don't know a math professor, try Tobias Dantzig's book *Number: The Language of Science*.) Meanwhile, see if you can figure out which two of these are primes: 19, 57, 81, 97.

A prime with at least 1,000 digits is known as a titanic prime. When that term was introduced in 1984, only 110 such primes were known. Now we know more than 1,000 of them. The Greeks understood the important concept about primes. Here it is again: Primes are like atoms, and the other whole numbers are what we think of as "molecules."

Answer: 19, 97

17 Archimedes' Claw

At last the Romans were reduced to such a state of alarm that if they saw so much as a length of rope or a piece of timber appear over the top of the wall, it was enough to make them cry out, "Look, Archimedes is aiming one of his machines at us!" and they would turn their backs and run.

—Plutarch (ca. 46–120 C.E.), Greek biographer, *Parallel Lives*

The position of Archimedes as the most creative and original mathematician of Antiquity has never been in question—indeed he is usually ranked with Newton and Gauss as one of the supreme mathematical geniuses of all time.

—Stuart Hollingdale (1910–), British author, *Makers of Mathematics*

It sounded as if he were bragging, although everyone knew that Archimedes (ar-kuh-MEE-deez) wasn't a braggart. But he did tell King Hiero II (ruler of the Greek city-state of Syracuse), "Give me somewhere to stand and I will move the earth."

This portrait of Archimedes was painted long after his lifetime. The Italian artist Giuseppe Nogari (1699–ca. 1763) depicts Archimedes with a compass, a tool that wasn't invented until the first century C.E.

Now that would be quite a feat! The Greek god Atlas was supposed to be holding the heavens on his shoulders—but hardly anyone took that tale seriously. They did pay attention to Archimedes (287–212 B.C.E.), who had discovered so many mathematical theorems, written so many scientific books and papers, and invented so many things, that hardly anyone could keep up with his accomplishments.

Archimedes especially loved geometry, which is all about shapes—flat shapes and solid shapes. He studied triangles, circles, ellipses, squares, rectangles, and other polygons (to name a few flat shapes). And he studied pyramids, cubes, cones, cylinders, spheres, and other polyhedrons (to name a few solid shapes). Then he figured out how to measure them all. (Euclid defined, compiled, and organized. Archimedes innovated. Both kinds of minds are needed.)

A POLYGON (meaning "many-angled" in Greek) has two dimensions (length and width) and three or more sides and angles. **A POLYHEDRON** is a solid, a three-dimensional (length, width, depth) figure with four or more faces.

Why would you want to measure the area of a circle? Or the volume of a sphere? Or of a cylinder? For starters, it would let you measure the water in a pipeline, the wood in a log, or the size of the universe. Those first two kinds of measurements had important practical uses in Archimedes' time, as they do now. But the universe? Could someone, even then, have been thinking of space travel? For Archimedes, with one of the most creative minds the world has known, nothing was beyond consideration.

In a book called *The Sand Reckoner*, Archimedes estimated

ARCHIMEDES' SOLIDS

Archimedes found 13 ways to make a polyhedron in which all the faces are regular polygons. For some solids, he just cut off the corners of one of Plato's five perfect solids (see page 99). The third shape, for example, is a truncated cube.

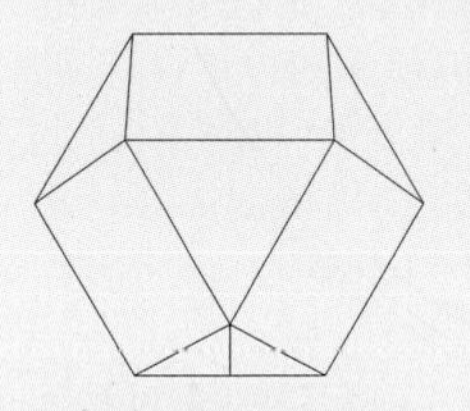

cuboctahedron:
8 triangles, 6 squares

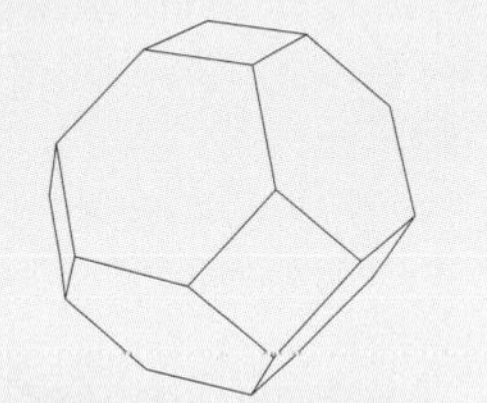

truncated octahedron:
6 squares, 8 hexagons

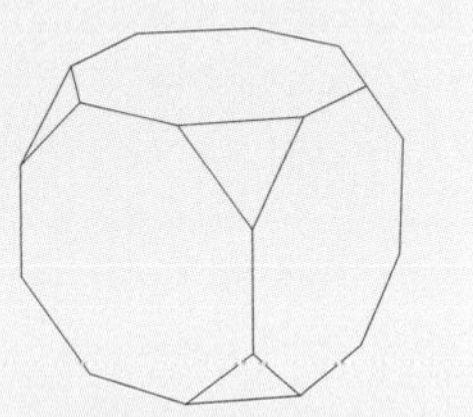

truncated cube:
8 triangles, 6 octagons

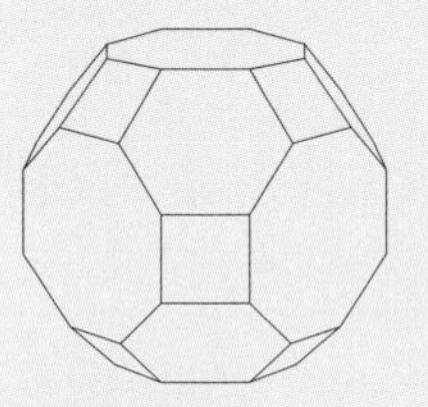

truncated cuboctahedron:
12 squares, 8 hexagons, 6 octagons

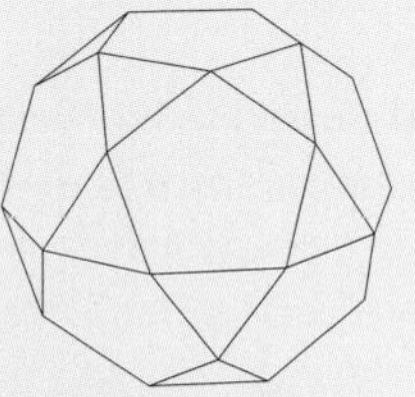

icosidodecahedron:
20 triangles, 12 pentagons

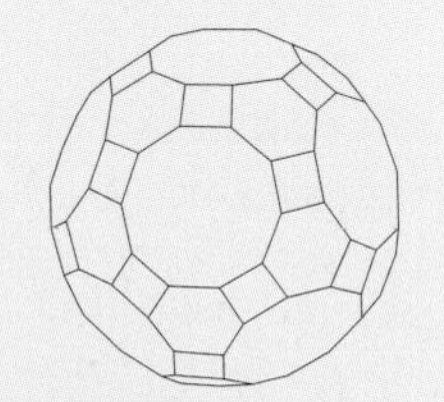

truncated icosidodecahedron:
30 squares, 20 hexagons, 12 decagons

Archimedes' Screws are still in use today. Seven of them pump wastewater in a treatment plant in Memphis, Tennessee. Each screw is 2.5 meters (8.2 feet) in diameter and can lift 75,000 liters (almost 20,000 gallons) per minute. (That's about 600 bathtubs full of water.)

Archimedes' Screw is an inclined (slanted) tube with a giant screw inside it. Turn the handle, and the screw revolves and lifts water to a higher level. The tool is named after Archimedes because he wrote about it in 236 B.C.E., but it's doubtful that he invented it.

the number of grains of sand needed to fill the universe. He was trying to prove two things: that numbers are infinite—you can never have a last number—and that it is possible to deal with really big numbers.

THINK BIG—AND BIGGER

Archimedes wanted to prove that no number is too big to measure. He began by creating his own big number. It was 10,000, and he called it a myriad. Then he said you could multiply a myriad by a myriad and have a myriad myriad, which is written today as 10^8 or 100,000,000.

Archimedes had a serious problem in dealing with these large numbers. The Greeks used alphabet letters for numbers. They didn't have digit symbols, as we do, and zero was missing. A myriad was an *M* with a little alpha (α) on top: $\overset{\alpha}{M}$. To say that Greek mathematicians were hampered is an understatement. It's amazing what they accomplished with what they had.

Archimedes used his system of very large numbers to estimate that the number of grains of sand in the visible world is 10^{63} (as we would write it). That's far too many, but his point was that nothing is too big to measure.

He may be right. We're still finding ever-bigger numbers. A favorite of mine is a googol: 10^{100} (1 with 100 zeroes after it), which is bigger than Archimedes' number of sand grains. The googol was named in 1955 by nine-year-old Milton Sirotta, the nephew of mathematician Edward Kasner. The biggest number unit named is a googolplex: 10^{googol} (1 followed by a googol of zeroes).

These big number words have found their way into everyday language. A *myriad* now means "an infinite number," too many to count. *Googol* with a spelling change (Google) is an Internet search engine many of us use. And here's a quotation from *The Record*, a newspaper in New Jersey: "In North Jersey, where the reservoirs are at 99.2 percent capacity and puddles have taken on a look of permanence, a googolplex of baby mosquitoes is being incubated." That means: If you go to North Jersey in a rainy year, you'd better bring mosquito repellent.

Archimedes sent off a letter to King Hiero II announcing his book. This time he said, "I will try to show you by means of geometrical proofs, which you will be able to follow, that, of the numbers named by me . . . some exceed . . . the number of the mass of sand equal in magnitude to the . . . universe."

Grains of sand to fill the whole universe? That would be some number, even though the universe to the Greeks was only what you can see with your eyes.

Archimedes was the son of an astronomer and, like his dad and Aristotle and most others of his time, he thought the stars and planets were attached to solid but transparent spheres. He built small planetariums in order to make the heavens understandable. Those models told him there was a problem with the concept of an Earth-centered universe. So, like Aristarchus of Samos (who was about 23 years older), Archimedes believed in the less-accepted idea that the Earth and planets revolve around the Sun. (We wouldn't know about Aristarchus if Archimedes hadn't written about him.)

Archimedes thought ideas were more important than things. As to practical science, the Greek historian Plutarch said that Archimedes believed engineering was "sordid and ignoble [as is] every sort of art that lends itself to mere use and profit." (That was Plato's idea. It was a tough influence to shake.)

Archimedes was fooling himself because every time he put his mind to it, he seemed to invent something that was useful—and usually profitable. You could call him an engineering genius—and not be wrong. The truth is, when pure thought and practicality are balanced, civilizations work at their best. And those two extremes were combined in this one amazing man. It doesn't happen often, in people or civilizations.

Archimedes lived in Syracuse, a city-state on the island of Sicily, in the middle of the Mediterranean Sea. (See the map on page 150.) When he was ready for what we'd call college, he sailed off to Alexandria. His teacher was Conon, whose teacher was Euclid. For the rest of his life, Archimedes stayed in touch with the scholars in Alexandria.

Here is what Plutarch says about Archimedes the writer. "It is not possible to find in all geometry more difficult and intricate questions, or more simple and lucid explanations. Some ascribe this to his natural genius; while others think that incredible effort and toil produced these, to all appearances, easy and unlaboured results."

Did Archimedes work hard at writing? I don't know, but it takes hard work for most people—like me—to produce clear and simple sentences. I have to write and rewrite.

ENGINEERING is about using the laws of science to build things—bridges, tunnels, temples, monuments, machines, and so on. If building useful things is SORDID ("dirty, vile") and IGNOBLE ("unworthy"), where does that leave Hero, the Alexandrian engineer who built mostly frivolous things?

A copper coin shows the head of Hiero II, who ruled Syracuse for 60 years. Plutarch describes Archimedes as a near relative of Hiero, a king who often sought Archimedes' advice on military and other matters.

That wasn't difficult. The Mediterranean world was linked by the sea. It made travel and communication easy—for traders, travelers, and armies on the move. And in Archimedes' time, two big armies were moving. The Romans (from Italy) and the Carthaginians (from Africa) were fighting for control of the sea-lapped lands.

Most of Sicily was ruled by Carthage. But prosperous Syracuse, at one end of the island, managed to stay independent. That wasn't easy. Carthage wanted Syracuse. So did Rome. Syracuse's location and wealth made it especially attractive to power seekers.

Politics didn't interest Archimedes. It was math and science that he cared about. But King Hiero II, who may have been his cousin, kept pestering him for help. So Archimedes invented some war machines, in case Syracuse needed to defend itself.

Then the king came to him with a personal problem. He had given a big hunk of gold to a jeweler to make a gold crown. But when he got the crown, King Hiero believed he

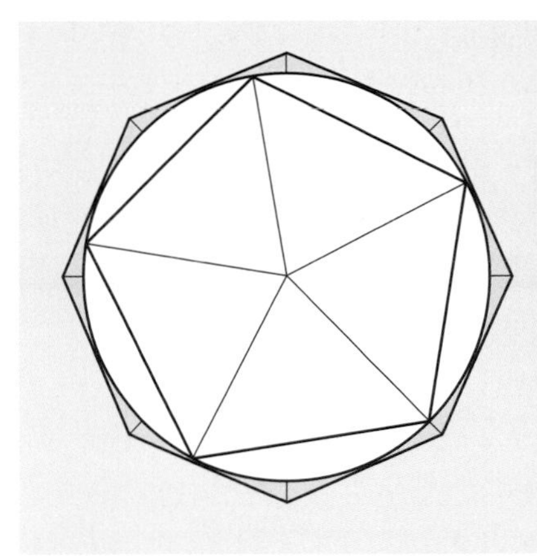

WHEN CLOSE COUNTS

How do you measure the curve of a circle? Archimedes worked out something called the method of exhaustion. It was a first step toward calculus, which he might have discovered if he had had decent number symbols to use. He came up with a value for π (pi) that was closer than any yet achieved (see page 71). He did that by drawing polygons inside and outside a circle. As the polygons were given more and more sides (this *was* exhausting), they came close to becoming a circle. The circumference of the circle—a measurement needed to calculate π—was in between the circumferences of the inner and outer polygons.

had been cheated. He thought the jeweler had mixed cheaper silver with the gold and kept some of the gold for himself. But no one knew how to tell a mixture of metals—an alloy—from a pure metal. And that included Archimedes.

He started to think about the problem. Soon he couldn't get it out of his mind. How can you tell a pure metal from an alloy? You can compare color or shine, and you can test hardness. The crown passed all those tests. A different approach was needed. King Hiero told Archimedes he was not to harm the crown. Archimedes thought about the problem when he was eating, when he was walking, and he must have dreamed about it when he was sleeping.

Density seemed a possible way to go. Gold is more dense—heavier for its size—than silver. So a pure gold crown would weigh more than a crown *of the same size* with both gold and silver in it. But this crown weighed the same as the chunk of gold the king had given the goldsmith. Another dead end.

If you're thinking scientifically, it helps to restate a problem. Archimedes may have reworded it this way: A gold-silver crown that weighs the same as a pure-gold crown *would have to be a bit bigger*; its volume would be greater. Was this crown bigger than the chunk of gold? Had the jeweler added extra silver to make its weight right? Measuring the exact volume of an irregular shape seemed impossible. How could it be done?

Archimedes was in the public bath when the answer came

To understand density, it helps to define mass and volume. MASS is the measure of the matter in an object. On the Moon, an object's mass (its quantity of matter) is the same as on Earth, but its weight is less because the Moon's gravity is weaker. On Earth, mass and weight are generally the same value. (The gravitational pull on a mountaintop is slightly less than in a valley, so even objects on Earth can vary in weight depending on their location.) VOLUME is the amount of space an object takes up. It's not exactly the same thing as size. Even if a Wiffle™ ball and a baseball are the same size, the hollow Wiffle™ ball has less volume. DENSITY is a ratio of mass to volume. It's expressed as mass (m) divided by volume (v), or $d = m/v$.

The woodcut at right (probably from the nineteenth century) imaginatively depicts Archimedes' "Eureka!" moment in the bathtub. In Archimedes' time, crowns were wreath shaped, like the one above.

Did Archimedes really run naked through the streets? One source is a Roman named Vitruvius, a military engineer in Africa for Julius Caesar. He tells of Archimedes' "Eureka!" moment in a 10-volume book on architecture and engineering. His work has a base in science that is unusual for books of any era.

to him. He was so excited that he jumped out of the bath and ran home through the streets of Syracuse—stark naked. He was too intent on his thoughts to take time to put on his clothes. (The Greeks didn't worry about nakedness as much as we do.) He kept shouting, "Eureka (yoo-REE-kuh)! Eureka!"—which means "I have it! I have it!"

When Archimedes had stepped into the full bath, his body displaced water. It splashed above the rim and onto the floor. That's what gave him his big insight. He realized that the water he displaced was equal in volume to the part of his body that was underwater.

Scientific breakthroughs are often made by seeing connections, and that's exactly what happened. Archimedes suddenly realized that his body and a chunk of gold must react the same way when put in water. **Any object—a man, a crown, or a piece of gold—when immersed in a liquid will displace a volume of liquid equal to the object's own volume.** Now all Archimedes had to do was to lower the crown into a bowl filled with water to the rim. Which he did. Then he measured the water that spilled over. He did the same thing with a piece of pure gold the same size as the one the king had given the jeweler.

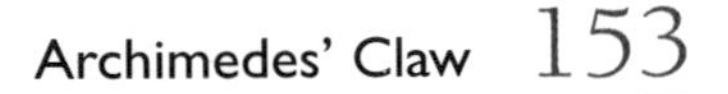

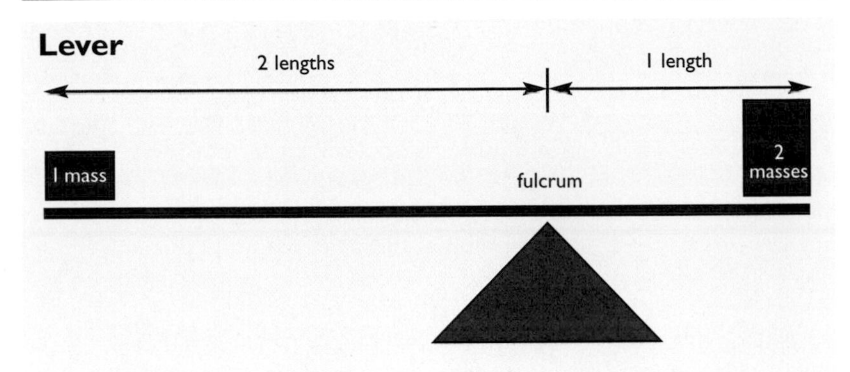

Archimedes' Law of the Lever led to the notion of center of gravity: For every object, there is a single point at which the force of gravity appears to act on that object. In this drawing, that point is where the lever rests on the fulcrum. The mass that is twice as heavy is half the distance to the fulcrum.

Did the king's crown displace the same amount of water as the pure gold? No. It displaced more water. The crown had a greater volume than the gold. It was padded with silver. The king had been cheated.

What about moving the Earth? Remember, Archimedes said that if he had somewhere to stand, he could move the Earth. To prove his point, Archimedes told the king to find something very heavy, and he would move it. The king picked his biggest ship and loaded it with cargo and sailors. The ship was in dry dock, so it couldn't glide through water, which would have made things easier. Archimedes had to lift it all by himself. He did. With one hand, or so the story goes.

In case you were wondering, the Earth's mass is just under 6 sextillion metric tons, or 6 followed by 21 zeroes. (A metric ton is equal to 1.1 standard U.S. tons.)

He worked out a combination of levers and pulleys and proved that with a lever that's long enough and balanced properly, you can lift anything, even the Earth. Archimedes was one of the world's first experts in mechanics.

In simple terms, his Law of the Lever, still a basic law in physics, says this: If you have a heavy person on one side of a seesaw and a light one on the other, you need to move the heavy person closer to the fulcrum and give the lightweight a lot of board. To balance, the product of weight times distance must be the same on both sides of the seesaw (see illustration above).

A LEVER is a rigid bar or beam that rests on a FULCRUM (fixed support). A PULLEY is a wheel-and-rope combination used to lift or pull objects. MECHANICS is the science of motion and machines (like levers and pulleys) that move things.

How about those war machines that Archimedes invented for King Hiero II? He hoped they would never have to be

A STOMACH GRABBER, NOW AND THEN

Guess who made the front page of *The New York Times* on Sunday, December 14, 2003? Yes, it was our friend Archimedes. How come? It seems that 2,200 years ago he came up with some puzzles that fit right into a vigorous branch of today's mathematical world. The big news is that some of his problems have been rediscovered after being lost for more than two millennia, and they are mind-bending, up-to-the minute, and fun too.

Here's the story: Archimedes wrote a treatise called "The Stomachion" (stoh-MAHK-yon), which means "stomach." It was lost long ago, except for a tiny bit of the introduction. No one could figure out what it was all about, so the work was mostly forgotten. Then a palimpsest (PAL-imp-sest) turned up in Baltimore, Maryland. (A palimpsest is a manuscript written over—like a painting done on top of an older one.) It seems that in the year 975, someone made a copy of Archimedes' work in Greek on parchment. In the thirteenth century, paper being expensive, some Christian monks tore that manuscript apart and washed its pages. Then they folded the pages in half, rebound them, and wrote out their prayers.

A Danish scholar discovered the palimpsest in Istanbul, Turkey, in 1906. He noticed faint mathematical formulas under the prayers and photographed most of the pages. Then the leather-bound book disappeared, to turn up in the 1970s in the hands of a French family. By this time it was in terrible condition—moldy, dirty, and ragged. Still, it was worth $2 million to an anonymous billionaire who bought it and then lent it to the Walters Art Museum in Baltimore.

Scholars—from Stanford to Johns Hopkins to Oxford—got to work. Ultraviolet light showed them both prayers and mathematics, but not clearly. Computer programs had to be written to separate the mathematics from the writing on top of it. There's a whole lot more to this detective story, but the scholarly experts did amazing work. This is what they found: "The Stomachion" deals with a robust branch of math called combinatorics.

Combinatorics? How many ways are there to make change for one dollar if you can use any number of coins of any denomination? Each way you find is a combinatorial solution. Combinatorics is everywhere. Loosely, it is the science of counting arrangements and possible combinations. Computers have energized the field and expanded its possibilities. If you get into combinatorics, you'll find it a stomach turner. That, at least, is the explanation some mathematicians have for Archimedes' title.

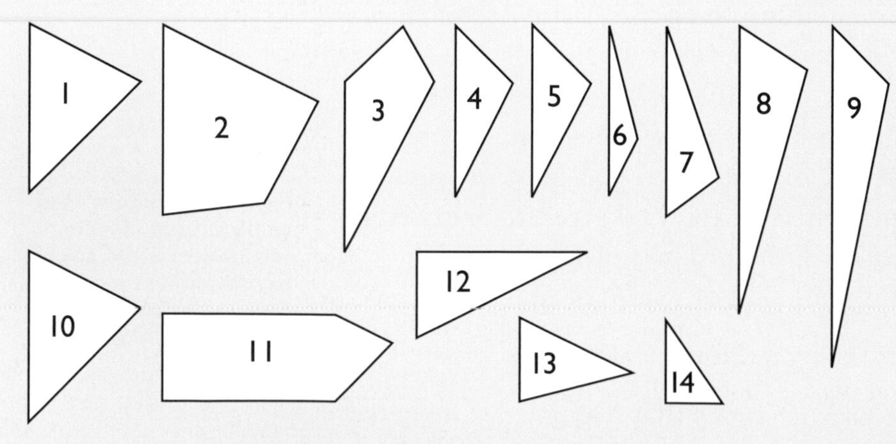

The 14 shapes in this illustration come straight from Archimedes' "Stomachion." The object of this puzzle is to put the pieces together to make a square.

Did sunlight and shields really set those Roman sails on fire? We have the word of Plutarch, a Greek historian writing more than 200 years later, but modern physicists are still debating the science and technology of Archimedes' mirrors.

used, but in 215 B.C.E. the Romans were on the march—and they wanted Syracuse. By this time Hiero was dead and his grandson, Hieronymus, made a big mistake in picking Carthage as the potential winner in the Second Punic War.

The Romans were confident. The powerful Roman legions (armies) had been squashing everyone who got in their way. Syracuse was a treasure to put in their chest. Its strategic location made it vital to any power that wanted control of the Mediterranean. Besides, the Romans hated the Carthaginians—and they worried that some people in Syracuse, like the new king, were getting too friendly with Carthage.

So Roman infantrymen—15,000 of them with shields and armor and fancy weaponry—were loaded onto a fleet of 60 ships and sent to capture Syracuse. The soldiers would have to climb the city's walls, but they were used to doing that. Theirs was the best army anywhere. They were prepared for almost anything—except the mind of a scientist.

Ancient catapults were powered by the tension of a wound-up rope. Soldiers used a winch and ratchet to pull the rope as taut as they could and to secure it in place. They placed a missile—a stone as heavy as a man—on the end of a wooden arm. As soon as they released the rope, the arm flung the missile in a long, high arc. The range of Archimedes' catapult was twice as far as a football field.

Archimedes is said to have arranged a large number of flat mirrors, or polished shields, that could be focused on ships approaching the island. The mirrors reflected the sun's rays and blinded the Roman sailors. The sum of all those small spots of sunlight was so intense that it set their cotton sails on fire. The ships turned back.

Meanwhile, the Romans weren't finished. They rebuilt their ships. This time they set sail on a cloudy day.

DIABOLICAL means "fiendish" or "wicked." It comes from the Greek word for devil.

Archimedes was ready for them with some diabolical machines. Giant catapults, like slingshots, hurled ballistic missiles (lethal stones) at the ships. That wasn't all. Cranes swung out from the city's walls, dropping huge boulders on those who got too near. At the same time, archers shot deadly arrows. Still, that wasn't the worst of it.

Combinatorics asks: How many possible ways can the 14 pieces on page 154 be put together to make a square? That's what Archimedes was trying to figure out. We still don't know his answer. (There are tears and gaps in his manuscript.) We do know what modern mathematicians have found: There are 17,151 ways to make a square from Archimedes' 14 shapes. Here are three of the solutions.

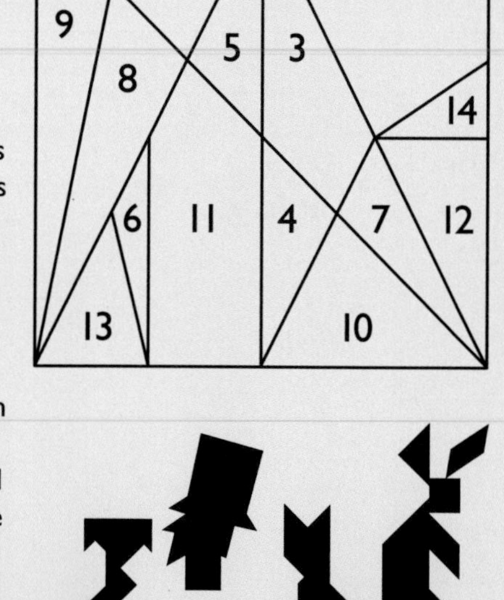

Do you recognize these characters from Lewis Carroll's *Alice in Wonderland*? Carroll made each one out of tangram puzzle pieces—five triangles, a square, and a rhomboid.

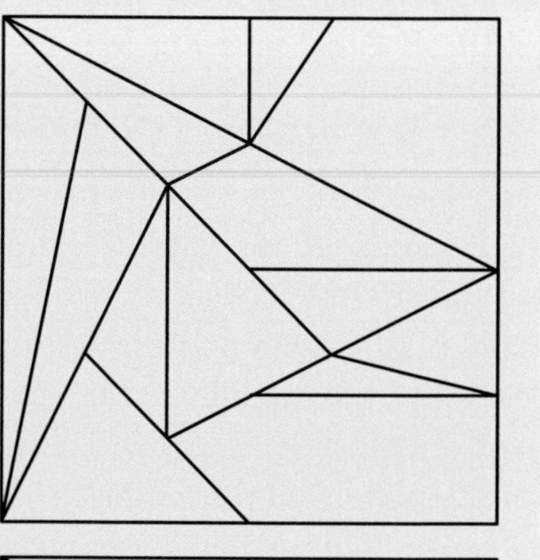

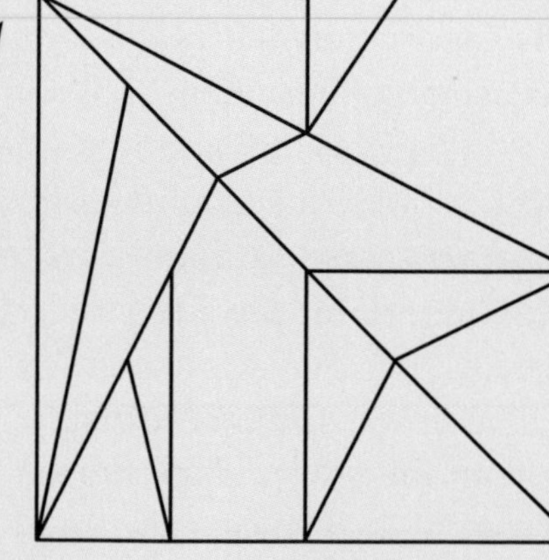

DEATH BY MATH

Despite Archimedes' war machines, the Romans finally besieged and captured Syracuse.

According to the story, Archimedes was concentrating on a mathematical problem, scratching numbers in the dirt at his feet, when a Roman soldier called out to him. Archimedes was so immersed in his work, he didn't respond. (He was famous for both his ability to concentrate and his absentmindedness.) The soldier ran a sword through his body.

A copy of a Roman mosaic shows Archimedes reaching for his chalkboard while an impatient soldier commands him to follow.

The Roman fleet was in for a horror beyond anything anyone could have imagined. It was about to meet Archimedes' Claw! Here is a description of what happened, written by the Greek historian Plutarch:

> *The ships, drawn by engines within, and whirled about, were dashed against steep rocks that stood jutting out under the walls, with great destruction of the soldiers that were aboard them. A ship was frequently lifted up to a great height in the air (a dreadful thing to behold) and was rolled to and fro, and kept swinging, until mariners [sailors] were all thrown out, when at length it was dashed against the rocks, or let fall.*

A 75-year-old scientist had stopped the world's greatest army with the power of his brain.

IS IT A CLAW OR A FLAW?

Is this what Archimedes' Claw looked like? Not a chance. Compare it to the engineer's precise and detailed diagrams of the weapon's mechanics (opposite). The painter of the piece below, Giulio Parigi, had a different goal. He relied on imagination rather than science to portray the drama and terror of a battle. His artistic claw, painted in 1599, is part of a wall mural in the Uffizi Gallery in Florence, Italy. Parigi also depicted Archimedes' mirrors (see page 174). If you visit the Uffizi, you'll find both murals in the Stanzino delle Matematiche, the Mathematics Chamber.

**id Archimedes** actually build "the claw" and those other military machines? Most contemporary records have been lost, and no one was taking photographs then. But, in addition to the words of the Greek historian Plutarch, we have those of Livy, a Roman historian. He described the Roman attack on Syracuse and was awed by Archimedes' achievement. And the Romans didn't like to give credit to their enemies.

Time often leads to exaggerations. Both Livy and Plutarch made up dialogue and didn't seem to worry about standards of accuracy, as modern historians do. Other reports came from terrified Roman soldiers who had even more reason to exaggerate.

The truth is, no one knows exactly how Archimedes' Claw worked or what it looked like. To engineers, that's an irresistible invitation to tinker. How big was the machine? What shape was the claw? How many pounds could the levers and pulleys lift? Who or what did the lifting? Is it even possible to build such a device as the historians described it and

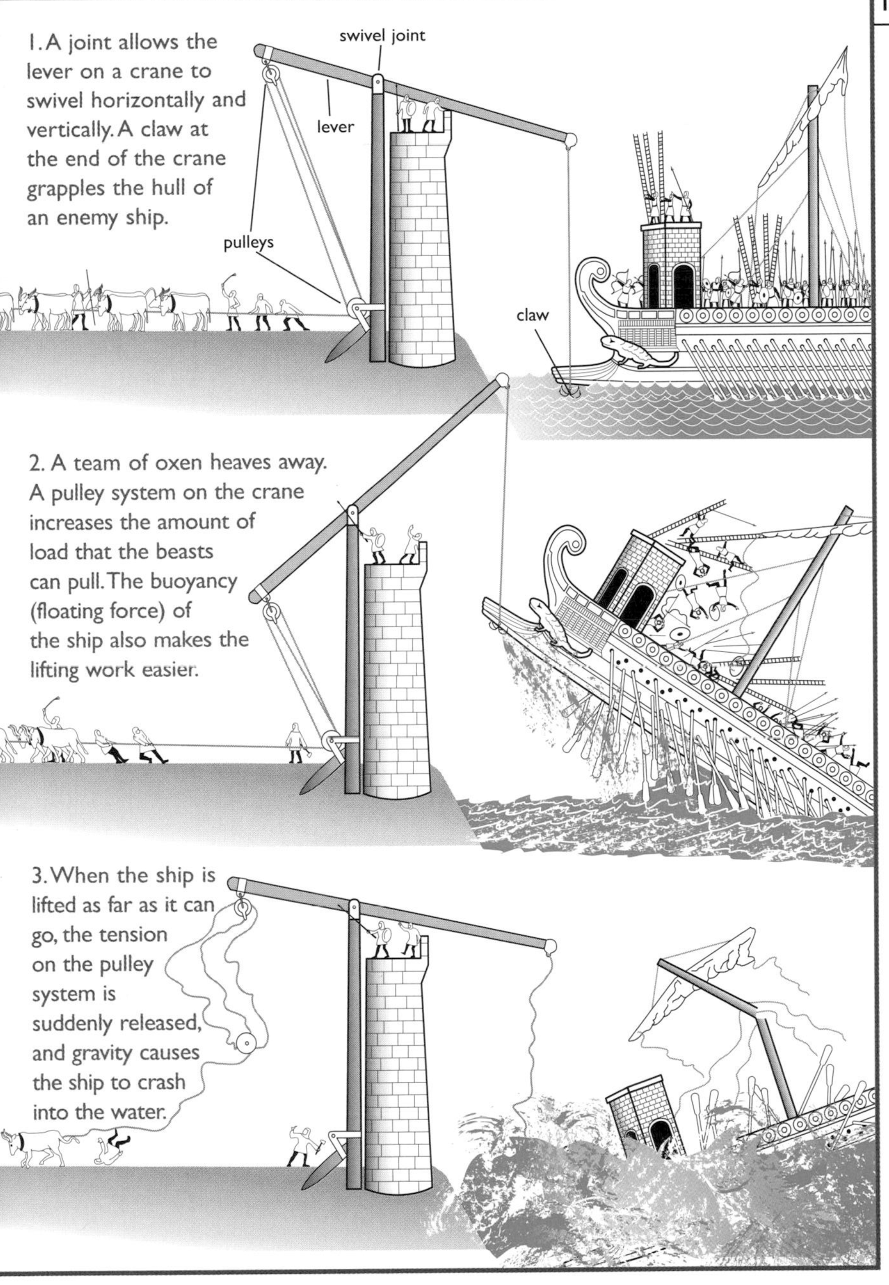

model their ideas to see what works.

Here's one modern engineer's vision of Archimedes' Claw. Would it get the job done? One flaw is that the ship has to float directly over the claw before it can be grabbed. Wouldn't the Roman soldiers see the rope and just row around it? Another challenge is speed. After upending a ship, how long would it take to reset the rope? Too much time would give Romans a chance to land and attack or get away. And what about materials? The claw's wood beams would have to be strong enough not to crack or warp—and be water resistant too in

follow the laws of physics?

Those laws haven't changed. Gravity works the same way today as it did in Archimedes' time. That means modern engineers can do what Archimedes did—scheme and sketch and create and

a seaside setting. The ropes had to stretch and snap—again and again—without breaking. Could they?

A good motto for engineers is "There's always a better way." With that attitude, someone is sure to find it.

18

Measuring the Earth

I am human and let nothing human be alien to me.

—Terence (ca. 185–ca. 159 B.C.E.), Roman slave who became a poet and playwright, *The Self-Tormentor*

[The Greeks showed us] that the universe is mathematically designed.... There is law and order in the universe and mathematics is the key to this order. Moreover, human reason can penetrate the plan and reveal the mathematical structure.

—Morris Kline (1908–1992), American math professor, *Mathematics: The Loss of Certainty*

Philosophy is not a theory but an activity.

—Ludwig J. J. Wittgenstein (1889–1951), Austrian philosopher, teacher, and gardener, *Tractatus Logico-philosophicus* (*"Treatise on Logical Philosophy"*)

Eratosthenes (ca. 275–ca. 195 B.C.E.), who was born in coastal North Africa (now Libya), went to school in Athens and then was called to Egypt by Ptolemy III, who asked him to become director of the library/museum/university at Alexandria. He was the perfect person for that job. Eratosthenes (er-uh-TOS-thuh-neez) was a great scholar himself, and he energized others. His nickname was "Beta,"

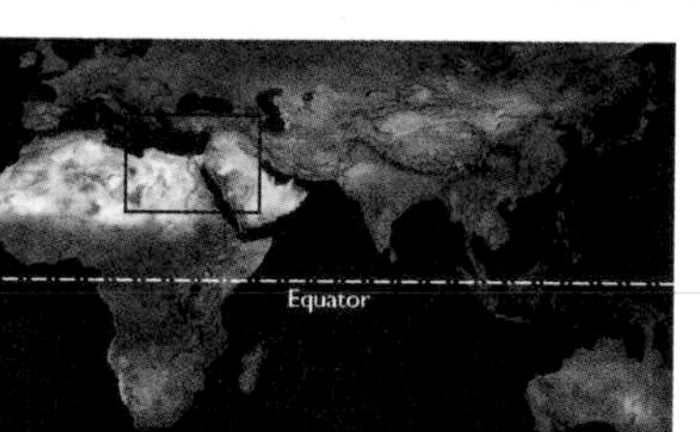

Eratosthenes was born in Cyrene, now in Libya, not to be confused with Syene, which is in southern Egypt. Cyrene was founded by emigrants from the Greek island-state of Thera in 631 B.C.E. In its day, it was a classy place to live. The Greeks started many colonies around the Mediterranean Sea. Croton was another colony.

When Cyrene came under Egyptian rule (from 323 B.C.E.), it had a heyday as a great intellectual center with a famous medical school. Eratosthenes was a renowned citizen (before moving to Athens and then Alexandria); so was the philosopher Aristippus. In 96 B.C.E., the Romans took over. The city prospered for the next several centuries until the Arabs conquered the region and the city was abandoned. Today it is Shahat, Libya.

It's exactly noon in the city of Aswân (formerly Syene) in southern Egypt. How can you tell? Noon on the summer solstice (June 20 or 21) is the only time the Sun is directly overhead and its rays can reach the bottom of this very deep well. Aswân is near the tropic of Cancer (which is 23.5°N), the latitude line that defines the northern limit of Earth's Tropical Zone.

which is the second letter in the Greek alphabet. Some people thought him second only to Aristotle in broad talent. We know Archimedes was one of his friends, because Archimedes dedicated a book to him. But today Eratosthenes is remembered mostly because of some measuring he did.

Eratosthenes discovered that at noon on the summer solstice (the longest day of the year), at Syene (an Egyptian city near modern Aswân on the southern Nile) the Sun's rays lit the very bottom of a deep well, and a stick cast no shadow at all.

That wasn't true at Alexandria, on the northern mouth of the Nile, where at the same time on the same day, the Sun's rays did cast a shadow.

Eratosthenes must have asked himself, "Why a shadow at Alexandria and no shadow at Syene?" Then he figured out that the Sun must be right overhead at Syene but not quite overhead at Alexandria, and he realized that he had hit

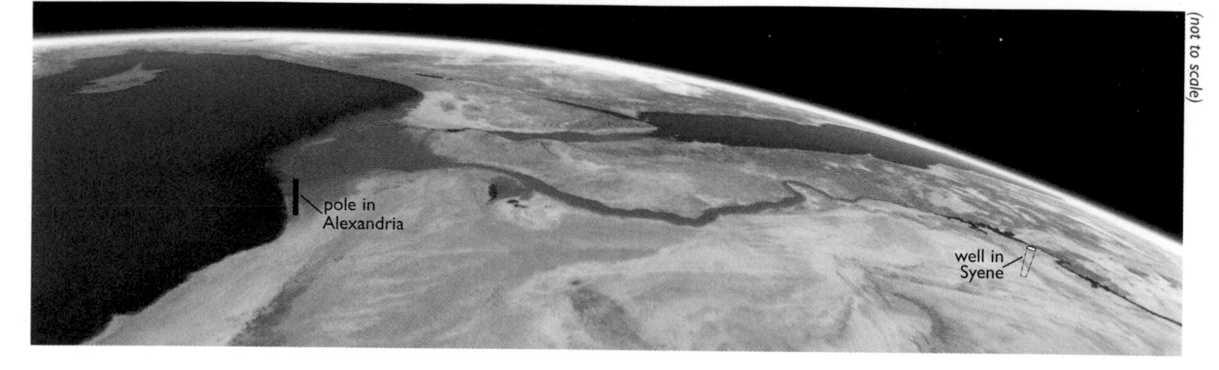

It's not a straight line from Alexandria to ancient Syene. Though the cities aren't far apart, the span of the Earth between them is curved. Because of that curve, sunlight strikes the cities at slightly different angles.

upon valuable information. He saw it as indicating that the Earth is curved. He wanted to prove that idea.

Eratosthenes is said to have put a stick in the ground in Alexandria and measured the angle of its shadow. It was 7.2°, which is 1/50 of a circle If the Earth is truly a globe—as he believed—then the distance between Alexandria and Syene is 1/50 of that globe. To find the size of the whole globe, all he had to do was measure the distance between the two cities. So he paid someone to walk from Alexandria to Syene, counting his steps, and then he multiplied that distance by 50. The answer was very close to the best modern measurements of the circumference of the Earth (see pages 164–165). That was more than two millennia

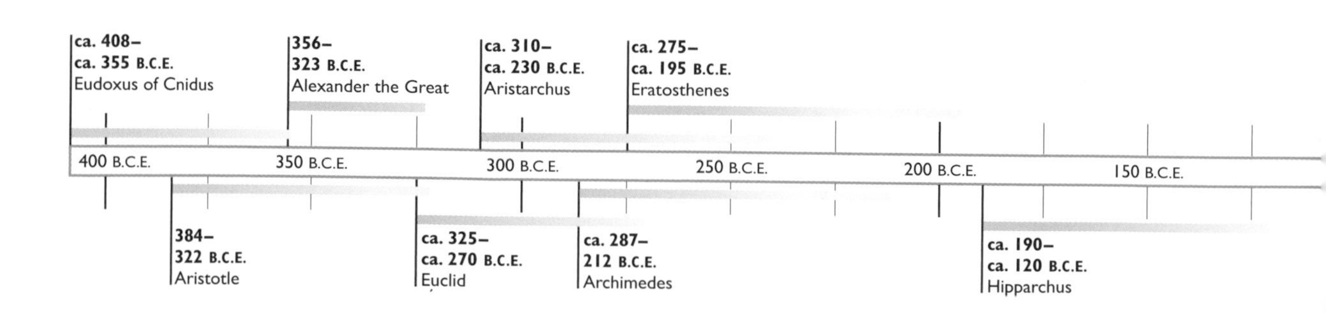

ago, and the only equipment Eratosthenes used was a stick in the ground, his brain, and a hired walker.

Given what he found to be the large size of the Earth and the small size of the known land, Eratosthenes surmised that there was a huge interconnected ocean. (That would be verified by the voyage of Magellan 18 centuries later.)

Most of Eratosthenes' works, like those of most of the ancient thinkers, have been lost. Much of what we know of him and his accomplishments comes from the written comments of others. And they tell us that he was a geographer, historian, and literary critic, as well as an astronomer. He was the first man, of whom we know, who was concerned with accurate dating. He set up a chronology that began history with the Trojan War. He devised a system for determining prime numbers that is still called the Eratosthenes' Sieve (see page 145.) But it was when he came up with that close-to-accurate measurement of the Earth that he demonstrated that the universe is understandable. Given some brainpower, we can figure out how it works. What would his contemporaries and future generations do with that insight? For a long time they just forgot all about it.

Eratosthenes is known as the Father of Geodesy (jee-OD-uh-see), the science of Earth measurement. Besides measuring the whole globe, he mapped the known world, from Britain to Ceylon and from the Caspian Sea to Ethiopia. He also made a star map with 675 stars, and he wrote a treatise on Greek comedy.

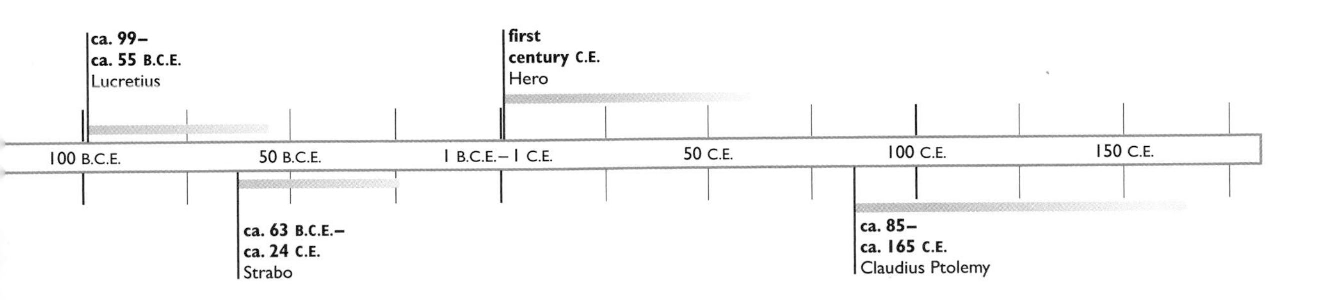

How Did Eratosthenes Come So Close?

Syene, near modern Aswân, Egypt, is close to the tropic of Cancer (23.5° N). At that line of latitude, at noon on the summer solstice (June 20 or 21), the Sun is directly overhead. Stand there and you'll have no shadow. Alexandria is farther north, so there the noon Sun is not quite overhead on the solstice. You, or a stick, will cast a shadow that's at about a 7° angle.

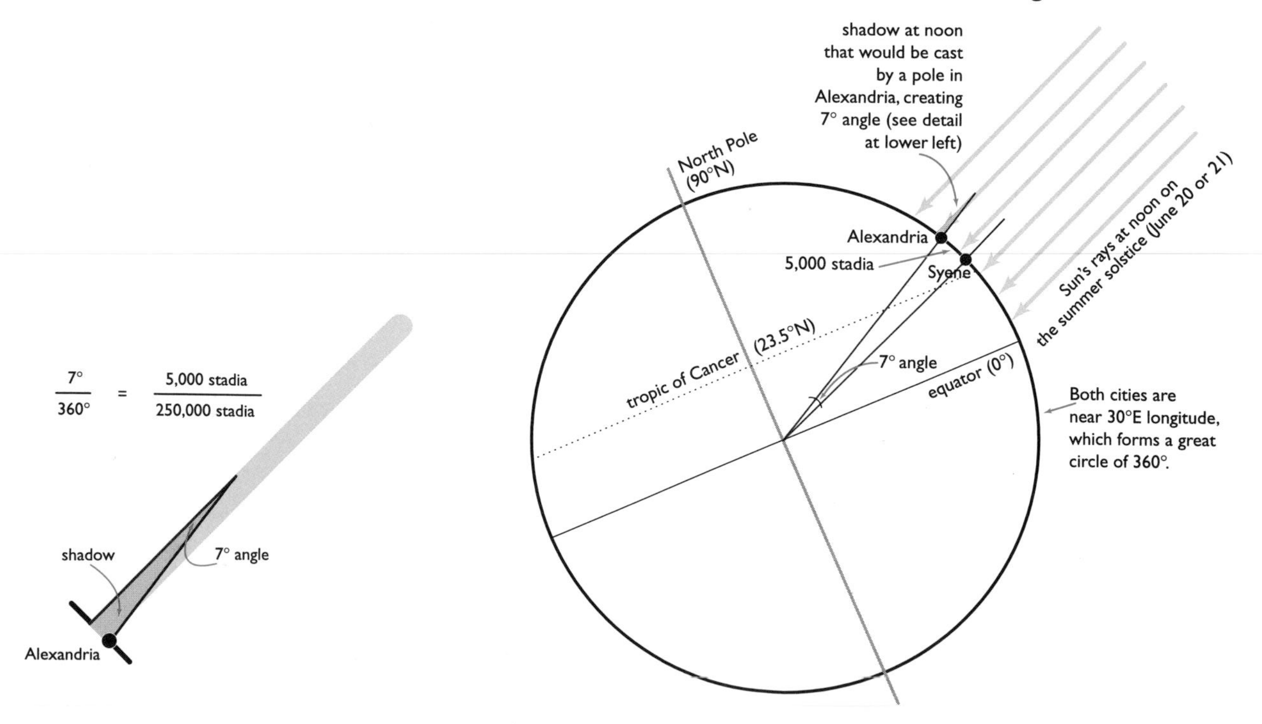

WHY WAS IT SO TOUGH?

Today's satellites can map and measure the Earth within inches of accuracy. It wasn't that easy for Eratosthenes. He had no way to measure long distances well, and even very careful step-counting is an uneven thing. The round number 5,000 stadia is a clue that Eratosthenes knew the result was an estimate.

Also, Alexandria isn't exactly due north of Aswân/Syene. It's a couple of degrees west of it. Aswân/Syene isn't on the tropic of Cancer. It's a little north of it. Eratosthenes couldn't know about those slightly off-kilter positions, which threw off his result a bit.

There's one more thing no one knew back then. Earth bulges at the middle and is a little flattened at the poles. That makes the meridians (the north-south great circles) a little shorter than the east-west Equator, which is 40,075 kilometers (24,901 miles) long.

Those tiny quirks probably wouldn't have mattered to Eratosthenes. He fudged by tacking on extra stadia to make the math easier, which shows he wasn't aiming for perfection. He just wanted a close estimate, and he did incredibly well.

Both cities lie near the same meridian, or longitude line, which is a north-south great circle (a circumference, really) around the Earth. Like all circles, the meridian has 360°. Since 7° is about 1/50 of 360°, the distance between the two cities (5,000 stadia) must be 1/50 of the distance around the meridian. By multiplying 5,000 by 50, Eratosthenes estimated Earth's circumference at 250,000 stadia. Actually, he fudged a little. He added 2,000 stadia to get a number that was easier to work with: 252,000. (It's evenly divisible by 60 and 360.)

So, you might be wondering by now, just how long was a stadium (singular for stadia)? The experts aren't quite sure, but they say it's between about 150 and 158 meters (164 and 173 yards). At 157 meters (172 yards), a popular choice, Eratosthenes figured Earth's circumference at about 39,250 kilometers (24,390 miles). That's amazingly close to today's measure of the north-south circumference, which is about 40,000 kilometers (24,855 miles).

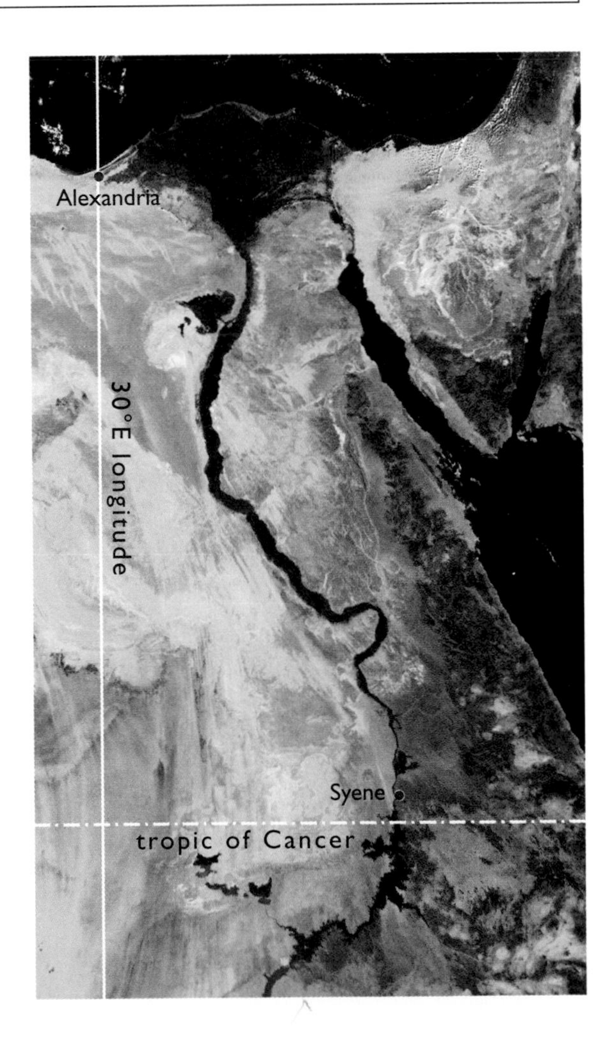

19 Rome Rules

The end of democracy is freedom; of oligarchy, wealth; of aristocracy, the maintenance of education and national institutions; of tyranny, the protection of the tyrant.

—Aristotle (384–322 B.C.E.), Greek philosopher and scientist, *Rhetoric*

Sovereignty in a state is thrown like a ball from kings to tyrants, from tyrants to aristocrats (or to the people at large), and finally to an oligarchy or to another tyrant. No single type of government lasts very long. This being the case, I regard monarchy as the best of the three basic types of government. But a moderate, mixed type of government, combining all three elements, is even better.

—Marcus Tullius Cicero (106–43 B.C.E.), Roman lawyer, orator, and statesman, *On the Republic*

So much were all things at Rome made to depend upon religion; they would not allow any contempt of the omens and the ancient rites, ...thinking it to be of more importance to the public safety that the magistrates should reverence the gods, than that they should overcome their enemies.

—Plutarch (ca. 46–ca. 120 C.E.), Greek historian, *Parallel Lives*

It was the Romans who accomplished what Alexander intended. Under the Roman Republic (509–31 B.C.E.), Rome grew from a small city-state to a mighty colossus that circled the Mediterranean Sea and went beyond. The republic was designed as a people's government, and it flourished impressively. Roman citizens elected the governing officials.

On this ancient Roman coin, Julius Caesar (left) looks back while his adopted son and great-nephew, Octavian, looks to the future.

Slaves, women, and conquered peoples didn't get a vote, so you can't really call them citizens. Still, Rome's representative government was a powerful idea and a big step toward broad, democratic, representative government. (When American patriots gathered in

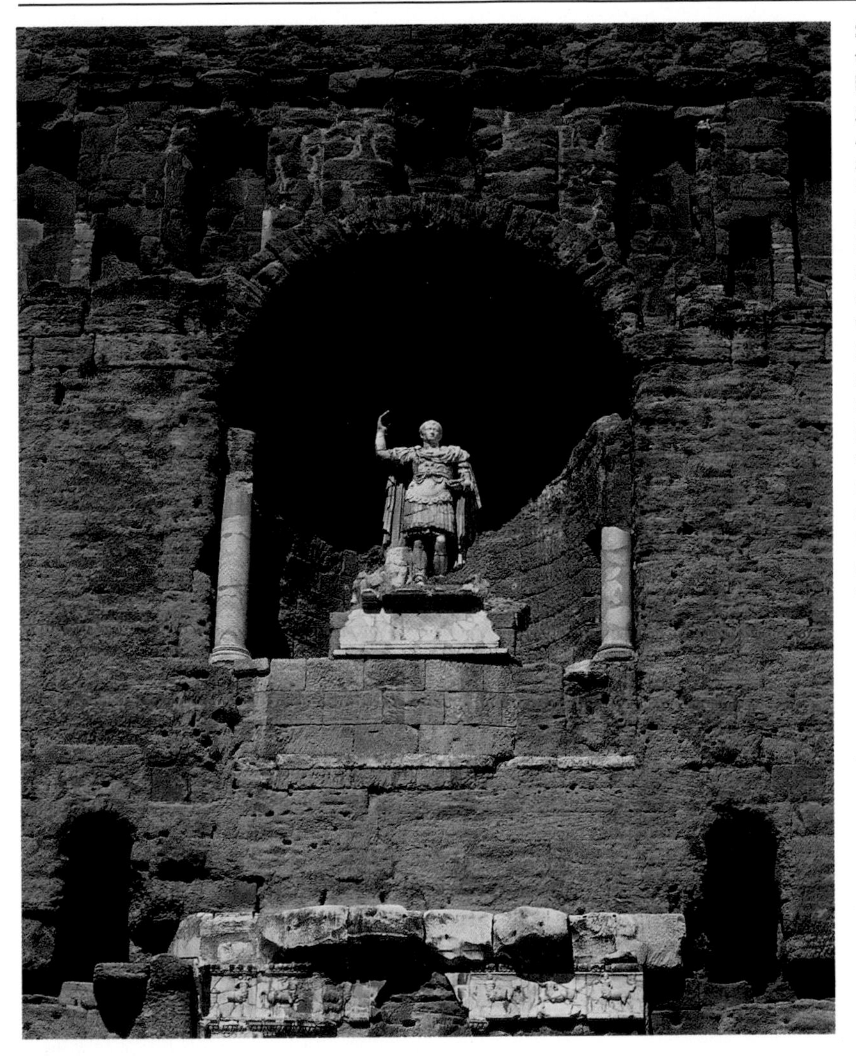

Sovereignty is the power to rule. **Oligarchy** is rule by a clique or small, tight group; *olig-* means "few." **Aristocracy** is rule by an elite. *Aristos* means "best" and -cracy is from *kratos*, meaning "power" in Greek. A **tyrant** is an all-powerful ruler (a king, queen, dictator, or emperor) who abuses that power. **Monarchy** is rule by a king, queen, or other royal leader. A **republic** is a government in which citizens elect the representatives who vote and act on their behalf. Rome had a limited republic; many people weren't citizens.

Emperor Augustus Caesar (Octavian) still greets visitors to one of the world's best-preserved Roman theaters, built some 2,000 years ago in Arausio (now Orange, France). The marble statue, twice as tall as a man, is perched high on a wall behind the open-air stage. The theater accommodates 7,000 patrons and is in use today.

Some Roman History

Julius Caesar (ca. 100–44 B.C.E.) was a great, charismatic Roman general and statesman at the time of the republic. He spent lavishly on public buildings, public athletic events, and wars (from Britain to Egypt). Caesar had big plans for more public works, for more libraries, and for codifying all of Roman law when, on the Ides of March (the fifteenth) in 44 B.C.E., he was murdered by Brutus and Cassius, who thought he was destroying republican freedom.

Julius Caesar's great-nephew, Octavian, succeeded him. He had to fight to do so. After becoming emperor of Rome, he was named Augustus ("sacred") Caesar in 27 B.C.E. (Remember: when you're in B.C.E., the numbers go backward. The smaller numbers, or dates, are more recent.)

A **colossus** is a giant statue or a "colossal" (huge) thing. **Augustus** means "sacred" or "venerable." You can see why Octavian wanted to take that name. The month of August is named for Augustus Caesar. The name *Caesar* led to the German *kaiser* and the Russian *tsar* (or Latin *czar*). **Charismatic** means "able to attract, charm, and influence others." It comes from the Greek root for "favor."

Philadelphia to write a republican constitution for their new nation, they turned to Rome for inspiration. Unfortunately, they copied the idea of nonvoting slaves as well as the idea of a voting citizenry.)

But the republican system, which had served the city-state well, fell apart as Rome grew huge. Its final decades were plagued by civil war. In 31 B.C.E., Octavian, a strong, energetic general, defeated Cleopatra and Mark Antony in a naval battle at Actium in the eastern Mediterranean Sea and became sole ruler of the Roman Empire. Octavian had pretended to be a republican, but four years later he assumed the title of Augustus; he was the first Roman emperor. He initiated a period of peace and prosperity and boasted that he found the city of Rome "built of brick, but left it built of marble." The poets Horace, Virgil, Ovid, and Livy thrived under Augustus. (Later, in the late seventeenth and early eighteenth centuries, England would call itself "Augustan" and its writers—Dryden, Pope, and Steele—would consciously imitate the wit and elegance of the Latin poets.)

An AQUEDUCT (AK-wi-duhkt) is a pipe or channel to carry water long distances.

A marvel of engineering, this ancient Roman aqueduct cuts through the center of Segovia, Spain. Built in ca. 50 C.E., it still delivers drinking water to the city. The Romans could not have imagined that cars and trucks would one day drive through its arches.

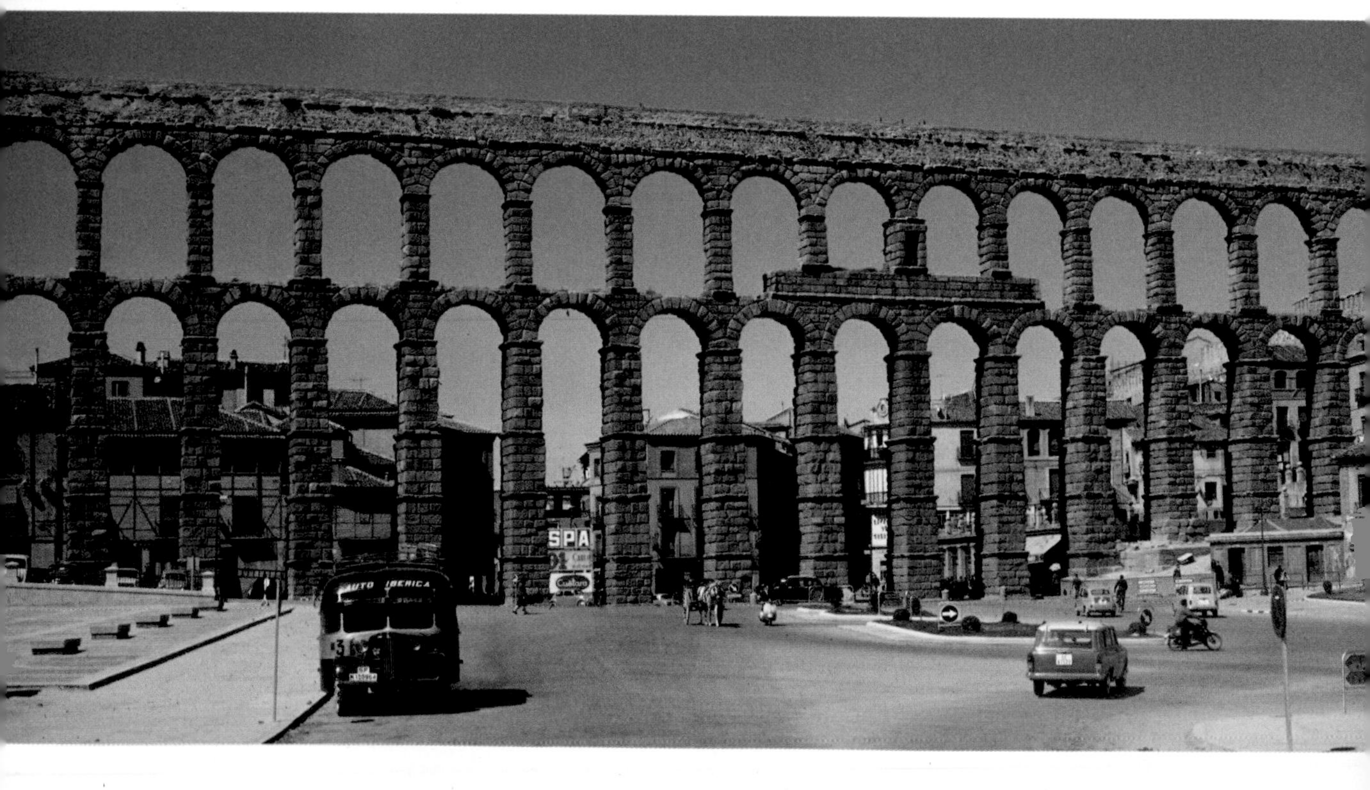

The Romans were different from the Greeks. They had their huge empire to control, so they used their energy to think about ways to govern and trade and fight. The Romans were pragmatists, which means "practical thinkers." They were astonishing engineers, bringing roads and waterworks and organized government to those they conquered. (They also collected taxes and slaves.) But pure science? And questions about why and how the universe works as it does? And technology that doesn't serve a practical purpose? The Romans didn't seem to care about those things. They didn't even bother to translate the works of most of the great Greek thinkers into the language they spoke, which was Latin. For the most part, they never understood the beauty of knowledge for its own sake. Or that you can never predict where learning may lead. Their leaders were too busy conquering and ruling—and gathering taxes from vanquished peoples.

Of course there were exceptions. Lucretius (ca. 99–ca. 55 B.C.E.), who lived during the Roman Republic, wrote about atoms (see page 92) and also studied lightning, thunder, sound, and light. But science without freedom doesn't go far, and as the empire grew, free thought diminished. The people Rome conquered had very little freedom.

Still, from England to the Mediterranean to Persia, Rome constructed state-of-the-art aqueducts, great buildings, and impressive public works. Roman ships policed the seas, making them safer than they had been. Roman rule kept much of the globe orderly and peaceful. Expansive Roman roads helped make control possible. But not everyone wanted to be controlled.

A Roman relief (below) from the time of Augustus Caesar (he died in 14 C.E.) depicts soldiers on a warship. The Romans changed the design of Greek ships to make them easier to handle—five men to an oar instead of a tricky three-tiered arrangement that required precise timing and skill.

Jerusalem was a conquered city, and its Jews chafed under Roman rule. One of those Jews was a young preacher named Joshua, who was born in Bethlehem (near Jerusalem) when Augustus Caesar was an old man. Later, his story would be written in Greek, and Greek for *Joshua* is Jesus.

Before Jesus was born, Julius Caesar's soldiers set fire to Alexandria's famous

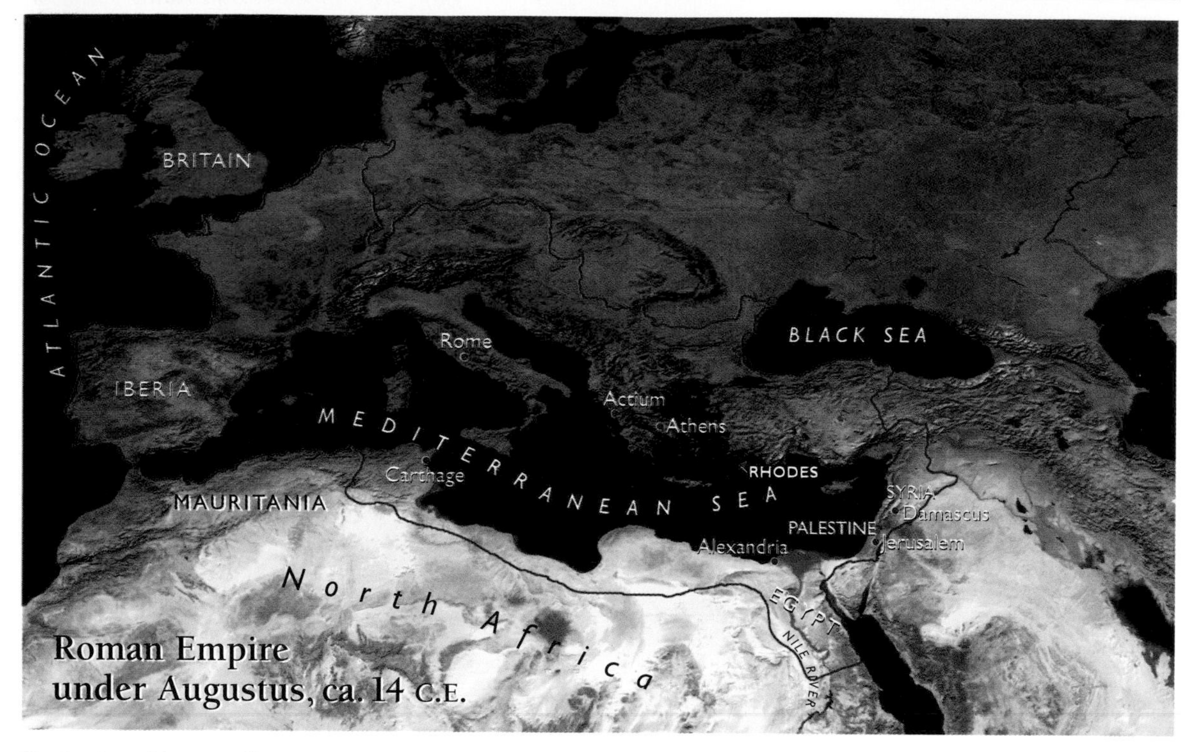

Roman Empire under Augustus, ca. 14 C.E.

During its golden age, Greece was made up of independent city-states—it had no unified central government. Many of those Greek city-states founded colonies. Alexander had a unified empire very briefly. The Romans had one for several hundred years.

library. According to the legend, the Romans torched an enemy fleet and the flames jumped to land and spread to the library. Many, but not all, of the books burned. That was in about 48 B.C.E. Alexandria remained a world center of learning and business, although Rome was now intent on taking its place.

The city of Rome was the hub of political power, but Alexander's city was still an intellectual hub, and it continued to prosper. It was a manufacturing and trading center, and its

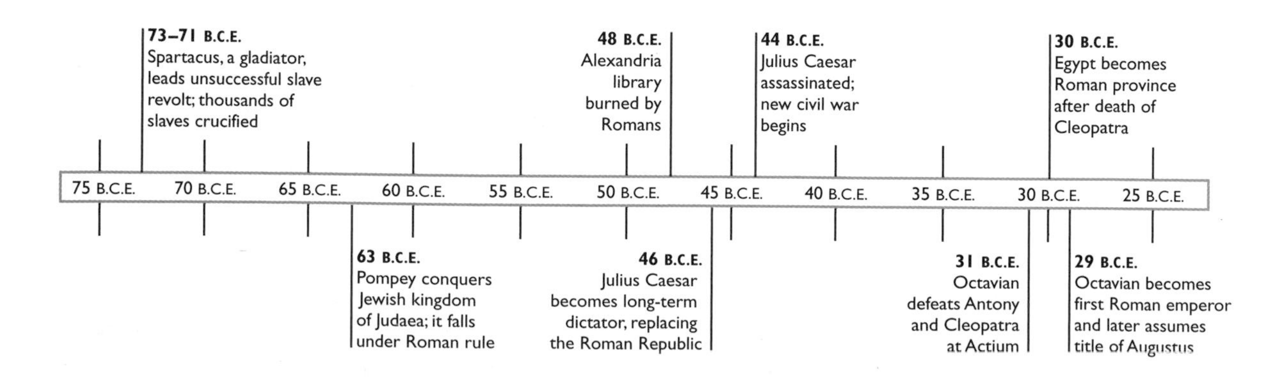

ideas and trade goods went back and forth from one end of the empire to the other—and beyond. Alexandria helped make Rome a cosmopolitan empire. The Greek geographer Strabo, who lived at the same time as Jesus Christ, wrote, "many merchants from Egypt sail to India each year." On the island of Samos, Augustus Caesar met with ambassadors sent by an Indian raja. It was an expansive time: Greece, Rome, and Egypt influenced Buddhist art, and vice versa.

Of China, the Romans knew hardly anything. On Strabo's map of the world you can see something called Seres, far to the east. It is China. Merchants from China had once followed a route to the fabulous city of Samarkand in Uzbekistan and on to Baghdad (in Iraq) and Damascus (in Syria). But Samarkand

RAJA (or RAJAH) is the Hindi word for king, prince, or other male ruler. It's akin to the Latin words *regis* ("regent") and *rex* ("king"), which come from the same Indo-European root (see page 173).

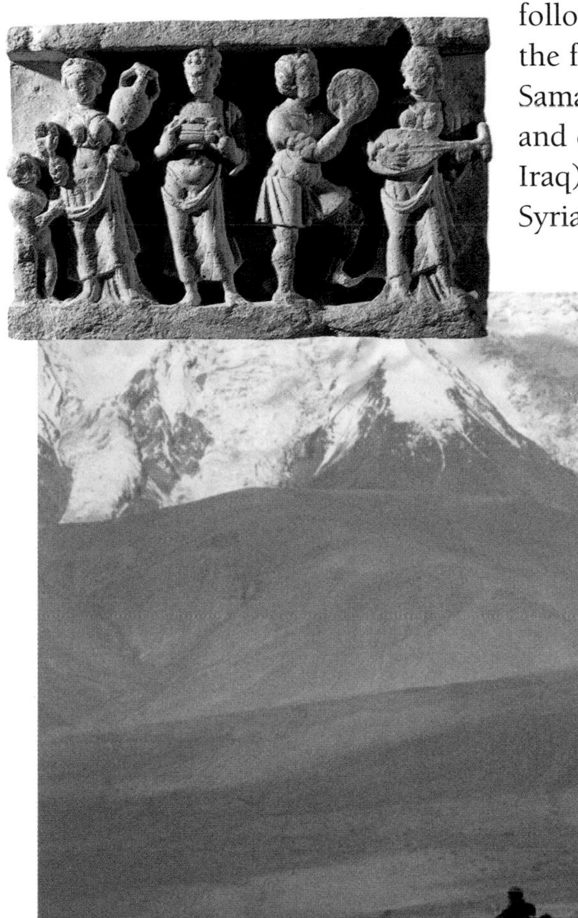

Goods, ideas, cultures, and people traded places along the Silk Road, the main land route between West and East. A frieze of musicians (left), made in Afghanistan, is a blend of Western Greco-Roman and Eastern Buddhist styles of art. (Compare it to the relief on page 169, which was made farther west.) On a stretch of the Silk Road in Kyrgyzstan in central Asia (below), modern merchants still use camels to cross the dry Pamir Plateau.

had been destroyed by Alexander in 329 B.C.E. For a while, a thin thread of travelers continued to bring silk across Asia to sell in the Syrian markets. By Augustus Caesar's time, the Silk Road was a dusty trail, made dangerous by marauding gangs, and there was almost no regular contact between the cultures of the Asian Far East and the Mediterranean West.

MARAUDING means "roving, raiding, ruining, and stealing." It comes from a French word meaning "rogue" or "vagabond."

The great Greek philosophers were long gone, but Greek science was not finished. Strabo, whose name means "squint eyed" and who surpassed Eratosthenes as a geographer, went up the Nile River in 24 B.C.E. and then, 20 years later, settled in Rome. Strabo had studied the ideas of Plato, Aristotle, and Eratosthenes. His writings tell us of those thinkers and of his own times.

Plato had mentioned that there was a lost continent somewhere beyond the known world. He called it Atlantis. Was there actually an Atlantis? That was a big question mark. (It still gets scientists arguing.) But it must have gotten Strabo thinking. He wrote, "It is conceivable that in the same temperate zone—that we inhabit—there are actually two inhabited worlds, perhaps even more, and particularly in the area of the parallel through Athens in the region of the Atlantic Ocean." Read that again and think about what he is saying. Another inhabited world on the same line of latitude as Athens? Was that a myth? Could it be that there were lands and people unknown to the Romans? (How do his words compare with those of people today who talk about other inhabited worlds?)

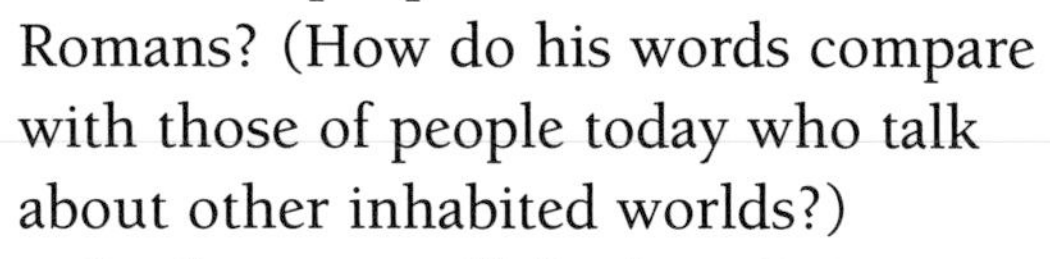

The Temperate Zones are between the tropic of Cancer and the Arctic Circle and between the tropic of Capricorn and the Antarctic Circle. Note that Athens, Greece, is at 38°N latitude (or "parallel," as Strabo calls it), a line that runs through the highly inhabited worlds of China, Japan, and the United States.

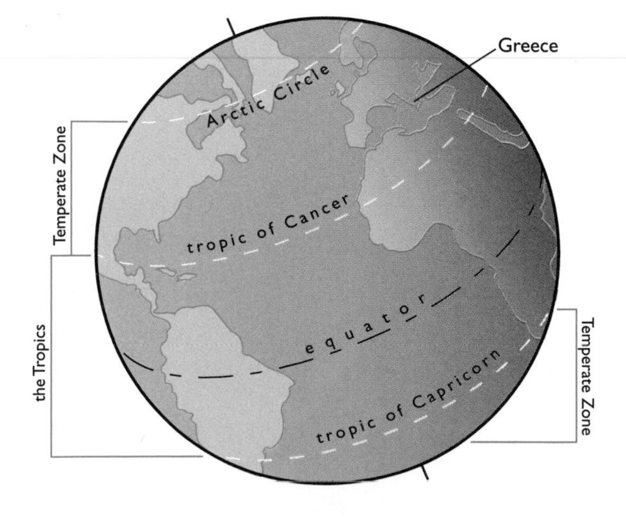

Strabo went still further. Citing Eratosthenes, he said, "The habitable world forms a complete circle, itself meeting itself; so that, if the immensity of the Atlantic Ocean did not prevent, we could sail from Iberia [Portugal and Spain] to India along one and the same parallel over the remainder of the circle."

Read that again. Check Strabo's dates. What is he suggesting?

ONE, TWO, BUCKLE THAT SHOE

The words in this book are mostly borrowed (if you're harsh, you can say "stolen"), but don't blame me. I didn't take them; English speakers have done that for centuries. We're still doing it. The English you and I speak, read, and write has far more imported words than native words. Latin and Greek rank as the top two word lenders. You'll find many of their roots in the word definitions on these pages. But English also has a sizable smorgasbord of Scandinavian and Germanic terms as well as a sprinkling of Arabic, Native American, Yiddish, and Asian words.

In the nineteenth century, linguists began comparing all these languages scientifically. One way was to make charts like the one at right. The linguists were genuinely surprised to discover that most European languages—including English—and a few Asian languages stem from the same tree. They called them Indo-European languages because the mother tongue, the language from which they all sprang, originated somewhere between Asia (*Indo-* refers to India) and Europe.

That mother tongue no longer wags, but dozens of offspring continue to branch and grow. In addition to *one* and *two*, interesting words to follow around the Indo-European family are *mother, father, brother,* and *sister.*

one, two . . . . . English
en, duo . . . . . . Ancient Greek
heis, dyo . . . . . Modern Greek
unus, duo. . . . . Latin
un, deux. . . . . . French
uno, dos. . . . . . Spanish
uno, due. . . . . . Italian
eins, zwei . . . . . German
odin, dva . . . . . Russian

A Bible story (Genesis 11) says our cacophony of languages is punishment for trying to build a tower to heaven—the Tower of Babel (shown in a 1563 painting by Pieter Brueghel the Elder).

20 Longitude and Latitude plus Two Greek Mapmakers

[The Lord] telleth the number of the stars; he calleth them all by their names.

—Psalm 147:4, Hebrew Bible (King James Version)

The stars about the lovely Moon hide their shining forms when it lights up the Earth at its fullest.

—Sappho (lived sixth century B.C.E.), Greek poet, *Fragment 4* ("Some say there are nine Muses, but they are wrong. Look at Sappho of Lesbos; she makes ten," wrote Plato.)

Hipparchus did a bold thing, that would be rash even for a god, namely to number the stars for his successors and to check off the constellations by name. For this he invented instruments by which...it could easily be discovered not only whether stars perish and are born, but also whether any of them change their positions or are moved and also whether they increase or decrease in magnitude. He left the heavens as a legacy to all humankind, if anyone be found who could claim that inheritance.

—Pliny the Elder (ca. 23–ca. 79 C.E.), Roman statesman and scholar, *Natural History*

It's easy to measure the Earth. Well, it did take tens of thousands of years to get it right—and satellites helped a lot—but Earth is made for mapping. It's punctuated with rivers and mountains and oceans and coastlines that are points of reference; they put things in place.

Mapping the night sky is something else. All you have are points of light—stars, mostly. That didn't stop Eudoxus (yoo-DAHK-sus) of Cnidus, another of those persistent Greeks, who, about 350 B.C.E., decided to map the sky. He realized that the heavens needed markers in order to create regions. So he put the northern polestar (which was Kochab back then, not Polaris—see page 62) in the middle of his map and drew imaginary lines fanning out from it, like

SKYLINES

This sky map (below), from a German encyclopedia called *Bilderatlas* (1860), uses celestial coordinates similar to those that Eudoxos created. At the center sits the north celestial pole, an imaginary North Pole in the sky. The North Pole on Earth is at 90°N latitude; its sky-high counterpart is at 90° declination (a term for celestial latitude). The first ring of declination is 80°; the second is 70°; and so on down to 0° declination, the biggest ring. That's the celestial equator, an imaginary circle above Earth's equator. Declinations south of 0° use negative numbers: –10°, –20°, etc.

The lines radiating from the celestial pole like spokes of a wheel are called right ascensions. Count them (below left). Like longitude lines, there are 24—one for each hour of the day. That's how they're labeled: 1 hour (h), 2 hour, etc., with spaces in between marked in minutes (m).

Using a spoke and a ring, you get a celestial coordinate—the point where the two lines meet. Here's a popular one:

Right ascension: 6^h45^m

Declination: –16°43′ (the foot symbol stands for minutes)

That's where you'll find Sirius ("the Dog Star"), the brightest star in the sky.

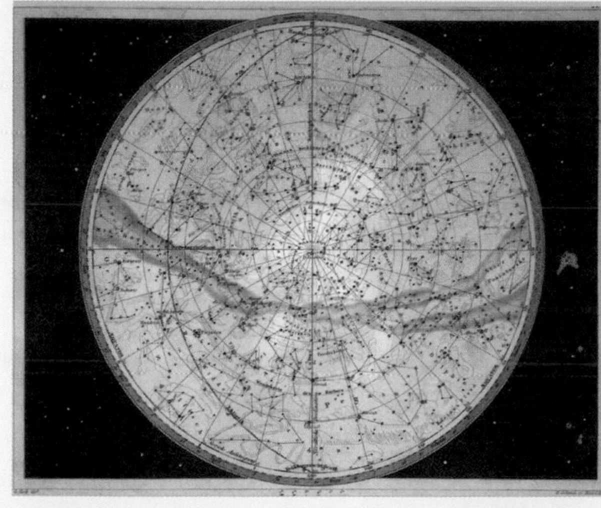

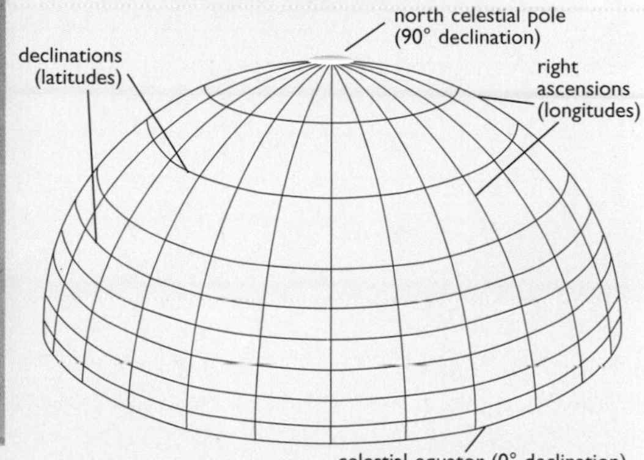

spokes of a wheel. Then, going back to the polestar and using it as a bull's eye, he drew concentric circles around it. The circles crossed the spoke lines to make a grid with coordinates (places where the lines meet). Those celestial spokes, which started as points and became fat and wide at the ends, would come to be called lines of longitude. The circles, evenly spaced, were lines of latitude, similar to those on a globe. Now on this imaginary sky grid, Eudoxus could pinpoint stars almost exactly. He thought he was fixing the heavens on his map and that the stars would stay in place forever.

About 200 years later, another determined Greek, Hipparchus (hi-PAR-kuhs), looked heavenward and saw a

Hipparchus is using a cross-staff to measure the altitude of the polestar, which is the same as the degree of latitude. (At 45°N latitude, the polestar is 45° above the horizon.) Later, astronomers figured out how to use the noon Sun to determine latitude—a trickier calculation, since the Sun's position changes over the course of a year.

bright star he had never seen before. It wasn't on Eudoxus's grid. Something was wrong. The stars were supposed to be unchanging. Where had this one come from? Perhaps it had just been missed in the past. But Hipparchus didn't really think so. He decided to remap the heavens, then if another star made a surprise appearance, he would be prepared and know if it was new. So he made a star map using Eudoxus's lines of latitude and longitude. Hipparchus placed close to 1,000 stars on his map.

He divided the stars into classes according to brightness (later called magnitude). The 20 brightest were of the first magnitude; stars of the second magnitude were slightly dimmer, and so on to the sixth magnitude, which is barely visible to the naked eye. With some modifications, we use that classification today.

In making his star map, Hipparchus compared the locations of stars he could see with those recorded by Eudoxus and earlier astronomers. He discovered that although, night after night, the stars appear to be fixed in position, over time they actually shift from west to east. But the shift is so slow, no one sees it during a normal lifetime; it takes about 100 years for the change to be

WHY DOES EVERY EQUINOX LOOK A LITTLE DIFFERENT?

Here's why: The Earth revolves around the Sun *a tad less* than once a year. Each March 21 (or thereabouts), our planet reaches the place in its orbit that is the vernal (spring) equinox. And each year, that place is a little bit short of last year's spot. Because Earth is in a different spot, the stars look as if they're in a different spot, too. It takes a century or so for someone standing on Earth to notice this tiny change, but given a thousand or more years, the yearly shortfalls seriously add up.

Four thousand years ago, on the day of the spring equinox, Earth was opposite the constellation Aries, the ram. For Hipparchus, some 2,000 years ago, the opposing constellation was Pisces, the fish. Now it's about to become Aquarius, the water carrier. (There's a hit song from the musical play *Hair* about the "dawning of the Age of Aquarius." Now you know what that catchy lyric means.)

If you know your zodiac, you realize that the constellations are going backward in relation to the calendar. The rate is about one constellation every 2,160 years. Because there are 12 zodiac constellations, it takes about 26,000 years for the Earth to end up in the same equinox position.

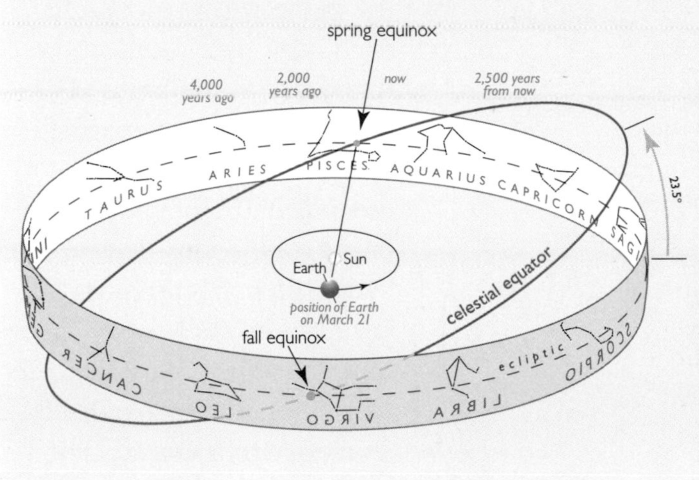

The 12 zodiac constellations don't really circle our solar system. Their stars aren't even grouped in a pattern—they're spread light years apart. It just looks that way to us. So this diagram isn't a map of space, and it's definitely not to scale. Think of it as a useful tool to picture why the stars appear to us as they do.

Plant yourself on the Earth. It's the spring equinox, March 21. If you're on the side of Earth facing away from the Sun, it's night; you see the constellation Virgo fixed among the stars. If you're on the side facing the Sun, it's daytime; you can't see any stars. Hidden from view, directly behind the Sun, lurks the constellation Pisces.

noticeable. Hipparchus didn't realize it, but at the rate that the stars shift, they complete a huge circle in our sky about every 26,700 years.

Other things change slowly, too. The equinox (the day when light and dark are equal) arrives a few seconds earlier each year. It's called the precession of the equinoxes (see above). More than 16 centuries after Hipparchus, a Polish stargazer named Copernicus would discover that this precession is caused by a slight wobble of the Earth on its axis.

Hipparchus, who is said to be the greatest of the Greek

Precede means "to go before" in Latin, so PRECESSION, the noun form of that verb, means "the act of going before." An OBSERVATORY is a place with instruments for making science observations—of the stars, the weather, pollution levels, and so on. A planetarium isn't an observatory. It's a model, or copy, of the night sky.

astronomers, set up an observatory on the island of Rhodes in the Aegean Sea; there he invented stargazing instruments that would be used by sky watchers for centuries to come. But his most ambitious achievement came when he worked out a complex scheme to explain the movements of the stars and, especially, of the planets. Remember, Aristotle had 54 layered transparent spheres holding the stars as they all moved in the same direction around a stationary Earth. But Aristotle never came up with a good explanation of the wandering movement of the planets.

Knowing what you know about the solar system, you might think this drawing of a planetary orbit looks really goofy. Yet if you are standing on Earth, these epicycles—loops within loops—provide a reasonable explanation of why planets sometimes move across the sky in an erratic path. Specifically, they are an imperfect attempt to account for retrograde motion (see page 113).

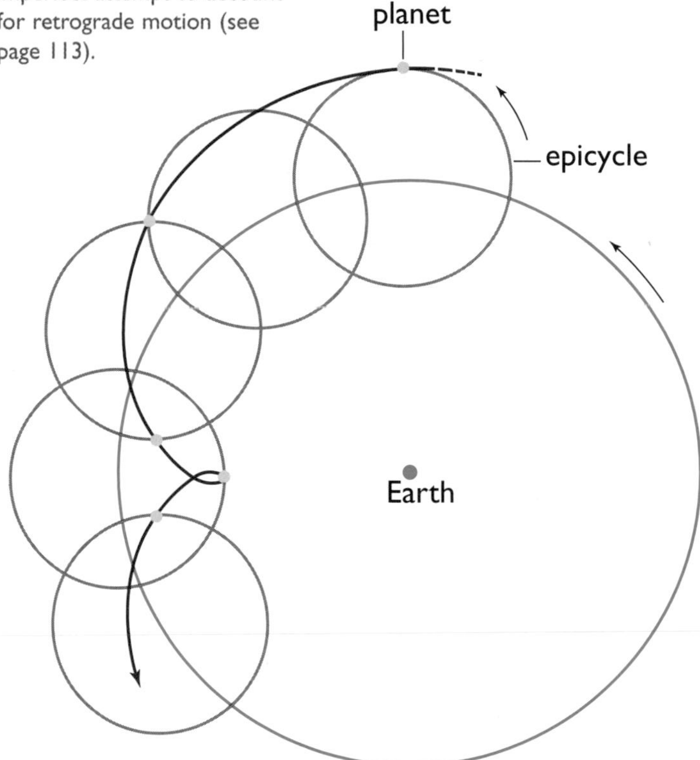

Hipparchus accepted Aristotle's basic idea but reduced the number of large spheres to seven. To handle the planets, Hipparchus had each orbiting a focal center that also orbited Earth. These planetary orbits were called epicycles. This very complicated scheme was based on the idea that orbits had to be perfectly round because the circle was thought to be the perfect shape. As it happens, orbits aren't perfectly round. The idea was wrong, but it would take a long, long time to figure that out.

Hipparchus knew that Aristarchus, who had taught and worked at Alexandria about 130 years before his time, had believed the universe was heliocentric (Sun-centered). The motion of the planets and Earth could be easily explained without epicycles, said Aristarchus, if they all moved around the Sun. But if that was true, Earth had to be moving, and that must have seemed bizarre to Hipparchus. It did to almost everyone else.

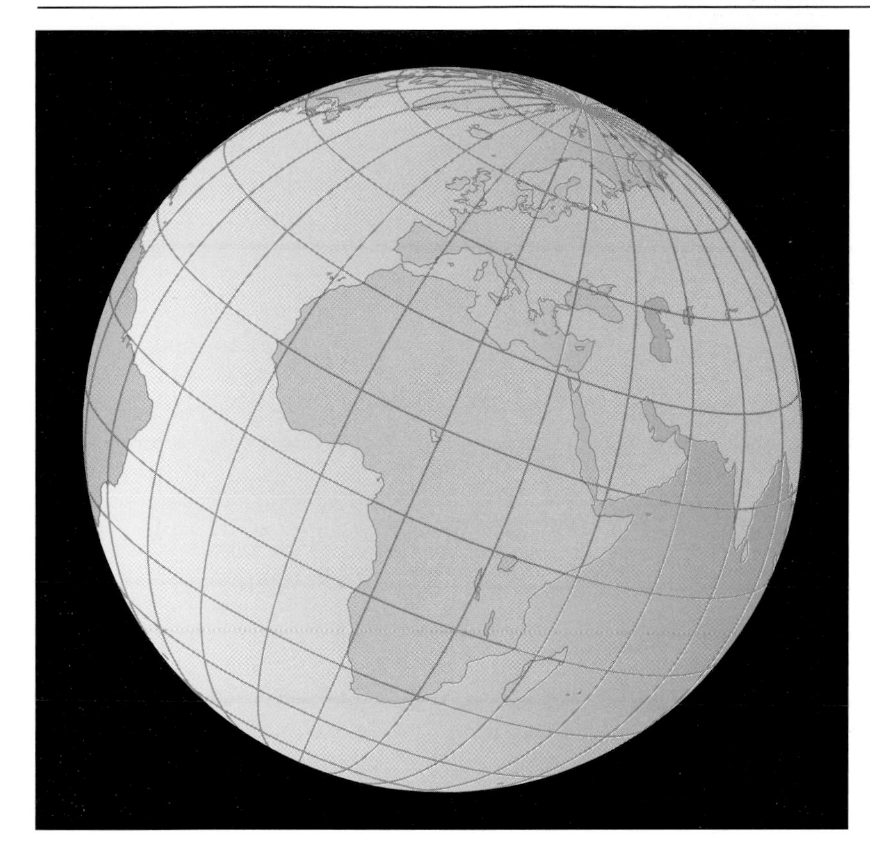

Lines of longitude on Earth meet at the North Pole—and at the South Pole too. Lines of latitude, looking like belts around the Earth—are parallel. They never meet. The equator is 0° latitude, and the North and South Poles are 90° latitude. Compare this Earth-bound mapping system to the celestial coordinate system on page 175.

Besides, Aristarchus had never worked out the mathematics of his solar-centered idea, so it was hard to take it seriously. Hipparchus was able to take his own model and provide the mathematics to make celestial prediction possible. For a long time it all seemed to work.

That's not all Hipparchus did. He is sometimes called the Father of Trigonometry. Trigonometry is about measuring the angles and sides of triangles (see page 180). Imagine you're at sea and lost, and then you sight a familiar star. By setting up an imaginary triangle from ship to star and comparing it with a known triangle from land to that same star, you can usually figure out your location.

Hipparchus did even more for those of us who want to know where we are. He took the lines of latitude and longitude from Eudoxus's sky maps and put them on a map of the land. That was around 129 B.C.E. We've been using those lines on Earth maps ever since.

WITH TRIGONOMETRY, YOU CAN KNOW IT ALL

There's more than one way to measure a triangle, but you already know that. Thales made a little triangle with the same angles as a huge one in order to find the distance of a ship (see page 43). Pythagoras did his bit with the three sides of a right triangle (see page 79). Hero said that if you know the three sides of any triangle, you can find its area (see page 132).

With trigonometry, you can know it all. (Well, all about triangles, anyway.) The word *trigonometry* means "three (*tri*) angle (*gōnia*) measure (*metron*)," and it's the ultimate math for measuring triangles of any size or shape.

Jacques Ozanam, a seventeenth-century mathematician, wrote in his book *Cours de Mathématiques* that he couldn't live without it: " 'tis by Trigonometry only that the Courses of the Phenomena and Changes which happen in the Universe can, with any certainty, be discovered . . . nor can anyone arrive at the knowledge of the Motions of the Celestial Bodies, but by that of the most simple Figures, which are Triangles. . . . The Usefulness of Trigonometry is so great, that it is in a manner impossible to live without it."

Trigonometry works because, as Ozanam points out, triangles are "the most simple Figures"—just three sides and three angles. If you change the length of one side or the size of one angle, you can't help but change other sides and angles too.

Try it: Draw some triangles on graph paper and see what happens when you stretch one side, for example. Right triangles (with one 90° angle) are especially easy to play with. You can also search the internet for a handy "triangle calculator," an online tool that lets you change a side or angle, and then it automatically recalculates the other values of the new triangle. Visual calculators (like the one at www.visualtrig.com) show the changing shape of the triangle, along with the changing numbers. Interactive ones—Java applets that let you grab and drag triangle points—are actually fun to play with.

What you'll discover if you explore triangles long enough is that the angles and sides come in fixed ratios. For example, for every 45-45-90° triangle (no matter how big or small), the two short sides are always equal length. For every 30-60-90° triangle, the hypotenuse is twice as long as the shortest side. (Triangles that have the same three angles—but are different in size—are called similar triangles.)

Trigonometry fans have brewed up table after table of these useful ratios. You just need to know one side and any other two pieces of a triangle; trigonometry will fill in the rest. In other words, you'll know it all.

In this German woodcut (ca. 1530), surveyors show how to use cross-staffs and trigonometry to measure the distance between any objects, in the sky or on land—a technique called triangulation.

WHAT'S THE POINT?

**point of reference** is any landmark that you use to judge the distance, direction, or location of a place. Street signs are obvious markers, but lots of familiar sights help you get around town.

If you're in the Northern Hemisphere, the Big Dipper is like the corner store—you can't miss it. It's a point of reference that looms large, night after dependable night. Its two end stars are always pointing at Polaris, the polestar (see page 63). The problem is, unlike a store, the Big Dipper takes a spin around the sky once every night. So, while you can tell where stars are in relation to each other, you need another way to describe an exact position in the sky. That's especially true for the Moon and planets, which pop up in lots of spots.

To locate a spot in a city, you can usually just say, "It's at the corner of This Street and That Street." Likewise, to pinpoint a star or

Any chance they could get, navigators hopped ashore to take their celestial measurements. On flat ground, gravity kept a plumb line (the string with the bob) at right angles to the horizon. A rolling, pitching ship made the line (and the navigator) sway.

to zenith

to star

90°

A quadrant is a quarter circle marked with degrees from 0 to 90. Its point of reference isn't the horizon (0°); it's the zenith, the point directly overhead, which is 90°. A quadrant is easy to use: Dangle the tool at the corner so that the free-swinging stick is vertical. The stick stays in position while you adjust the quarter circle with your other hand so that the right edge points to a star. The stick marks the star's degree of altitude.

planet, you need two "streets"—two points of reference. One point tells you altitude—how high in the sky it is above the horizon. The other tells you which way to look—north, east, south, west, or some spot in between. Another word for that is azimuth, which comes from the Arabic for "the way."

Over the millennia, astronomers, like Eudoxus, have mapped stars using a bunch of different coordinate systems. (Coordinates are overlapping points on a grid.) In case you're curious, here's the royal road (the easy way) to geometric star charting (below and opposite page).

ALTITUDE: 90° OF SEPARATION

No matter where you're standing on Earth, no matter which way you're facing, there's an imaginary point directly above your head. It's called a zenith, and it's at right angles, 90°, to your horizon. Everything that you can see in the sky is somewhere between the horizon, call it 0°, and the zenith. All you have to do is measure its altitude in degrees. It's easy, if you just want a rough idea. Hold your hand at arm's length and follow the guide at right.

Of course, there are big hands and little hands, long arms and short arms. To take a better reading, you need an instrument with degree markings. On paper, that instrument would be a compass or a protractor, each with a degree marker shaped like a semicircle. An astrolabe is just a protractor for the sky (see photo, opposite page).

There's another little hitch: Altitude changes with latitude. The farther north you are, the higher Polaris is in the sky. So your measurement, no matter how accurate, only works for your latitude.

To have a universal coordinate system, you can't use the horizon as your base. You need a plane of reference that works for everyone—like the equator or the ecliptic (the Sun's path in the sky). That's where the math gets sticky, but if you're still curious, look up "equatorial coordinates" on the Internet or in an astronomy book.

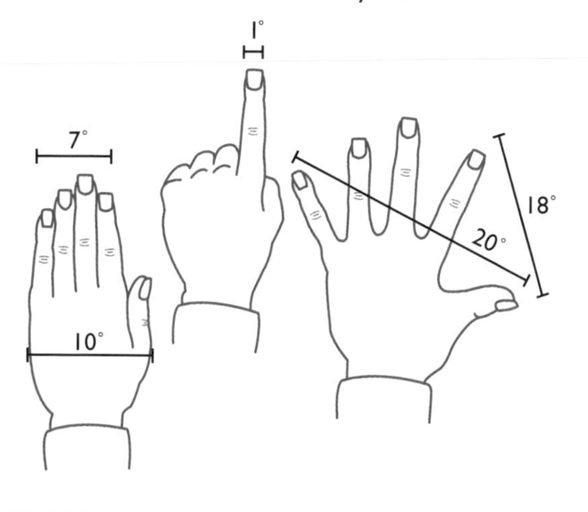

Azimuth: A Big Circle All Around You

Once you know a star's altitude, you can angle your head at that height, slowly spin around, and search the sky in a big circle. Somewhere along that circle, known as the azimuth, you'll find your star.

It's easy to tell a friend, "Look west about 20 degrees above the horizon if you want to see Venus at nine o'clock tonight." But what if Venus isn't due west? If you're in the Northern Hemisphere, it's usually more like southwest or west-southwest.

There's a better way, and once again, geometry provides it. Since the azimuth is a circle, you can divide it into 360°. North is 0°, east is 90°, south is 180°, and west is 270°. So to those in the know, you can now say, "Venus is 20 degrees high at 230 degrees on the azimuth."

As with altitude, this system isn't foolproof. Earth is spinning around, so the azimuth changes constantly. To solve this problem, you need to use celestial longitude lines called right ascensions (see page 175).

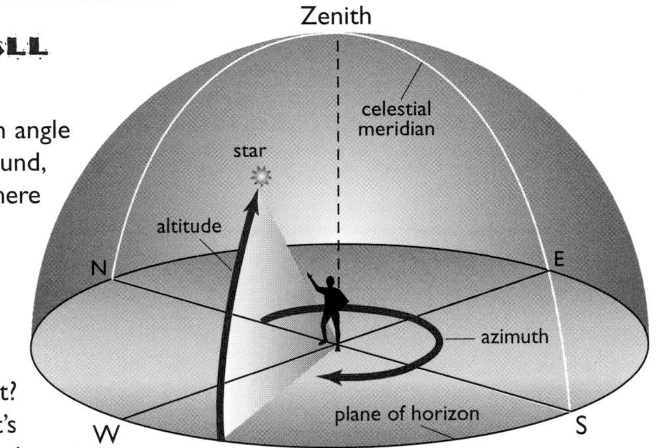

Don't confuse the zenith with the north celestial pole (page 175). The zenith is much more personal. It's the point directly above your head—and a different spot for someone standing elsewhere on Earth. (The north celestial pole is always straight above the North Pole.) Likewise, the horizon on this diagram isn't the equator or even the celestial equator. It's your personal horizon, the line where sky and land meet all around you.

This seventeenth-century astrolabe has all that a mariner needs for finding latitude at sea. An astrolabe is an ancient tool with a double pointer, called an alidade, that spins 360° around. One person dangles the tool by the ring so that the vertical bar lines up with the zenith and the horizontal bar with the horizon. A friend holds it in place (not easy on a rolling ship) while pointing the alidade at a star or the Sun. Astronomical astrolabes include extra parts for charting stars (see page 207).

21 The Greatest

I know I am mortal by nature...but when I trace at my pleasure the windings to and fro of heavenly bodies I no longer touch the earth with my feet: I stand in the presence of Zeus himself and take my fill of ambrosia, food of the gods.

—Claudius Ptolemy (ca. 85–ca. 165 C.E.), Alexandrian mathematician, astronomer, and geographer, *Almagest ("The Greatest")*

We live in a galaxy that is about one hundred thousand light-years across and is slowly rotating; the stars in its spiral arms orbit around its center about once every several hundred million years. Our sun is just an ordinary, average-sized, yellow star, near the inner edge of one of the spiral arms. We have certainly come a long way since Aristotle and Ptolemy, when we thought the earth was the center of the universe!

—Stephen Hawking (1942–), British physicist, *A Brief History of Time*

Claudius Ptolemy (TOL-uh-mee) was known as the world's greatest astronomer, geographer, and mathematician. In the second century C.E., everyone in the Alexandrian world believed that. There would be a long interlude when Ptolemy was forgotten, but then he would bounce back stronger than ever. During the Middle Ages and the Renaissance, Ptolemy was revered. People believed he was the greatest in 1492, when Columbus sailed the ocean blue and happened upon what was to him the New World. So put Ptolemy in your head as another of the most influential authors of all time. He wrote huge books that organized all the known math, geography, and

It is impossible for a man to begin to learn that which he thinks that he knows.

—Epictetus (ca. 50–ca. 138 C.E.), Roman slave who became a philosopher, *The Discourses*

Ptolemy studies an armillary sphere in this wood panel painted by Flemish artist Joos van Ghent (ca. 1435–ca. 1480). The panel is part of a series commissioned for the palace of the duke of Urbino, in Italy.

astronomy of his day.

This Ptolemy was not an emperor or a general, as were the Ptolemys who ruled Egypt. We hardly know anything about his personal life. We do know that he was born in North Africa and was part of the Greek-speaking Hellenistic world. Most authorities believe that he was Greek, but a few think he may have been Egyptian. He lived and studied in Alexandria a century or so after Strabo and more than three centuries after Eratosthenes.

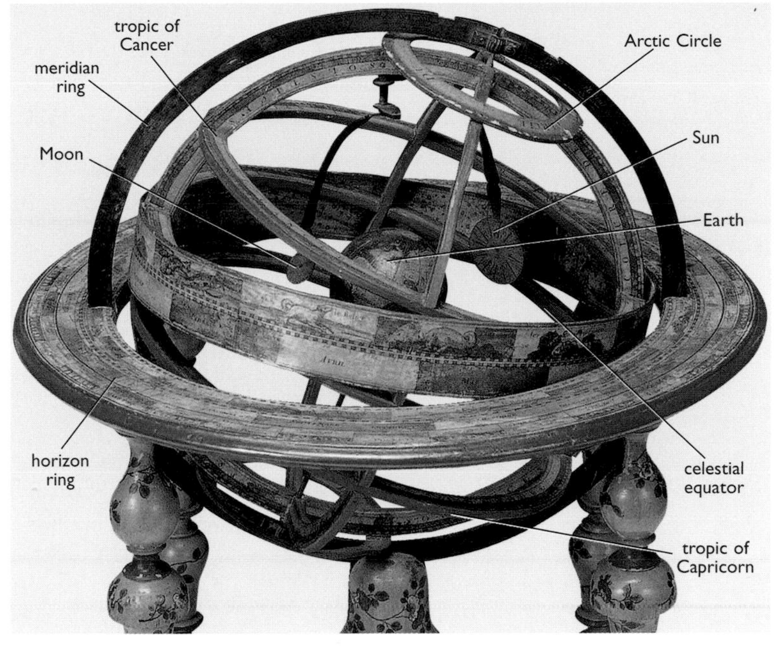

An armillary sphere is a see-through celestial sphere, a model of the sky as it appears from Earth. The tilted Earth is caged in the middle, while rings carrying the Sun and Moon rotate around it. The planets are missing because Ptolemaic (Earth-centered) spheres can't model their erratic paths across our sky. In the sixteenth century, Copernican (Sun-centered) armillary spheres with planets began to appear in Europe.

When he considered the heavens, Ptolemy rejected those thinkers. He looked to Hipparchus, who studied at Alexandria in the second century B.C.E. Hipparchus had followed and improved Aristotle. All three (Aristotle, Hipparchus, Ptolemy) came to the conclusion that a round Earth *is* at the center of the universe and that it stands stone-still while everything else in the heavens revolve around it.

Remember, Aristarchus had figured out that Earth goes around the Sun. But that idea—of a heliocentric universe—was never very popular. The Earth whirling about in space? That seemed absurd. Ptolemy rejected the notion and followed where Hipparchus and Aristotle had led.

He based his great work on their geocentric mistake. That wasn't Ptolemy's only mistake. He also believed that Earth is mostly dry land rather than ocean. (That was another idea that went back to Aristotle.) He thought the stars and planets orbited in perfect circles. (That was Plato's perfection idea.) And he thought Earth much smaller than it actually is. When Ptolemy calculated, Earth came out 30 percent smaller than what Eratosthenes had figured it to be—and what it really is.

You're looking at the world upside down (north is at the bottom), the way mapmaker al-Idrisi saw it in the twelfth century. The well-traveled Moroccan gathered geographic data from sources in the East and West, both new and ancient (including Ptolemy).

Seeing Stars (and Planets Too)

The ancients could see five planets with the naked eye: Mercury, Venus, Mars, Jupiter, and Saturn. They could see thousands of stars, perhaps as many as 9,000. Naming them and marking their coordinates on a grid was painstaking but to dedicated scholars, very worthwhile.

By the third century B.C.E., Chinese astronomers had cataloged more than 800 stars. About 100 years later, Hipparchus, a Greek astronomer, cataloged 850 stars and set up his 6-point scale for magnitude (apparent brightness). In the second century C.E., Ptolemy added more than 170 stars to that list. He also named 48 constellations, which still appear on star maps today.

What the Greeks Saw

Magnitude 1 (brightest):	20 stars
Magnitude 2:	50 stars
Magnitude 3:	150 stars
Magnitude 4:	450 stars
Magnitude 5:	1,350 stars
Magnitude 6 (barely visible):	4,000 or so stars

The telescope, which came along in the seventeenth century, changed everything. Stargazers could now begin to glimpse the vastness of the universe. They saw tens of thousands more stars, although no one had an idea that there are a multitude of galaxies and that we are not the center of it all. In 1781, another planet, Uranus, emerged from the darkness, followed by Neptune in 1846.

In the last half of the nineteenth century, photography began to allow us to take pictures of the sky and study it in detail. The images revealed a multitude of moons, distant nebulae (gas clouds), and other galaxies. At the Harvard College Observatory, a team of women called "computers" glued their eyes to photographic plates, counting dense fields of stars. As they called out facts and figures, "recorders" added each star to a growing list, later named the *Henry Draper Catalogue*. (Henry Draper, a physician and amateur

Besides that, Ptolemy made Asia huge, stretching it far beyond where it actually is. Later, all those errors would make Christopher Columbus think that Asia was just across an ocean that wasn't terribly big.

Remember (it's in the previous chapter, so you don't have to go far) the Hipparchus/Aristotle model of the cosmos? Earth, in the center of the universe, is circled by transparent moving spheres that have planets and stars attached to them. It was an ingenious way to explain the fixed movement of the stars across the sky. It also seemed to explain why the stars don't fall from the heavens.

When it came to the problem of the wandering planets, Hipparchus had come up with a complicated system of epicycles (see page 178) to explain their motion. Ptolemy

Ptolemy said the Earth couldn't be moving, because if it were, people would be tossed around and birds would fall from the trees. A ball thrown in the air would come down in a different place than it went up. It wasn't a foolish observation. Think about it: If Earth does move, why *can't* we feel it? More than 14 centuries after Ptolemy, a scientist named Galileo would finally answer that.

photographer, is said to have taken the first picture of a star's spectrum. Those nineteenth-century women got little credit for their work.)

We now believe that our galaxy, the Milky Way, has 200 to 400 *billion* stars—maybe more. And the Hubble space telescope and other powerful eyes in the sky have found many, many more galaxies. "Computing" stars has just begun.

Among the thousands of stars visible since ancient times, a handful have become celebrities. The North Star, a favorite among mariners, is famous. But when it comes to starring in art, mythology, and astronomical maps, it's hard to top the 12 zodiac constellations. Above, they're featured players in a ceiling fresco by Italian artist Taddeo Zuccaro (1529–1566). A Catholic cardinal—Alessandro Farnese—chose this starry theme for his palace. You'll also find zodiac art in churches and public buildings throughout Europe.

Ptolemy's influence on mapmaking stretched well into the fifteenth century. A German map published in 1486 echoes his *Geographia*, including the idea that most of the globe is covered by land. Today, we know the opposite is true: oceans occupy more than two-thirds of the planet's surface.

worked hard to improve and refine Hipparchus's work. It seemed the best explanation there was, so most people thought it correct. (It wasn't.) Knowledge is a step-by-step process. A step may have rotten boards that seem to be sound. Eventually, those boards get found and replaced, but sometimes they provide support before they fall apart. Ptolemy's steps seemed to make sense. They stood in place for about 1500 years, and they were essential to the development of science.

Ptolemy wasn't much of an original theorist, but he was a solid thinker who compiled and organized and extended the work of the past and left a base for future scientific study. His great contribution was in explaining the world mathematically. Those who followed would be compelled to do the same thing.

Ptolemy wrote massive volumes on science, geography, and mathematics. And his ideas seemed to work. With his model and his mathematics, you could predict the motions of the Sun, stars, and planets. You'd be close enough so that farmers, sailors, and teachers could use the results—which they did. Today we're apt to look back at Ptolemy and forget how important he was for centuries and centuries and centuries. Those who turned to Ptolemy were rejecting superstition and magic and using a solid work of scholarship.

His monumental book was called *Megale mathematike syntaxis ("Great Mathematical Composition")*. Sometimes it was just called *Megiste ("The Greatest")*. It mapped and charted the visible stars, going still further than Hipparchus. Ptolemy's star charts became enormously helpful to anyone trying to navigate a ship at sea. As to his writings on mathematics, math historian Carl B. Boyer calls them "by far the most influential and significant trigonometric work of all antiquity." Today, no one is sure how much of it was Hipparchus's work and how much was Ptolemy's.

Measuring the altitude of stars above the horizon to find latitude was easy (see page 182). Finding longitude in a ship at sea was close to impossible without an accurate clock. Ships' motion made all the known timepieces inaccurate. Read the book *Longitude* by Dava Sobel for details.

But don't think of Ptolemy as a copycat. He was a corrector and improver. Yes, he got most of his ideas from others, but he worked hard to extend them. When he wrote an atlas, the *Geographia*, he did follow Hipparchus by putting grids with lines of latitude and longitude on its 27 maps. Then he introduced many previously unknown places using information from sailors' reports. In China, cartographers, who had no idea the world was round, also developed a grid system for maps. The Chinese maps showed details far better than maps made in the West, but they were limited by the flat-world idea.

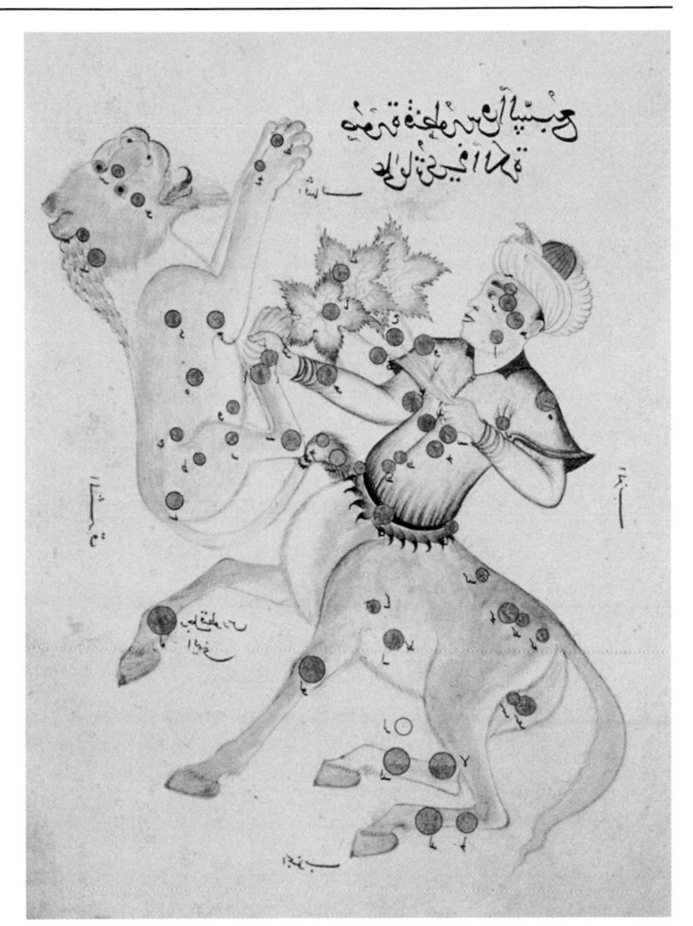

This illustration, a copy from a tenth-century Arabic manuscript called *Treatise on the Fixed Stars,* by al-Sufi, combines two constellations derived from Ptolemy. The constellation Sagittarius, in the form of a centaur (half man, half horse), battles Leo, a lionlike beast.

By the time Ptolemy finished writing his books, things were in turmoil in the world centered at Alexandria. Rome was now dominant, and the Romans weren't terribly interested in pursuing scholarly ideas. They never even translated Ptolemy's books into their Latin language.

Arab scholars were interested. They translated most of his books into Arabic. They took the *Megiste*, put the Arabic "*al*" in front of it (it means "the") and got *al-majusti* , which comes to us as *Almagest*, meaning "the greatest." And for a long, long time, that's what scientific thinkers thought it was. The Arabs probably saved Ptolemy from being lost forever. Although, for a while, hardly anyone in Europe seemed to care.

That's because Ptolemy and Aristotle and the rest of Greek science were going into a deep freeze. And question asking? It would soon be just about out of style.

22 A Saint Who Was No Scientist

An unjust peace is better than a just war.
—Cicero (106–43 B.C.E.), Roman orator and statesman, *Letters to Atticus*

But, Rome, 't is thine alone, with awful sway,
To rule mankind, and make the world obey,
Disposing peace and war by thy own majestic way;
To tame the proud, the fetter'd slave to free:
These are imperial arts, and worthy thee.
—Virgil (70–19 B.C.E.), Roman poet, *The Aeneid*

Rome had, for men of that time, seemed the organizing principle of all human history. Rome gone, what sense was to be made of the world?
—Garry Wills (1934–), American historian and professor, *Saint Augustine*

After some impressive centuries of rule and a time of peace and harmony unusual in human history, the vast and mighty Roman Empire was in trouble. Historians will give you a whole lot of reasons for its decline:

- poor leadership
- economic woes
- urban (city) problems connected to size and rapid growth
- nasty political fighting
- crime

BARBARIC originally described anyone who stammered or babbled. To the Greeks and Romans, it referred to someone from another country. To Christians, non-Christians were barbaric. Today, the word describes anyone who is crude, rough, or primitive.

Whatever the cause, when wild, uncivilized, barbaric tribes began attacking, the empire started to collapse. It didn't happen overnight; it was a long process. Those

A carving on a third-century C.E. sarcophagus (stone coffin) captures the chaos of a close battle between Romans and barbarians. The Roman soldiers look superior—clean-shaven with helmets and armor. The barbarians, in this case Germans, appear crazed with their messy beards and long hair. Guess which culture made the sarcophagus.

barbarians—bands of warring hoodlums—came from the north, and kept coming. They were being pushed out of their lands by still fiercer hordes who had fought on horseback all the way across Asia from Mongolia. Wherever they went, these barbarian tribes brought disaster, destruction, energy, and change.

While the Roman Empire was tumbling, a new religion was spreading like a grass fire (which means fast). The new religion was Christianity. It preached love and brotherhood and brought a vision of a world to come. It was a message of love and hope. It spoke to those who were facing confusion, violence, and wrenching change in what had once been civilized, secure communities. Christianity led believers inward, to a quiet, spiritual life. At a time when the known world was often unbearably brutal, Christians focused on the world of the soul and the promise of eternal life.

"Blessed are the poor in spirit: for theirs is the kingdom of heaven," said Matthew in the New Testament.

But an otherworldly philosophy often wasn't good news to scientific thinkers who focus on the world about them. "To discuss the nature and position of the Earth does not help us

Baptism, to a Christian, is the washing away of sins. Above, Jesus Christ undergoes the baptism ritual in the Jordan River. This sixth-century ceiling mosaic decorates a church in Ravenna, Italy.

A PAGAN is a person who believes in more than one god—or in no god. The word started out meaning "villager" or "civilian" in Latin and applied to anyone living in the country. Sometime after the Roman Empire became Christianized—with the exception of remote rural areas—*pagan* took on a strong religious tone. Another word for a pagan is a HEATHEN. Pagans and heathens sometimes committed HERESY, the then-serious crime of being antireligious.

in our hope of the life to come. For us, curiosity is no longer necessary," wrote Tertullian, an early Christian.

In 313 C.E., Emperor Constantine made Christianity a tolerated and lawful religion of the Roman Empire. Christians, who had been struggling outcasts, began to hold political as well as spiritual power. Their ideas on science now mattered. Should priests and bishops encourage Christians to study the natural world? Or should they urge Christians to concentrate on saving their souls? Studying science and mathematics meant studying the Greeks. But the Greeks were being called pagans. If the Greeks were wrong about religion, as the Christians believed, could they be right about scientific matters? This was a serious, troubling dilemma.

Greek scientists had taught that the Earth is round. But if the world is round, then some people must be hanging upside down, said the priests. "Can anyone be so foolish as to believe that there are men whose feet are higher than their heads, or places where things may be hanging downwards, trees growing backwards, or rain falling upwards?" wrote the much-loved African priest and author Lactantius. (He was tutor to Emperor Constantine's son.) The side of the Earth opposite Europe (and below the equator) was called the antipodes (an-TIP-uh-deez). How could anyone live on the antipodes? The Bible's stories seem to make it an impossibility. Belief in a habitable round Earth was often seen as proof of heresy. (Hardly anyone understood that there is no top and bottom nor up and down in space.)

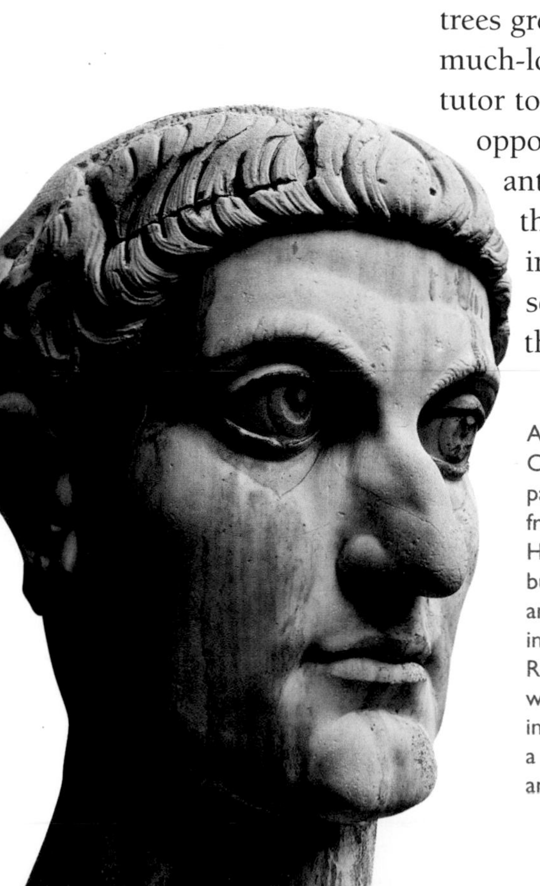

Although Roman Emperor Constantine was born a pagan, he freed Christians from the threat of persecution. He ordered churches to be built throughout his empire, and Christianity soon influenced all aspects of Roman life. This marble head was part of a colossal statue in the Basilica of Constantine, a government building in the ancient Roman Forum.

Then there was the problem of astrology. If everything is prewritten in the stars, as the popular Roman astrologists believed, how could God leave people free to make their own choices? (That power to make choices is known as "free will.") Historian Daniel J. Boorstin wrote, "The very struggle to become a Christian—to abandon pagan superstition for Christian free will—seemed to be a struggle against astrology."

It was a difficult time for thinking people. There was no technology—no microscopes or telescopes—to test scientific theories. Original thinking in the sciences didn't seem to lead anywhere. Question asking just for the sake of learning—the Greeks' great gift to all of us—began to seem pointless.

To further complicate things, there was a language problem. The barbarians couldn't speak Latin or Greek. They couldn't read at all. It would take hundreds of years for today's languages to evolve and more years for books to be written in modern languages. It would take just as long for the Huns, Goths, Franks, Vandals, and other northern tribes to meld with the Mediterranean peoples and create new cultures.

In the meantime, at the end of the fourth century, chaos and confusion reigned. No one knew where the Mediterranean world was going, and not knowing creates fear and often leads to fearful behavior. In 391, a big fire was lit at the library in Alexandria when the Roman emperor Theodosius the Great attacked the city. The teachers at the library—the best of them—weren't going to be intimidated. They kept right on teaching.

Hypatia (hy-PAY-shuh) was thought to be the very best. A professor of astronomy and mathematics, she was said to be wise, of high character, and beautiful too. The Christian bishop Synesius wrote in praise that her lectures drew students from all over the Greek-speaking world. But another bishop, Cyril, said her lectures were pagan.

Cyril was fighting an Egyptian leader, Orestes, for political

We know that Hypatia of Alexandria, shown here in a terra-cotta, was a brilliant scientist, mathematician, and philosopher, mostly because of the writings of her best-known student, Synesius. Hypatia was stoned to death in 415 C.E.

Imagine digging a hole from your house straight through the Earth. You'd emerge at your ANTIPODAL POINT on the opposite side of the globe. From most places in the United States, that would be the Indian Ocean. The word ANTIPODES means "the opposite side of the Earth." It is also an informal name for Australia (the land "down under") and New Zealand, which are opposite England on a globe. (If the first maps had been drawn in Australia, would Europe be "down under?") At the time of the Roman Empire, Europeans believed they lived on the top of the Earth, so the land south of the equator was called "the antipodes." It seemed like the Earth's bottom. In the Earth that Cosmas described, no one was thought to live below the equator. If they did, their feet would be higher than their heads (or so people thought). Everyone laughed at the idea of upside-down Antipodeans.

power; Hypatia supported Orestes. That was a mistake. Cyril became archbishop of Alexandria in 412. He expelled the city's Jews in 415. He had a worse fate in mind for Hypatia.

A CANON is a set of religious rules, laws, holy books, or holy people. So **CANONIZING** is about adding someone to the canon of saints, which only happens after death.

One man who worried and thought and wrote during these turbulent times helped keep the Greek love of learning from a final burial. He became a Christian and was eventually canonized as a Catholic saint. But he was no scientist. He thought that the idea of people living on the underside of a round Earth was crazy and said so. Still, without him, Christianity might not have been open to the modern science that would come eventually.

His name is Augustine (AW-guh-steen), or Aurelius Augustinus in Latin. He was born in 354 C.E. in Numidia (now Algeria, Africa), when it was part of the Roman Empire.

The Colosseum was a giant open-to-the-sky amphitheater in Rome, built in 72 C.E. (Actually, it took some years to complete it.) For 400 years it was the site of gladiators' battles. Today, a COLISEUM is an arena or stadium for sports and entertainment. An AMPHITHEATER is an oval or round stage or arena surrounded by tiers of seats.

But the great Roman Empire—with its towns, coliseums, aqueducts, gymnasia, amphitheaters, workshops, and temples—was doomed. Religion wasn't the issue. Rome had permitted Christianity since 313, and many of the barbarians were now Christians. The issue was that the political system and the way of life were collapsing. In 410 the Christian city of Rome fell to the Christian Visigoth Alaric I. Rome had dominated the Mediterranean world for more than five centuries. Jerome, a Latin scholar and church father living in Bethlehem, wrote, "The light of the whole world is extinguished. . . . If Rome can perish, what can be safe?"

TIME TO THINK

In *The Confessions*, St. Augustine wrote, "What, then, is time? If no one asks me, I know. If I wish to explain it to him who asks, I do not know." Philosophers and scientists are still trying to define time, but it's a slippery thing, and the way we think about it keeps changing.

St. Augustine reflects in a 1480 fresco by Sandro Botticelli.

Not much, as it turned out. In 415, a mob got aroused and yet another fire was lit in Alexandria. Soon, many of the library's greatest treasures were blazing. Works of the ancients (like Democritus, Epicurus, Pythagoras, Archimedes, and Eratosthenes) were lost forever. Then the mob went really wild. Hypatia was brutally killed, some say the crowd used pieces of broken pottery and stones. Archbishop Cyril is

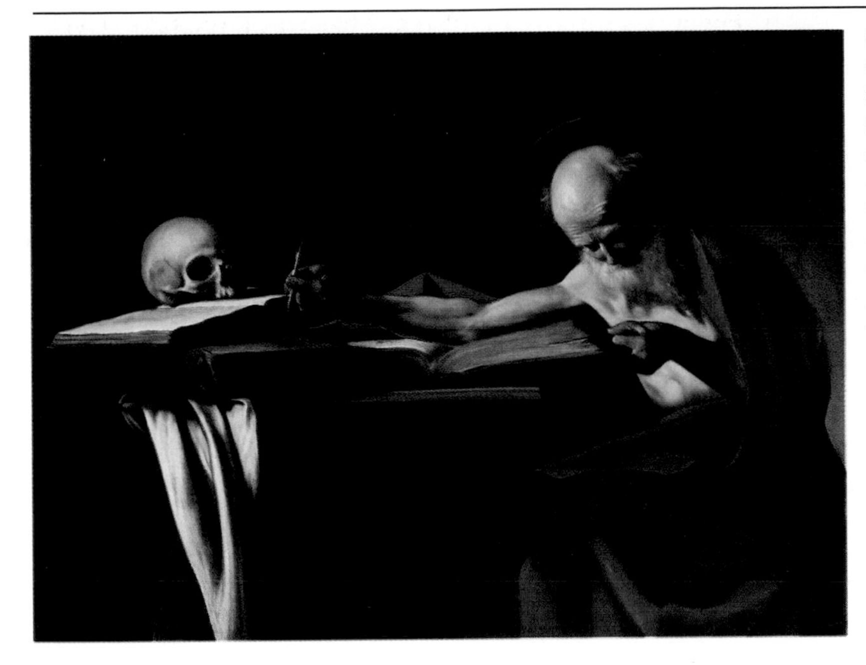

In 405 C.E., Jerome completed the first translation of the Bible from Hebrew and Old Latin into standard Latin. It was called the Latin Vulgate Bible. (*Vulgata* means "ordinary," not "crude," as its descendant *vulgar* means today.) Jerome was a fiery guy who didn't like Augustine and said so. Both were learned and eloquent, and both became saints after they died. St. Jerome was a popular subject of Renaissance art. This painting by Michelangelo da Caravaggio (1571–1610) depicts the saint bent over bound and printed books that didn't exist in Jerome's lifetime.

believed to have incited them. Free thinking and free speaking had become life-threatening pastimes.

Vandals laid siege to the fortified port city of Hippo (now Annaba, Algeria). Augustine was the bishop of Hippo. It was a modest, outback kind of town—there were 700 Christian bishops in Africa—so it wasn't Augustine's job that made him important. It was his writing. He used words well and wrote at least 93 books and hundreds and hundreds of sermons and letters. Here's a sample: "To handle words with words is to interweave them—like interlaced fingers: rubbing them together makes it hard to tell, except by each finger on its own, which is doing the itching and which the scratching."

His two most famous books are *The Confessions*, said to be the first literary autobiography, and *City of God*, in which he substitutes God's realm for earthly ones. Augustine puts love and faith in God ahead of all else. But he is also concerned with practical matters. In a work called *The Teacher*, he wrote, "Unfettered inquisitiveness, it is clear, teaches better than do intimidating assignments." "Unfettered inquisitiveness"? "Intimidating assignments"? His message to teachers seems to be: Let your students ask questions and search out answers rather than overwhelm them with assigned schoolwork.

The VANDALS were peoples from northeast Germany who, in the fifth century, ravaged Gaul (France), Spain, North Africa, and then Rome itself. Wherever they went, they destroyed artworks, literature, public buildings, and people's lives. Today the word *vandalism* is used to describe acts of wanton destruction.

The VISIGOTHS, or GOTHS, were another Germanic tribe that invaded and devastated Europe in the third to fifth centuries. When Romans called someone a Goth, it was an insult, meaning "rude and crude." Later, *Gothic* became the name of a style of art and architecture. Today, the word describes a subculture of people who like dark and scary things.

A handwritten, hand-painted edition of St. Augustine's book *City of God* from fifteenth-century Italy features Rome as the title city and the author sitting inside a big letter *G* (at bottom).

This is what Augustine wrote about science:

> *It is not necessary to probe into the nature of things, as was done by those whom the Greeks call* physici; *nor need we be in alarm lest the Christian should be ignorant of the force and number of the elements—the motion, and order, and eclipses of the heavenly bodies; the form of the heavens; the species and the natures of animals, plants, stones, fountains, rivers, mountains; about chronology and distances; the signs of coming storms; and a thousand other things which those philosophers either have found out, or think they have found out.... It is enough for the Christian to believe that the only cause of all created things, whether heavenly or earthly... is the goodness of the Creator, the one true God.*

In everyday language, that means there is no need for Christians to concern themselves with the details of scientific inquiry—as the Greeks did. All that is needed is an understanding that God created all things.

The Greek PHYSICI are studiers of all of nature. Today's *physici*—we call them physicists—are scientists who study energy and matter, which they will tell you is all of nature.

Now that wasn't exactly encouraging to the scientific thinkers. But, unlike many of his fellow priests, Augustine didn't see anything wrong with studying Greek philosophy. Of the Greeks, it was Plato he most admired. "Among the disciples of Socrates, Plato was the one who shone with a glory which far excelled that of the others, and who not unjustly eclipsed them all," he wrote in *City of God.*

Because of Augustine, Christian thought had a big measure of Plato's ideas added to it. (There was no conflict. Christianity deals with spiritual and historical ideas, not philosophical and scientific ones.) Christians began to think in terms of ideal forms and timeless perfection.

Augustine actually studied Aristotle's ideas on motion and may have been the first to understand that Aristotle didn't have it right when he said an arrow flies because air pushes

Life of St. Jerome: Vision of St. Augustine (1504) by Venetian artist Vittore Carpaccio shows Augustine in an idealized Renaissance study, where both faith and science are practiced. Volumes by pagan Greeks are mixed with Catholic prayer books. An armillary sphere (upper right) offsets a religious altar (center). Legend has it that Augustine was writing a letter to Jerome, who then appeared and announced his own imminent death and ascent to heaven.

it. (Motion—how things move—is a central concern of science. It was Galileo and Isaac Newton who helped us understand what really makes an arrow fly.)

The year after Augustine died (430), the town of Hippo surrendered to the Vandals. Rome was finished there. Imagine yourself living in those times. Everything you know is threatened. The wisest man of your time is gone. The best way to cope with the destruction, said Augustine's followers, was to forget earthly things, forget science, and concentrate on God and the world to come. In the world they knew, the barbarians had won.

The Vandals gave Rome a major sacking in 455. The end of the Roman Empire is usually dated as 476, when the last emperor was deposed. But it was really gone before that.

Do You Believe the Polyhistor? Well, Lots of People Did

In the mid-1970s, a teenager named Wayne Douglas Barlowe began drawing aliens: humanoids, insectoids, reptilelike beings, and others too strange to name in earthly terms. Unlike the Polyhistor, he wasn't trying to fool anyone into believing they existed. He clearly labeled each creature, including this Triped, as a character in a science-fiction story. Yet he also designed his aliens with a grain of scientific truth to make sure their bodies were well adapted to their home planets. Tall, thin bodies with little muscle make sense on a planet with weak gravity, for example. (What does the Triped's body suggest to you about its planet?) The book, *Barlowe's Guide to Extraterrestrials*, featuring the artist's alien menagerie, is a wonderful meeting of literature, art, and science.

Caius Julius Solinus, who lived and worked about 250 C.E., was sometimes called the Polyhistor, or the "Teller of Tales." His writing was filled with tales—tall tales—although a lot of people thought they were the truth. Solinus had an active imagination. He was also a plagiarist. He stole straight from a Roman author named Pliny, who also presented make-believe as truth. It didn't matter; they were both believed. St. Augustine read Solinus's stories and pictures and took them seriously. To be fair, scholarly standards were not well defined, and the line between fiction and nonfiction was wavy.

Solinus described and drew inhabitants of distant lands. They included dogheaded people who barked, ants the size of big dogs, people with eight toes and four eyes, people with one foot so large it could serve as an umbrella, and snakes that drank milk from cows. His book *Collectanea rerum memorabilium* (*"Collection of Memorable Things"*) was a best-seller in his time and for centuries to come.

Scholarly standards are rules that professors and other professional researchers follow to make sure their work is accurate—citing sources in footnotes, for example. **Plagiarism**, stealing someone else's writing and passing it off as your own, is definitely against the rules.

Sea monsters and other strange creatures inhabit *Cosmographia universalis* ("*Universal Cosmography*," 1544), a geography book by Sebastian Münster, an influential German geographer and religious scholar.

"The stories and fabulous images that Solinus detailed enlivened Christian maps right down to the Age of Discovery," said historian Daniel J. Boorstin. It's what we now call fantasy, like tales of alien abductions that some people (quite a few people) seem to believe today. Fantasy can be fun to read and to write, just as long as you're clear about the difference between it and reality.

This so-called *Bocca della Verità* ("*Mouth of Truth*") dates from the fourth century C.E. The mouth supposedly snapped shut on anyone who put a hand into it and swore falsely.

Searching for Truth: Words from Some Experts

"I would ask you to be thinking of the truth and not of Socrates: agree with me, if I seem to you to be speaking the truth; or if not, withstand me might and main, that I may not deceive you as well as myself in my enthusiasm, and like the bee, leave my sting in you before I die."
—Plato, *Phaedo*

"The man who loves truth, and is truthful where nothing is at stake, will still more be truthful when something is at stake. . . . The states by virtue of which the soul possesses truth . . . are five in number; i.e., art, scientific knowledge, practical wisdom, philosophic wisdom, intuitive reason; we do not include judgement and opinion because in these we may be mistaken."
—Aristotle, *Nicomachean Ethics*

"A liar is not believed even though he tell the truth."
—Cicero, *On Divination*

"A liar ought to have a good memory."
—Quintilian, *Institutio Oratoria*

"I have met many who wished to deceive, but not one who wished to be deceived."
—Augustine, *The Confessions*

"A small error in the beginning is a great one in the end."
—Thomas Aquinas, *On Being and Essence*

23 No Joke—the Earth Is Pancake Flat!

Neither do men put new wine into old bottles: else the bottles break, and the wine runneth out, and the bottles perish: but they put new wine into new bottles, and both are preserved.
—Matthew 9:17, Holy Bible (King James Version)

The books were delivered to the flames; and all who should presume to [create] such writings, or to profess such opinions, were devoted to an ignominious [shameful] death.
—Edward Gibbon (1737–1794), British historian, *The History of the Decline and Fall of the Roman Empire*

In 529, the Christian Roman emperor Justinian closed the last of the traditional Greek schools. In Athens, the academy that Plato had founded *almost 900 years earlier* was shut down. Hardly anyone in the Christian world read the Greek philosophers anymore; they were labeled pagans (nonbelievers). Some of Plato's ideas did get incorporated in Augustine's philosophy, but the science of Aristotle and Ptolemy was now off-limits. Those scientific ideas were thought to be dangerous. Since most ordinary people were illiterate, they had to believe what others told them.

An ABBEY is a monastery (monk's home) or convent (nun's home) run by a religious leader called an abbot (male) or abbess (female).
ACCESSIBLE—easy to get to—is the opposite of CLOISTERED, which means "shut off from the world." A CLOISTER is a quiet, secluded place.

The Greek thinkers would soon be almost forgotten, except by a very few scholars—such as a learned Roman statesman named Boethius (boh-EE-thee-uhs), who translated Aristotle's writings on logic (but not science) into Latin and kept them from disappearing altogether.

Something else happened in that year, 529. It was an important symbol of the change that was underway. The first Benedictine abbey was founded at Monte Cassino, Italy. Plato's Academy had open gardens and walks and urban

The first abbey founded by St. Benedict (Monte Cassino, in 529) started a European trend toward keeping knowledge behind closed doors. About 900 years later, painter Luca Signorelli told the saint's life story in pictures—a series of frescoes at a Benedictine monastery in Italy. Here, St. Benedict is predicting doom for his new abbey. In fact, it was destroyed and rebuilt four times: in 577 by the Lombards (Germanic invaders of Italy), in 883 by the Saracens (Muslim raiders), in 1349 by an earthquake, and in 1944 during World War II.

buildings. It was in the middle of town. The abbey was on a mountaintop. Intellectual life was being sent off to tucked-away, cloistered monasteries.

At Alexandria the book burning wasn't over. Now there was a dynamic new force in the world—Islam. Like others with deep beliefs, some Muslims felt a need to destroy the old in order to replace it with something new. So in 642, when Islamic armies conquered Alexandria, they too expressed their power and rage by torching scrolls and codices. By this time, few Greek scientific writings were left anywhere. Those that did survive were saved by Arab and Byzantine scholars. They rescued what they could and took

BYZANTINE (BIZ-uhn-teen) refers to the Eastern part of the Roman Empire after it was divided in 395. Its religion was Eastern Orthodox Christianity. BYZANTIUM (by-ZAN-tee-uhm), its capital, was renamed Constantinople (after Constantine I, a Christian Roman emperor) and later renamed again as Istanbul after the Byzantine Empire fell to the Ottoman Turks in 1453.

Boethius (ca. 475–524 C.E.) was a Roman born into a family of wealth and influence. About the year 500, he was made court minister (like a secretary of state) to the Gothic king Theodoric the Great. Boethius read Greek and managed to keep some classical learning alive. He attempted to influence Theodoric to be a sound, fair ruler and to leave Roman law intact. But Theodoric was an absolute monarch, and in his old age he became suspicious, cruel, and nasty. He turned on Boethius and had him executed. In this illuminated manuscript, Philip IV (seated), the king of France from 1285 to 1314, extends his hand as Jean de Meun (kneeling), a French poet, offers him Boethius's book *The Consolation of Philosophy*. The poet helped translate the work, preserving and spreading Greek thought for future generations.

them from Alexandria to centers of learning elsewhere. Finally, earthquakes destroyed the lighthouse and most of the remains of ancient Alexandria. After that, Greek science was mostly forgotten in the West.

What happened to Ptolemy's great work and that round-Earth idea the Greeks had developed? A sixth-century monk named Cosmas wrote a 12-volume *Topographia Christiana*, which was meant to replace it. The first volume of his work has this title: *Against those who, while wishing to profess Christianity, think and imagine like the pagans that the heaven is spherical.* The Earth, said Cosmas, is rectangular—twice as long as it is wide—and surrounded by an ocean, which is surrounded by a second Earth: Adam's Paradise. The sky is like the roof of a tent with stars and planets pushed by angels.

Did people believe Cosmas? Yes, most did. But it didn't seem to matter much. Hardly anyone was concerned with the science of the universe. Its moment had passed. No other culture anywhere had done what the Greeks had done; but now Alexandria was becoming a forgotten town. Science and question asking were out of fashion.

The Sumerian idea of the universe returned—this time richly embroidered with descriptions of sea monsters and bizarre land creatures. It stayed around for almost 1,000 years. (To give yourself an idea of 1,000 years, it's twice as long ago as Columbus's voyage to America.)

Historians call the time between the fall of the Roman Empire (476) and the beginning of the Renaissance (early fifteenth century) the Middle Ages. Some divide that medieval era and call the first of it (from 476 to about 1000) the Dark Ages. Some call those years the Great Interruption. Whatever name you choose, it describes a time when most of Europe took a big step backward. Life became coarse and primitive, not just in comparison to our life today, but in contrast to other world civilizations of the time, such as the T'ang and Sung dynasties in China, the Eastern Orthodox Byzantine Empire centered in Constantinople (now Istanbul, Turkey), and the Islamic Empire (led by the culture-loving Abbasid rulers).

In Europe, for a few monks, there *was* a challenging life of the mind. Aristotle was responsible for that. His ideas on logic set rules for orderly discussions and arguments. Using Aristotle's logic, the monks learned to think analytically. They examined abstract ideas. That skill will be very useful in later centuries when science makes a comeback.

In the meantime, arguments about religion were so intense that one of them split the Catholic Church into two branches: Roman and Eastern Orthodox. Today it seems they were arguing

New Voices

After the fall of Rome (476), new voices began speaking out in new languages. Here is one of them, from *Beowulf*, an Anglo-Saxon epic poem written in the first half of the eighth century.

To the sound of the harp
the singer chanted
Lays he had learned, of
long ago;
How the Almighty had
made the earth,
Wonder-bright lands,
washed by the ocean;
How he set, triumphant,
sun and moon
To lighten all men that live
on the earth.

Logic is the art of thinking through ideas in a rational, orderly way. In logic, an argument doesn't mean a fight or disagreement. It's a way of reasoning by using logical statements to support an idea. An argument for a round Earth is "The Earth's shadow on the Moon is curved." A nonlogical argument over religion often is a disagreement that leads to a fight. A large number of the world's wars are about religion.

AN ILLUMINATING BOOK

Before the mid-fifteenth century, books were written by hand or printed from individually cut wooden blocks—so there weren't many books. They were called manuscripts, from two Latin words—*manus* ("hand") and *scriptus* ("written")—and often beautifully illuminated (illustrated).

This miniature painting (right) is part of a fifteenth-century chronicle of the life of Charlemagne, a Frankish king who conquered most of the European Christian world. (The name *France* comes from the Franks, a Germanic people.) Charlemagne (in the purple robe and crown) is discussing his conquest of Spain.

Charlemagne was crowned Holy Roman Emperor on Christmas Day in 800. After years of war, he devoted himself to organizing his vast empire, promoting education, and building plush palaces and awe-inspiring churches.

about tiny differences of ideas—but they didn't think so.

While this was going on, the tradition of scientific question asking shifted to Arab lands, where it stayed alive for some five or six centuries and helped create a golden age of Islam. (More on this in the next chapter.) But then Islamic religious leaders cracked down and, about the year 1100, Islamic science came to a crashing halt.

Science was already close to dead in the rest of Europe.

In China, technology was making enormous strides. Astronomers and mapmakers there were way ahead of those left in Europe. They were charting regions of the Earth exactly, although they still hadn't figured out that Earth was round. Ships' rudders, the compass, wheelbarrows, canals with locks, paper, printing, gunpowder, stirrups, and harnesses—those were just a few of the ideas and inventions that were eventually carried across the Silk Road from the East to the West. But neither the Chinese nor the Japanese nor the Mayans nor any other culture anywhere matched the Greeks when it came to asking questions about the universe. Without the right questions, you don't get the right answers.

The world's earliest known printed book comes from China. It's a wood-block printing of the Diamond Sutra, a sacred Buddhist text. (That's the Buddha in the center, surrounded by disciples.) The book is dated "on the 13th of the 4th moon of the 9th year of Xiantong," which corresponds to May 11, 868.

You don't even get energetic discussions. These *were* dark ages for science and dreadful years for most Europeans—especially for those with intelligence and imagination.

If anyone, then, had suggested that Europe was going to develop a science-based culture that would be a beacon to most of the world—well, that hardly seemed possible. By the year 1000, the once-famous Roman Forum had become a pasture for cattle. Squatters were living in Rome's Colosseum and hanging laundry from its windows. Alexandria? You could ask anyone: Its glory days were in the mostly forgotten past.

A FORUM is a public square in the center of a city where courts and gatherings are held. It can also be any open discussion of ideas.

WERE THE DARK AGES DARK?

Some historians argue that the Dark Ages weren't all that dark. You can research and decide for yourself. Here are a few ninth-century snapshots to get you started:

This was a time to watch out for Viking raiders. Danes attacked the east coast of England; Norsemen hit Ireland and Scotland. Vikings founded Dublin, destroyed Hamburg, and went deep into Russia.

In 881, Charles the Fat (Charlemagne's great-grandson) was named emperor of the Franks. But when Vikings laid siege to Paris, Charles did nothing. He was deposed and followed to the throne by Charles the Simple, who was worse.

China was politically troubled with the T'ang empire in decline. Zen Buddhism, on the rise, encouraged thoughtfulness and a focus on self-knowledge.

The conflict between the Christian pope (in Rome) and the Christian patriarch (in Constantinople) got heated. Each excommunicated the other. In 867, Basil I became Byzantine emperor and the empire surged forward.

In the Arab caliphate, al-Khwārizmī and al-Battani, great astronomers, improved on Ptolemy.

24 Don't Worry—the Round Earth Is Back!

The universe is as close as the veins in our necks.
—The Qur'an (Koran), Islam's holy book

For I ask any one, had he rather joy in truth, or in falsehood?
—Augustine (354–430), Christian saint, *The Confessions*

It is through reason that we are men. For if we turn our backs on the amazing rational beauty of the universe we live in, we should indeed deserve to be driven there from, like a guest unappreciative of the house into which he has been received.
—Adelard of Bath (ca. 1075–ca. 1160), English cleric and translator of scientific writings from Arabic into Latin

Pope Sylvester II, who reigned during the millennial year 1000, was sometimes called the Magician Pope. It was meant by some to be a compliment, for he was a versatile thinker with many talents. But others worried that he might actually be the devil in human guise. After all, he was the first French pope, and to the Italians who had dominated the church, that was reason enough for alarm. Besides, his mind often took him to fields well beyond religious studies. He built his own wind organ, and he played it with verve. He also built a planetarium with wooden spheres

Guise means "appearance"; it shares a root with *disguise*.

Pope Sylvester II is given credit for Christianizing Poland and Hungary. Here's the so-called Holy Crown that he gave to Hungary's first Christian king, Szent István (St. Stephen). Pope Sylvester was also a mentor and teacher to the German kings Otto II and III, both of whom became Holy Roman Emperors.

to plot the movement of the stars and planets. He collected ancient manuscripts. And he seems to have been an expert on the astrolabe, a navigation device used to calculate the position of heavenly bodies.

As a young man, Sylvester—whose name then was Gerbert of Aurillac—went off to Islamic Spain, where he studied Arabic math and science. While he was there, he read the ancient Greeks—in Arabic! The pope-to-be read Plato and Aristotle and even the love poems of Ovid, a first-century Roman. (You can see why this Frenchman worried some people.) In Spain, he discovered the abacus. Try multiplying MCXX by CXCVI. Can you manage? That's what people in Rome had to do if they wanted to deal with numbers. One Roman scholar said there were no calculations possible beyond MMMMMMMMM. (Even with Roman numerals, the ancients had gone way beyond that, but their achievements were forgotten.) For Gerbert/Sylvester and other thinkers, the abacus must have seemed an exciting high-tech tool (much like the computer seemed a few decades ago).

"Just as conventional calculations are swallowed up by the modern microchip, so the mechanism of the abacus obliterated the need to write out figures, speeding calculation in a magical fashion," writes Robert Lacey and Danny Danziger in a book titled *The Year 1000*.

When he was 39, Gerbert was the star of an all-day public debate sponsored by the German king Otto II. Scholars and students traveled from all over Europe for

Compare this astronomical astrolabe (made in 1215), with its interchangeable disks, to the simple mariner's astrolabe on page 183. The disks are guides to the locations of zodiac constellations, the ecliptic (the Sun's path in the sky), stars such as Sirius and Spica, and other celestial landmarks.

Some thought Sylvester II was the Antichrist. It was widely rumored when he died that he had made a pact with the devil and requested that his body be cut in pieces so the devil could reclaim it. In 1648, to put the tale to rest, his body was exhumed (his tomb was dug up). The skeleton was whole.

MUSIC IN THE AIR

Hero of Alexandria invented whistles and trumpets that worked by forcing air through tubes of different shapes (see page 134). A wind organ—also called a pneumatic (air-driven) pipe organ—is a fancy version of the same technology. It has anywhere from a handful to hundreds of air pipes that open and close at the push of a key or pedal. To play it with verve, as Pope Sylvester II did, means "to play with energy or spirit."

THE SECOND DIGITAL COMPUTERS

The name *abacus* may have come from *avaq*, the Hebrew word for "dust." The earliest of these calculating devices were probably boards dusted with sand, on which numbers were traced. The abacus most of us know has beads on wires, or pebbles in slots, where they can be lined up in columns of ones, tens, and hundreds, which makes adding and subtracting easy. (And since multiplication and division are just adding and subtracting done many times over, the abacus makes them easy, too.)

The Babylonians are thought to have invented the abacus, although some claim it was the Egyptians, and others, the Chinese. We know the devices were in use in Egypt in 500 B.C.E. For some unknown reason, they had just about disappeared from the West when Pope Sylvester reintroduced them around the year 1000. He thought the abacus an Eastern novelty. He knew it was used in China (and called a *suan pan*) and in Japan (known as a *soroban*), as well as in Indian and Arabic lands. Despite the pope's enthusiasm, it took a while (actually many centuries) for the abacus and the numbering system it helped spawn (which put numbers in columns) to catch on in Europe.

This Roman "pocket abacus," which is made of bronze, is roughly 2,000 years old.

Science and math writer Isaac Asimov calls the abacus the "second digital computer"—the first being the fingers on a hand. Now often a plaything for children, the abacus can be a powerful computing tool.

Roman-Numeral Refresher
I = 1
V = 5
X = 10
L = 50
C = 100
D = 500
M = 1,000

these events. Gerbert argued that physics is a branch of mathematics, won the debate, and was soon court mathematician and advisor to the powerful Otto. So, along with a high-powered mind, he now had political influence.

He was already a controversial figure. When he came home from Spain, he told friends and colleagues about Arabic mathematics and about Greek thinkers. A few other Christian scholars set off for Spain—although outsiders weren't welcomed. They were more likely to bring trouble than anything else. Most Spaniards thought they had little to learn from other European peoples.

Like Christian emperors, Islamic CALIPHS were both religious and political leaders. The separation of church and state is an American innovation.

The Spanish city of Córdoba (COR-duh-buh), for instance, had 80,000 shops, 1,600 mosques, 900 public baths, and 70 libraries. La Mezquita, a tenth-century mosque, held 32,000 worshipers. A caliph's palace included hundreds of buildings and 4,000 marble columns. Córdoba's fleet of merchant ships,

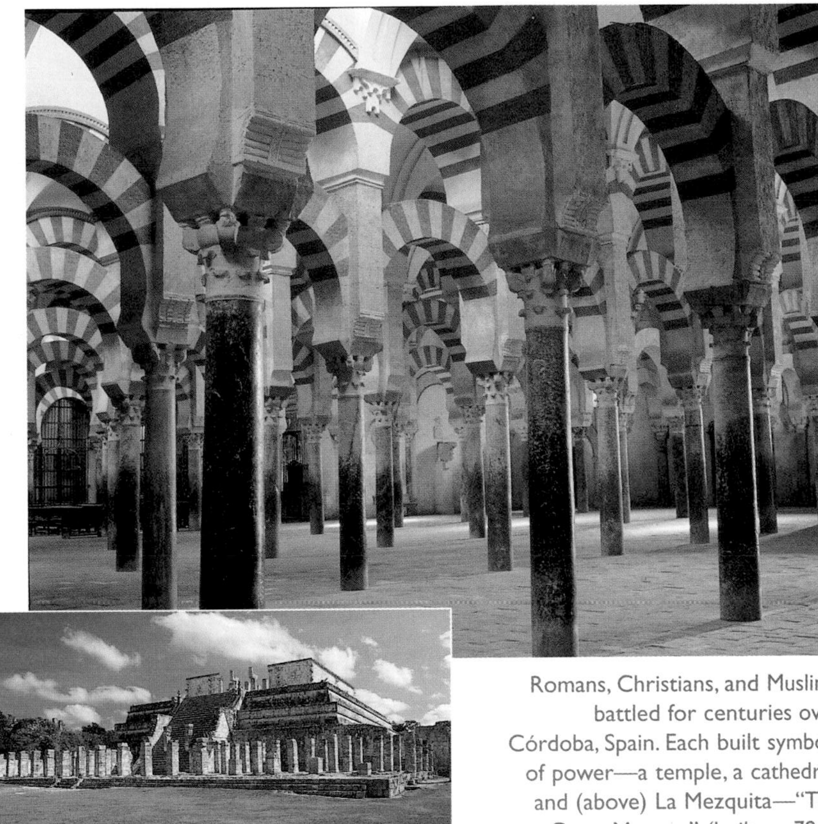

Romans, Christians, and Muslims battled for centuries over Córdoba, Spain. Each built symbols of power—a temple, a cathedral, and (above) La Mezquita—"The Great Mosque" (built ca. 784–ca. 987)—with its striped arches and 800-plus pillars. Around the same time, Mayans advertised their supremacy with the Group of the Thousand Columns (left) at Chichén Itzá (now in Mexico).

based at the Mediterranean port of Almería (al-muh-REE-uh), was huge. No city in Italy was anything like Córdoba. (Chichén Itzá, a Mayan city on Mexico's Yucatán Peninsula, might have impressed the Spaniards if they'd known of it.)

If we could zoom ourselves to the year 1000, we would find that Córdoba is a center of scholarship and trade, as well as of Islamic study. Christians and Jews are allowed to follow their own practices as long as their laws don't conflict with Islamic custom. The city, like much of Spain, has a prosperous, self-confident feel, although there are problems. Slaves keep plotting and rebelling. And the Christians can be troublesome. They are anxious to reconquer Spain and convert the infidels—Muslims and Jews. So Christians are tolerated but not encouraged.

THE REIGN IN SPAIN

Spanish history is complicated but fascinating. Because Spain is a gateway to Europe and Africa, everyone seemed to want the place. The earliest fossils show Neanderthals living there. Then came Tartessians, Iberians, Greeks, Celts, Visigoths (Germanic tribes), Jews, Romans, and Christians.

Maybe it was that hodgepodge that led to civil war. Whatever caused it, Visigoths and Christians were fighting it out when the Visigoths invited an Islamic army to come and help them. That was in 711. The Muslims, who were on a conquering spree, stayed. That led to a "golden" era in Spain when Jews, Christians, and Muslims worked together and produced great works of art and scholarship. Spain was prosperous. But Christians fought back, very slowly reconquering Spanish territory. The end of Islamic rule came when Granada was reconquered in that pivotal year: 1492.

Around 1110, an English philosopher named Adelard of Bath disguises himself as a Muslim student and heads for Spain. (Bath is a resort city in southwest England known for its mineral baths. The Romans loved the place.) Adelard has studied and taught in France and traveled to Italy, Syria, and Palestine. He speaks Arabic, so he has no problem pretending to be a Muslim. He is anxious to learn all he can in Spain's great schools. One of the first books he encounters is Euclid's *The Elements*.

In a world where hardly anyone can multiply or divide big numbers, Euclid is astonishing. Adelard is swept away by *The Elements*; it makes numbers clear and useful.

For out of old fields, as men sayth,
Cometh all this new corn from year to year,
And out of old books, in good faith,
Cometh all this new science that men hear.

—Geoffrey Chaucer (ca. 1340–1400), English writer, *The Parliament of Fowls* (translated from Middle English, an early form of the language)

Adelard translates Euclid from Arabic into Latin, giving Europe an enormous gift. He helps make it the most influential book on mathematics ever. (It will be the basis of almost all thinking about geometry until the nineteenth century.) Adelard also learns about the Greek theory of atoms and writes about it, and by the twelfth century, the idea of small basic particles is being taken seriously again.

Thanks to the influential Roman philosopher Boethius, Aristotle's writings on logic have been known for centuries. Now, with the Dark Ages receding, his works on science get translated and begin to be studied by a few Christian scholars. At the same time, three philosophers from the Arabic world write commentaries on Aristotle—clarifying and extending his thinking. They are Maimonides, who is Jewish; Avicenna, who is Persian; and Averroës, who is Spanish. Their observations

1,000 Years

This timeline spans 900 years, nearly a millennium. It seems miraculous that ideas could be passed down over so many centuries, across so many borders, and in so many languages. Consider that we're closer in time to Thomas Aquinas (far right), a medieval thinker, than he was to St. Augustine (beginning of timeline).

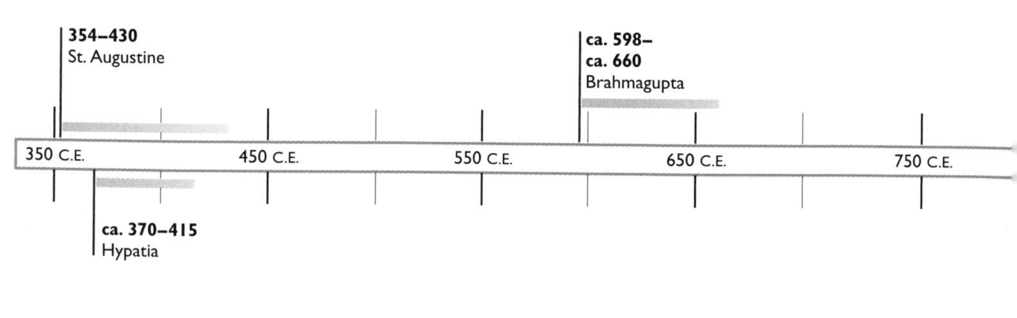

A BRAINY TRIPLE TREAT

If you are looking for someone to research and write about, Averroës, Avicenna, and Maimonides are all worth considering.

Averroës (uh-VEHR-oh-eez) was a famous Muslim philosopher who explained the Qur'an in terms of Aristotle, which was controversial in the Arabic world. So were the ideas he brought to Christians. Finally, Pope Leo X condemned those Christians who followed Averroës' philosophy—but the philosopher's ideas left an impact on Christianity. Averroës also wrote a book on medicine, and that book was widely used for centuries.

Avicenna (av-i-SEN-uh), another Arabic philosopher, was a physician to the vizier (high Islamic official) in Persia. He mixed Aristotle's ideas with Plato's, adapting them to his time. He was a prolific writer (he produced some 300 works) on a variety of subjects. One of his books—a compendium on diseases and treatments—became the most influential book on medicine written during the Middle Ages. And a poem he wrote, on healing, used to be memorized by medical students in Islamic lands.

Maimonides (my-MON-i-deez) was a rabbi who was born in Córdoba, Spain, in 1135. He became very influential in both the worlds of Islam and Judaism. He moved to Cairo, Egypt, where he was physician to the Islamic ruler Saladin. (Now Saladin is a fascinating person, too—a warrior, leader, and visionary!) Maimonides wrote many commentaries on the Hebrew Bible. His most famous book is called *Guide for the Perplexed.*

The Mishneh Torah, a code of Jewish law by Maimonides, built on Aristotle's ideas. That was an uncommon and unpopular move—but not a surprising one for a rabbi who straddled many cultures.

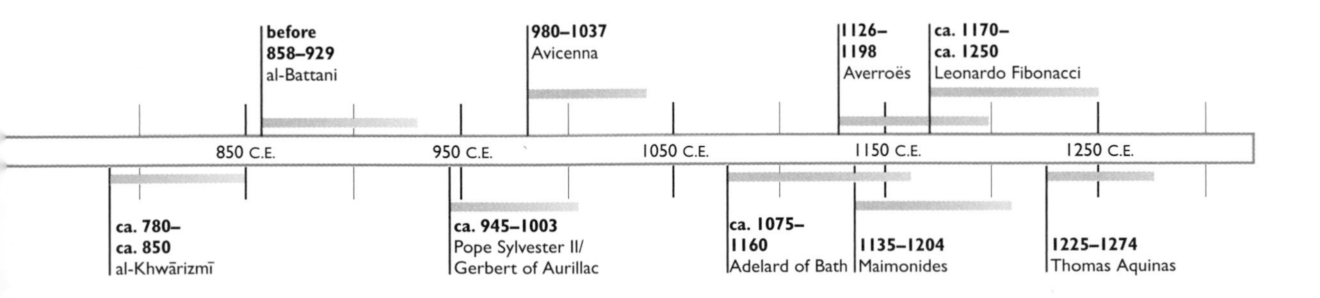

These two examples of religious art (right) were made in neighboring centuries, and yet they're worlds apart when it comes to reality. In the twelfth-century piece (far right), monks climb a heavenly ladder under an unnatural golden sky. The good monks greet Jesus Christ at the top; the bad ones fall off and are escorted to hell by black devils. In the thirteenth century (near right), St. Francis looks like he's in someone's backyard as he talks to birds gathered under a simple tree.

The early Gothic Chartres Cathedral (below) in northern France was built and rebuilt (after a fire) in the twelfth century. Its grotesque gargoyles (inset) are firmly planted in mythology, not reality. (Even so, these hideous beasts do serve two practical purposes—acting as spouts to throw off rainwater and attracting tourists.)

give thinkers much to discuss. After centuries of intellectual stagnation, this is all astonishing. Imagine living in the dark and suddenly lights are turned on.

Painting and sculpture begin to change, too. Early medieval paintings set religious figures and icons in a beautiful but unreal

gold background. The new paintings still have a religious focus, but real-looking plants and animals and people are also included. These natural subjects can be found in sculptures on the splendid Gothic cathedrals rising in the thirteenth century. The artistry of these churches mirrors the expanding ideas.

Some say much of the ferment can be traced to the new schools and universities. One of the first of them is built in the twelfth century at Chartres (SHAR-truh), France. The great cathedral houses a school where professors begin to teach about nature and science. Universities are soon founded in most large towns across Europe.

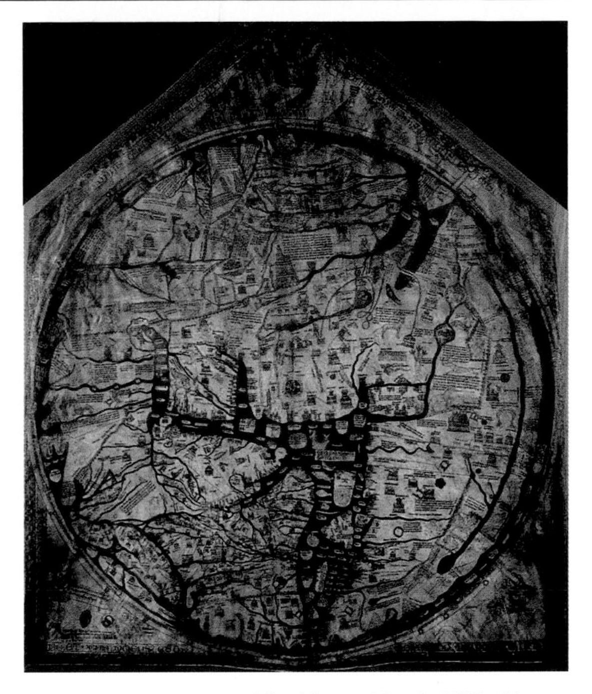

The Mappa Mundi ("World Map," ca. 1300) in Hereford, England, is the largest medieval map still around. The circle is 1.3 meters (52 inches) in diameter, with Jerusalem at its center and ocean all around. It's a T-O map: The T within the O is formed by the Nile and Don Rivers, which outline the three known continents—Europe, Asia, and Africa. Bizarre beasts lurk in distant lands, according to the map's drawings and descriptions. The Phanesii, for example, have huge ears for wrapping themselves against the cold. A Sciapod shades himself from the sun with his one giant foot.

Professors need books to teach, which leads to a demand for scholarly works. Scribes are put to work copying the new translations of the ancient works. Some scholars actually begin thinking for themselves. Still, ideas change slowly.

In a book called *Life in a Medieval City*, authors Joseph and Frances Gies describe the general level of scientific learning:

> *The pupil at the cathedral school absorbs relatively little true scientific knowledge. He may be given a smattering of natural history from the popular encyclopedias of the Dark Ages, based on Pliny and other Roman sources. [That means the Polyhistor—see page 198.] He may learn, for example, that ostriches eat iron, that elephants fear only dragons and mice, that hyenas change their sex at will, that weasels conceive by the ear and deliver by the mouth.*

It may all sound foolish to us today, but it is more than their parents and grandparents learned—and it makes some young learners very inquisitive (curious).

A few schools (very few) are teaching a new system of

THERE'S THAT COMET AGAIN

Raoul Glaber (ca. 990–ca. 1050), a monk from Burgundy (in France), tells of a scary comet that lit the skies in 989. That was probably just before he was born, but comets were taken so seriously that tales were told of it for many years. It was the comet we now call Halley's comet. Glaber wrote:

> *It appeared in the month of September, not long after nightfall, and remained visible for nearly three months. It shone so brightly that its light seemed to fill the greater part of the sky, then it vanished at cock's crow. But whether it is a new star which God launches into space, or whether He merely increases the normal brightness of another star, only He can decide.... What appears established with the greatest degree of certainty is that this phenomenon in the sky never appears to men without being the sure sign of some mysterious and terrible event. And indeed, a fire soon consumed the church of St. Michael the Archangel [Mont-Saint-Michel], built on a promontory in the ocean which has always been the object of special veneration [adoration] throughout the whole world.*

The comet returned 77 years later, in 1066, as it would about every 77 years after that. In 1066, the Battle of Hastings ended Anglo-Saxon rule in England. The conquerors were from Normandy in northern France, where the famous *Bayeux Tapestry* hangs, telling in pictures the history of a comet and a war.

Do you see Halley's comet? It looks like a shuttlecock flying over a castle tower in this scene from the *Bayeux Tapestry*. To the right of the tower, King Harold is being warned that this comet is bad news. For him, it was—Harold soon died in battle—but not for William the Conqueror, the victor.

numbers that has come from India by way of the Arabs and Spaniards and is being taken up by business people. But zero, a key to those Arabic numbers, is still almost unknown in the Europe of the twelfth and thirteenth centuries.

When it comes to geography, European maps show Earth with three continents: Africa, Europe, Asia (see map on previous page). All the world's known land—those three continents—is surrounded by the "Ocean Sea." That ocean is

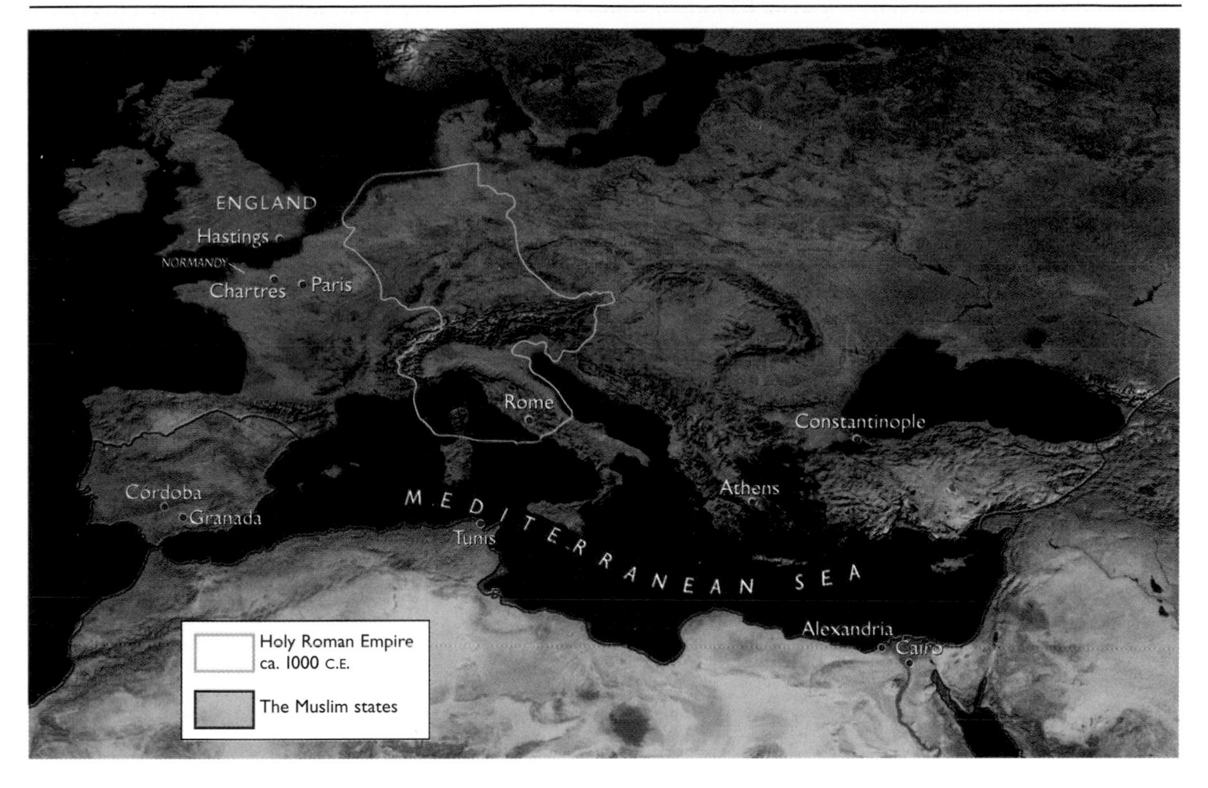

terrifying because no one has sailed across it and told the tale; sea monsters are thought to abound.

Aristotle's round world is being taken seriously at most of the universities, but no one with a ship is dealing with the idea. Around the year 1000, Leif Eriksson, a Viking explorer, sailed west from Greenland and discovered a new land that he thought promising for growing grapes. He called it Vinland. But, even a century later, hardly anyone elsewhere knows of that voyage, and no one understands its significance.

A few prophets are suggesting that there may be unknown islands and even great lands in the sea. Some are even reading the Greek geographer Strabo. Could aliens be living on those lands? What might they look like? Maps have illustrations showing humanlike creatures with dog heads or bizarre shapes—still reflecting the Polyhistor's lively imagination.

Those who are designing the magnificent Gothic cathedrals are the technological thinkers of the day. To do their job, they must know mathematics and geology and engineering—and be artists too. Their work is breathtakingly gorgeous.

A fourteenth-century miniature painting is split 50-50. Saracens (Muslim soldiers) in white headdresses with red shields (at left) battle Christian crusaders (at right) mostly hidden behind armor and blue, lion-crested shields.

The Christian Crusades would dominate political and religious life in the Western world during much of the twelfth and thirteenth centuries. It would spawn stories of heroism and knighthood and adventure. The actuality was very different. Too often the Crusades were ill-fated, ignorant, avaricious, out-of-control rampages. The Christian knights did begin to roll back the Islamic Empire, which controlled much of the Mediterranean and Eastern trade. A competitive European trading and shipping industry would evolve, bringing great wealth to the continent and energizing a class of seafarers who, a few centuries later, would head out to unknown lands elsewhere on the globe.

But most people in the new millennium are neither architects nor scholars nor priests nor sailors nor merchants. Most people are peasants. And for them, Europe is still a place of walled towns and walled thinking. New ideas are feared. Perhaps there is reason to fear them. Ideas are like viruses—they spread. The "new" Greek and Arabic ideas will eventually act like dynamite charges under the old walls.

For centuries, the Bible has seemed to hold all the answers—not just answers about God and faith, but also answers about the natural world and the universe. Now things are getting more complicated.

When Christian forces conquer Islamic Spain (Córdoba falls in 1236; Granada in 1492), soldiers and other travelers discover Islamic culture. They learn about Arabic math. They meet Greek art and science. It is all new to them. It gets them thinking. They have a lot to tell their friends when they get home.

Around the World in Two Medieval Centuries

The five big towers of Angkor Wat in Cambodia stand for the peaks of Mount Meru, the center of the Hindu universe.

In the twelfth century (the 1100s), the world population is nearing 330 million people. China is the most populous nation (it still is today) with more than 100 million people. In Cambodia, the Hindu temple at Angkor Wat is built with a 65-meter (71-yard) central tower and a moat with a 6.4-kilometer (4-mile) circumference. Here are a few more snapshots of that interesting time.

1107: A multicolor printing process is invented in China. It makes counterfeiting (copying) money much more difficult.

1174: A campanile (bell tower) is begun next to the cathedral in Pisa, Italy. The land beneath it is unstable, and the tower soon begins to lean.

ca. 1150: In South America, the Inca city of Cuzco (now in Peru) is founded by Manco Capac. Dazzling architectural structures are soon under construction.

1212: A Children's Crusade, led by a 12-year-old French boy, Stephen of Cloyes, attracts tens of thousands of European teenagers. With the best of intentions, they set out to free the Holy Land from what are called "infidels" (Muslims). Most of the youngsters die of hunger or disease; some are sold into slavery; none reach their goal; and only a few return home. The story of the Pied Piper of Hamlin is said to be based on this sad, awful, absurd enterprise.

1227: When the Mongol ruler Genghis Khan dies, he has amassed the largest empire the world has ever seen. Far larger than Alexander the Great's empire, the Mongol Empire stretches from the Caspian Sea to Korea. The Mongol conquering spree is accompanied by horrendous violence and murder of civilian populations. When Islamic Baghdad falls to the Mongols, most of its population—perhaps a million people—is murdered.

1281: Kublai Khan (grandson of Genghis), having conquered much of the world he knows, assembles an invasion force intended to subdue Japan. A fleet with more than 100,000 warriors sets out for that island. Then a typhoon strikes. Most of the ships sink. The survivors who manage to swim ashore are quickly killed by sword-wielding Japanese soldiers called samurai.

Genghis Khan is about to run over another enemy in a Persian illustrated history of the world (ca. 1400).

25 Absolute Zero

Socrates: **And all arithmetic and calculation have to do with number?**
Glaucon: **Yes.**
Socrates: **And they appear to lead the mind towards truth?**
Glaucon: **Yes, in a very remarkable manner.**
Socrates: **Then this is knowledge of the kind for which we are seeking.**
—Plato (ca. 427–347 B.C.E.), Greek philosopher, *The Republic*

The point about zero is that we do not need to use it in the operations of daily life. No one goes out to buy zero fish.
—Alfred North Whitehead (1861–1947), English mathematician and philosopher, *An Introduction to Mathematics*

In this Flemish tapestry, "Lady Arithmetic" teaches adults multiplication and division—tough but necessary skills for sixteenth-century professionals.

Do you want to be a mathematical genius? Just time-warp yourself back to the Middle Ages, or almost any time before the seventeenth century, and you will be able to do mathematics that only geniuses—or at least very, very bright people—can do. I guarantee it. If you've made it to third grade or beyond, you can do the kind of arithmetic that would have made your ancestors gasp with wonder at your brain.

"I cannot yet cast account either with penne or counters," wrote Michel de Montaigne (1533–1592) in a famous book titled *Essays*. To "cast account" means to do arithmetic! "Counters" means an abacus. Montaigne and most of his contemporaries couldn't do the arithmetic you can do easily.

Imagine yourself in a schoolroom in medieval Europe, and

you'll see your classmates struggling to do arithmetic with those cumbersome Roman numerals. Besides that, they've never heard of zero. Without zero as a helper, math is really tough. But zero isn't the only mathematical tool they are missing. They haven't learned the concept of *numerical place.*

> **Zero was a long time in establishing itself as a number in its own right; it was hard for mathematicians to accept that 'nothing' could be 'something'.... Negative numbers were also long denied legitimacy in mathematics.... The Greeks considered geometry the only acceptable form of mathematics, and since distance cannot be negative, they had no use for negative numbers.**
>
> —Jan Gullberg, *Mathematics from the Birth of Numbers*

Numerical place? It's the idea of a ones column and a tens column and a hundreds column, and so on. The fancy term for it is *positional notation.* Try multiplying 139 by 62. Or 49 by 8. Without number columns, you will have to add 139 sixty-two times. Or 49 eight times. That's the kind of calculating many Europeans were still doing just four centuries ago.

With Hindu-Arabic numerals, zero, and positional notation, ordinary people could do arithmetic. Multiplication and division were easy! And with zero as a number, mathematicians had a new tool in their toolbox. They could

INFINITE WISDOM

The Greeks rejected the idea of infinity along with zero. They found the idea of endlessness troublesome. But infinity fit easily into Eastern philosophy. In 628, Brahmagupta wrote of infinity, defining it as the opposite of zero. It was more than 1,000 years later, in 1656, when an English mathematician, John Wallis, first used the symbol for infinity ∞.

This Indian relief from the fifth century C.E. shows the god Vishnu, often called "the infinite ocean of the universe," sleeping on the endless Sesha, the serpent of infinity.

The Bakshali Manuscript is an ancient Indian math text written on bark—though only scraps remain. Zero is represented by a dot. The numerals are somewhat similar to those in the middle row of the box on the opposite page. Here's an algebra problem from the Bakshali: One person has 9 horses, a second has 7 well-bred horses, and a third has 10 camels. Each person gives 1 animal to each of the other two, making them equally rich. How much does each type of animal cost (in any money unit)? (Answer on opposite page.)

The Sanskrit name for ZERO was *sunya* (it means "empty"). In Arabic, that turned into *sifr*. In Latin, it got changed to *zephirum*. And from that root comes *zero* and also *cipher*.

do complex calculations that had been impossible before. "In the history of culture, the discovery of zero will always stand out as one of the greatest single achievements of the human race," says math historian Tobias Dantzig.

So how did we get from Roman numerals, no zero, and no positional notation to our present system?

The idea germinated in India and came to flower in the Arab world. We don't know exactly when Indian thinkers developed a whole new number concept—it took several centuries; what's important is that they did it. In the year 499 a Hindu mathematician named Aryabhata (ar-yuh-BUHT-uh) published a famous book on math and astronomy in verse! India was in its "classic period," when art, sculpture, and poetry reached a pinnacle (at the same time, in Rome the empire was falling apart). Aryabhata used nine digits and zero in a base-10 system. (Remember, the Babylonians

ZERO INFORMATION

1. Suppose you multiply a bunch of numbers together and the answer is zero. That tells you that at least one of those numbers is zero.
2. You can't divide by zero. Ever.

had used base 60.)

An Indian scholar by the name of Brahmagupta (brah-muh-GUP-tuh) wrote a mathematical treatise in 628 that borrowed some algebraic equations known in ancient Greece. But Brahmagupta did something the Greeks had never done: he solved equations using zero. He understood that zero can be more than just a placeholder; it can be a number itself. With zero as a number, negative numbers suddenly made sense. Brahmagupta used them too.

The Indian thinkers took three ideas that had been used elsewhere and made a powerful connection between them. The ideas were positional notation, zero, and the base-10 (or decimal) system. When that trio fell into place, they had come up with a way to use just 10 digits to represent all possible numbers. They had created a democratic number system; almost anyone could handle its basics.

Someone had to tell the rest of the world about that accomplishment. Arab mathematicians did so. In the eighth

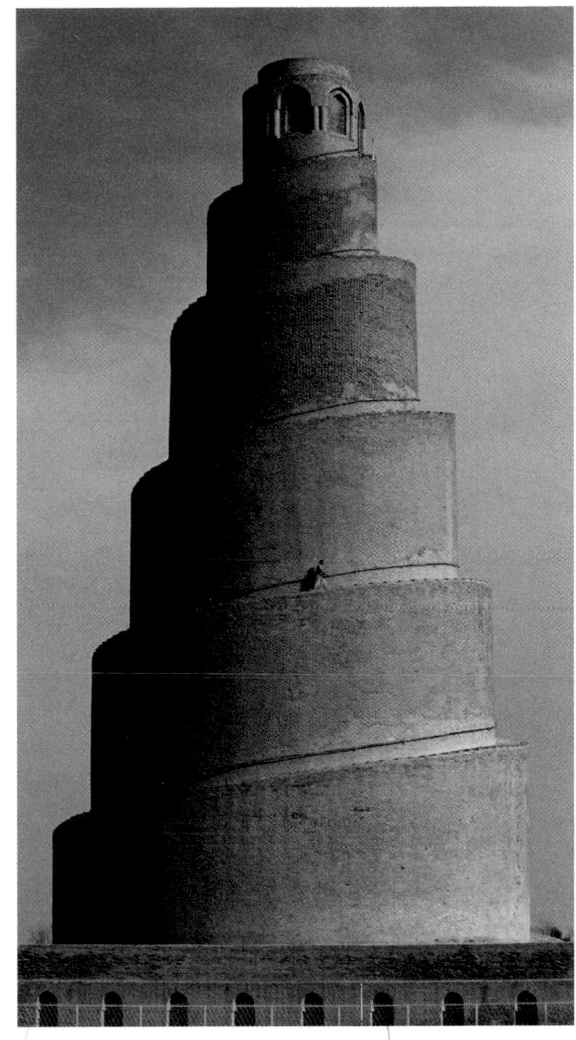

The Malwiyya (above) is a spiral minaret belonging to one of the largest mosques in the world. The Great Mosque of Samarra (now in Iraq) was built ca. 850 by Caliph al-Mutawakkil.

GO FIGURE

Keep in mind:
Hindu-Islamic = religions
Indian-Arabic = cultures

Our numerals are called Hindu-Arabic, but that's a bit like comparing apples and oranges. It should be Hindu-Islamic or Indian-Arabic. But it isn't. Go figure—but not with Roman numerals.

I II III IV V VI VII VIII IX X

٠٩٨٧٦٥٤٣٢١

1 2 3 4 5 6 7 8 9 10

When it comes to arithmetic, the Roman numerals I through 10 (top) are a headache to use. Arabic numerals (middle, read from right to left) and modern Hindu-Arabic numbers (bottom), make math easier.

Answer: Each person ends up with 262 money units' worth of animals; a horse = 28, a well-bred horse = 42, a camel = 24.

Harun was the caliph described in *Thousand and One Nights* (which is often called *Arabian Nights*), a book in which Scheherazade avoids a sentence of death by telling the all-powerful ruler unfinished stories of Aladdin, Ali Baba, and Sinbad the Sailor. On the right is a seventeenth-century Indian portrait of Caliph Harun as a young man.

century, the Islamic world was having its own classic period. Harun al-Rashid (hah-ROON ahl-rah-SHEED) was caliph (Islamic ruler) of an empire that stretched from the Mediterranean to India, with Baghdad as its capital.

Harun, who was an absolute monarch, could do what he wanted, and he wanted to bring arts and sciences to his realm. He was just getting into high gear when he died, in 809, and his two sons fought to succeed him. One of his sons, al-Mamun (ahl-mah-MOON), won the fight and went even further than his father. In 830 he founded an academy in Baghdad called the House of Wisdom, an intellectual center that included an observatory and a major library.

Baghdad became a new Alexandria and the world's richest city. It was home to Arabs, Indians, Greeks, Jews, Christians, Muslims, and pagans. Sailing ships from distant places, like Zanzibar and China, lined up at palm-shaded docks along the river Euphrates. In the middle of the city, the sumptuous Golden Palace and the Great Mosque were surrounded by three rings of walls. Outside the walls were fabulous gardens, plazas, and bazaars where you could buy cinnamon from Sumatra, cloves from Africa, silk from India, and local spinach (unknown in Europe). You could also buy slaves.

The word ALGORITHM is said to come from the name al-Khwārizmī (but a bit mispronounced). An algorithm is a step-by-step problem-solving procedure in math.

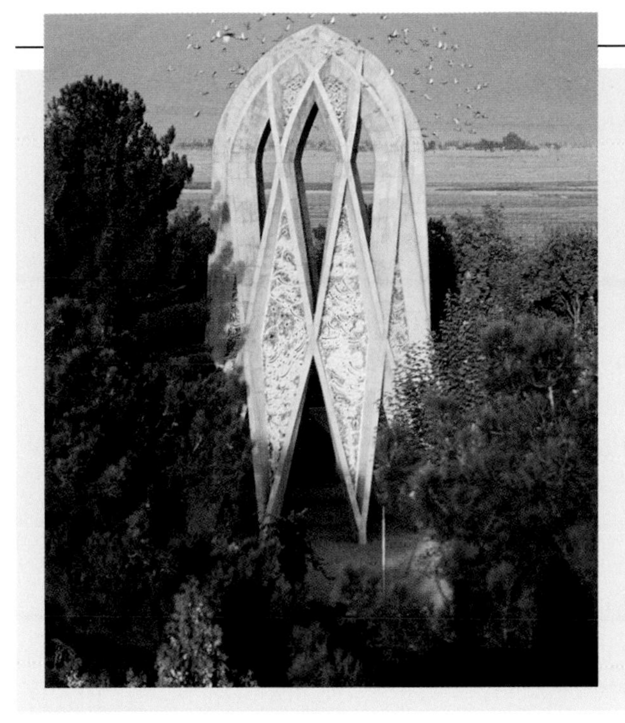

DON'T CHANGE A WORD

Omar Khayyam is known in the Persian world as the greatest mathematician and astronomer of his day. In the West he is famous for his four-line poems—called *rubaiyats*—that deal with life's pleasures and pains. (In the original Persian, the quatrains rhyme.)

The Moving Finger writes; and, having writ,
Moves on: nor all your Piety nor Wit
Shall lure it back to cancel half a Line,
Nor all your Tears wash out a Word of it.

The modern monument that houses the tomb of Omar Khayyam incorporates arabesque motifs and interlaced diamonds.

Al-Mamum was said to have had a dream in which he spoke to Aristotle. After that, he was determined to have Arabic translations made of all the Greek works he could find. (We can thank him for saving many treasured writings we now enjoy.)

Meanwhile, Arabs opened a House of Wisdom in Fez (Morocco) in 859, and another in Cairo (Egypt) in 972. In Córdoba (Spain), Muslim and Jewish scholars worked together at a fine library keeping the ideas of the ancients alive.

One of the ninth-century superstar scholars at the House of Wisdom in Baghdad was a mathematician named Muhammad ibn-Musa al-Khwārizmī (ahl-KWAHR-iz-mee). He wrote more than a dozen books and is thought to be among the greatest mathematicians of all time. Al-Khwārizmī used those Hindu digits, put numbers in columns, wrote of zero, and calculated with algebraic equations.

Two centuries later, Omar Khayyam (ky-AM), a renowned Persian mathematician, poet, and astronomer, wrote of the "majority of people who . . . confuse the true with the false, and . . . do not use what they know of the sciences except for base and material purposes." By the eleventh century, the

WHAT'S IN A YEAR?

Omar Khayyam measured the length of the year as 365.24219858156 days. He was attempting to reform the calendar, and that was an amazingly close figure. What he didn't realize is that the length of the year (measured by Earth's voyage around the Sun) changes during a normal lifetime. It changes in the sixth decimal place—that's a tiny amount—but over centuries it does make a difference.

Calendar reform involves politics as well as science; Khayyam's calendar was not implemented.

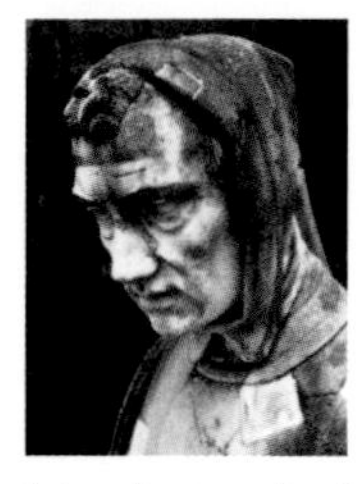

Leonardo Pisano (left) was nicknamed Fibonacci. His book *Liber Abaci* begins by introducing the base-10 numerals of the Indians—the digits 1–9 plus a zero. Yet while he was telling everyone else to use that decimal system, most of the calculations in the book are done with the Babylonian base 60. *Liber Abaci* is sometimes translated as *"Book of the Abacus"* and sometimes *"Book of Calculation."*

Islamic golden age was in decline. Assassinations and turmoil had led to a loss of freedom. Scientific progress rarely happens in an unfree atmosphere.

This story now shifts to Italy, where Leonardo Pisano becomes an important player. (He's not to be confused with Leonardo da Vinci or Leonardo DiCaprio.) Pisano got his last name because he came from Pisa, Italy. But one nickname didn't seem to be enough. Today, Leonardo is known by a writing nickname, Fibonacci (fee-buh-NAH-chee), from the Latin *filius Bonacci*, which means "son of Bonacci." Here are some words he wrote in 1202:

> *My father was a public scribe of Bejaia [Algeria], where he worked for his country in Customs, defending the interests of Pisan merchants who made their fortune there. He made me learn how to use the abacus when I was still a child because he saw how I would benefit from this in later life. In this way I learned the art of counting using the nine Indian figures . . . as follows: 9, 8, 7, 6, 5, 4, 3, 2, 1. . . . With these nine numerals, and with this sign 0, called* zephirum *in Arabic, one writes all the numbers one wishes.*

No human investigation can be called real science if it cannot be demonstrated mathematically.
—Leonardo da Vinci, *Treatise on Painting*

Fibonacci's father had ties to Italian merchants and North African Muslims. He understood the value of the new numbers. So it made sense to him to have his son tutored by mathematicians in Algeria. Fibonacci discovered a scholarly book on number notation; it was written by the Muslim mathematician al-Khwārizmī. It opened a new world to him.

When Fibonacci returned to Italy, he wrote his own book, *Liber Abaci*. It made a strong case for the use of Hindu-Arabic numerals and for zero. That zero idea, that nothing can really be something, was mind-blowing for the few who got it. His book made a big impact—on a small number of scholars. It would still be centuries before the average person used the new numbers. (This was before printing presses; books had to be copied by hand.)

Hindu literature indicates that the idea of zero may have been discovered even before the birth of Christ, but no actual zeroes have been found in any inscriptions before the ninth century. That's not true of American culture. In what is now Central America; the Mayans were using zeroes as placeholders.

Change was also slow because European nations were embarked on military and religious expeditions, or Crusades, to the Holy Lands in the eleventh, twelfth, and thirteenth centuries. The idea was to capture Jerusalem and other biblical lands from Muslims and put them under Christian control.

So it wasn't a great time to introduce Hindu-Arabic numbers to the West. Hindu-Arabic numbers were called seditious and unchristian and were actually outlawed in some places in Europe. Which meant the spread of this great idea was snail-paced.

Seditious? Here are some synonyms: rebellious, mutinous, disobedient, insurrectional. How can numbers be seditious? Well, they can lead to ideas and behavior that are dangerous to established ways.

There was an exception to this. Merchants from Genoa, Pisa, Venice, Milan, and Florence were carrying on a profitable trade with the Arab world. The Italians were experimenting with merchant capitalism, and anything that could make it work appealed to them. They were a bit like the Ionian merchants almost 2,000 years earlier, whose successes had helped encourage and support scientific thinking. Hindu-Arabic numbers were spectacularly efficient and easy to use, so the practical-minded and informed noticed and used them. They became a secret code for those in the know, especially for those savvy merchants. Others had to wait. Finally, by the eighteenth century, word had spread about the ease of the new numbers. After that, hardly anyone in Europe used abaci or finger-and-toe arithmetic. They were calculating as you do today.

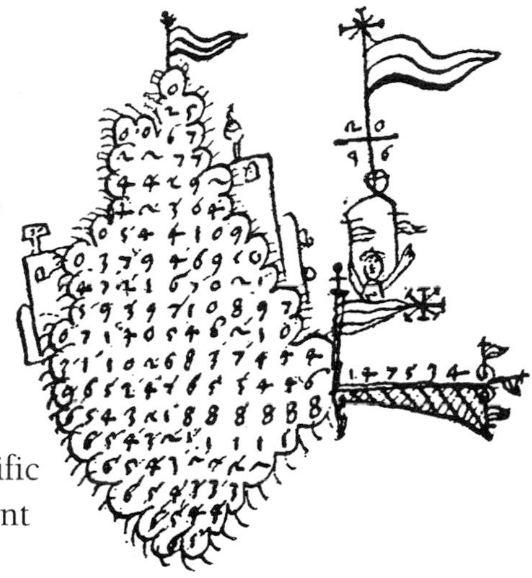

Most of us use long division to divide two big numbers, but above is an ancient method that's still taught in a few Arabic schools. It's called galley division—possibly because the calculations end up in the shape of a ship. The six-digit answer (on the right) sits on a dock; the two rows to the left of the dock contain the original two numbers being divided. Galley division was probably invented in India, but al-Khwārizmī, the ninth-century Muslim mathematician, publicized it. It was widely used in Europe until the early sixteenth century.

In the trading port of Venice, Italy (left, ca. 1650), ambitious merchants welcomed Hindu-Arabic numbers.

MR. FIBONACCI'S NUMBERS

Seeds, leaves, petals, buds, fruit sections—look for Fibonacci numbers in all kinds of plant parts. These petals number 3 (trillium), 5 (morning glory), 8 (bloodroot), 13 (black-eyed Susan), and 21 (daisy).

Fibonacci is famous for a special sequence of numbers that he may have discovered. (We know he wrote about the sequence and made it famous.) Here are the first 13 numbers of the Fibonacci series: 1, 1, 2, 3, 5, 8, 13, 21, 34, 55, 89, 144, 233.... Can you see the pattern? Add a number to its neighbor and you get the next number in the sequence.

Right away, mathematicians were fascinated by this series. They found that if you take any Fibonacci number and divide it by the previous number, something interesting evolves. For example: $8/5 = 1.6$; $13/8 = 1.625$; $21/13 = 1.6153...$. Do you see where these numbers are heading? If you keep going with the sequence of Fibonacci ratios, they will get closer and closer to the unattainable (because it is irrational) Golden Ratio, ϕ, which is 1.61803... (see pages 84–85).

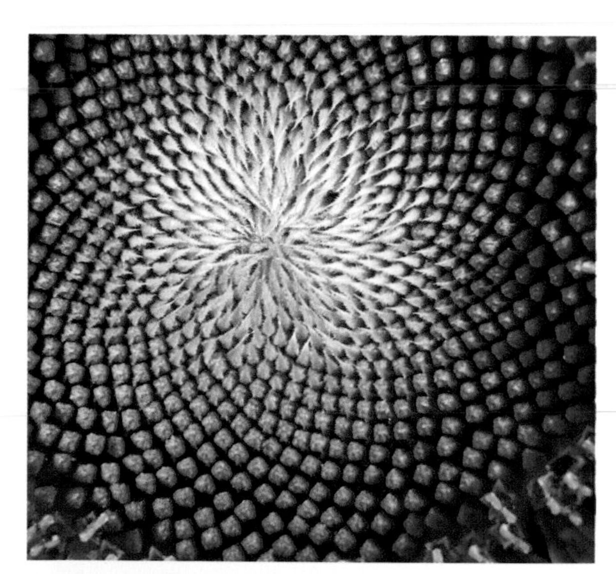

Sunflower seeds grow in a crisscrossing spiral. The number of seeds from center to edge, along either left or right spiral, is usually a Fibonacci number—34, 55, 89, etc. The angle formed by any two seeds is 137.5°—the Golden Ratio's reciprocal (.61803...) multiplied by 360° and then subtracted from 360. This arrangement packs in the most seeds. Even numbers and angles would leave bigger spaces.

The Greeks, as far back as Pythagoras, noticed that the Golden Ratio appears in nature far more often than chance would allow. You can see it in the diamond-shaped scales of a pineapple, with eight rows sloping to the left and 13 to the right. It seems that

Finger and hand bones are generally in Fibonacci proportions—2 (tip), 3 (middle), 5 (base), 8 (hand) units. Are yours? (They're easier to measure if you bend your fingers.) An intriguing science question is Why? Is it a coincidence or accident of nature? Or is there some physical advantage to these lengths?

many plants have a Fibonacci number of growing points (petals, leaves, seeds, branches, etc.). Daisies can have 34, 55, or sometimes even 89 petals. Why? Could the arrangement of leaves in a Fibonacci pattern be related to the amount of light each leaf receives? (If the ratio were even, with one leaf above another, wouldn't the leaves shade each other?)

Some coniferous trees show Fibonacci numbers in the bumps on their trunks. And palm trees show the numbers in the rings on their trunks. Why?

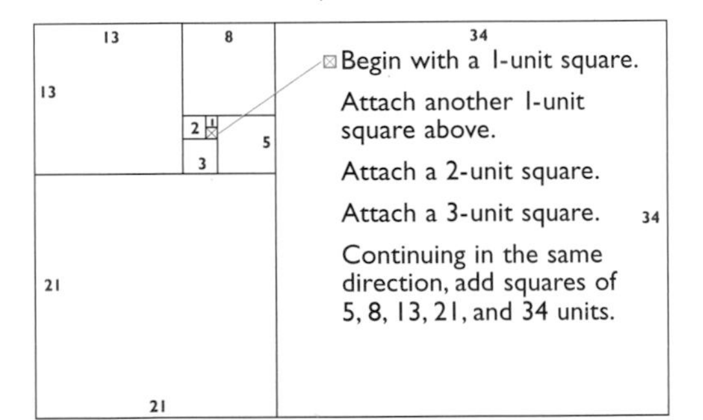

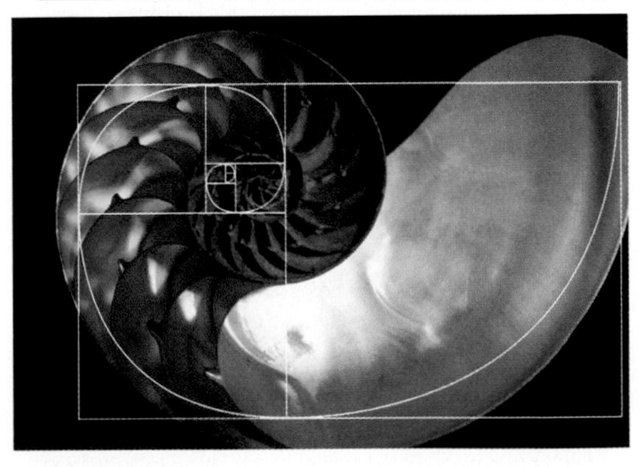

Nature doesn't say, "Hey, these are nice numbers; let's use them." That Golden Ratio of Fibonacci numbers seems to represent an efficient pattern of growth. It appears as part of a natural process. That is why the spirals are sometimes imperfect. The plant is *not* responding to mathematical rules but to nature's needs.

The Greeks had no idea Fibonacci was going to appear with a sequence of numbers that would help define the Golden Ratio, but they had that ratio, $1/\phi$, and they found it beautiful. It turns up often in their art and architecture (see page 85).

In our everyday world, the dimensions of a standard playing card are a Fibonacci ratio. So is the octave on a piano keyboard, with eight white keys and five black keys. Phi also occurs often in geometry. It is the ratio of a side of a regular pentagon to its diagonal.

Start looking for those Golden Ratios. They're all around you. A good book on the subject is *Fascinating Fibonaccis*, by Trudi Hammel Garland.

The nautilus starts as a small sea creature in a tiny chamber. As the nautilus grows, it adds rooms to its home, chamber after chamber. Each chamber is larger than the preceding one to accommodate its growing occupant. The pattern of growth is very close to the Golden Ratio—but not always exact.

26

An "Ox" Who Bellowed

Since there is a twofold way of acquiring knowledge—by discovery and by being taught—the way of discovery is the higher, and the way of being taught is secondary.

—Thomas Aquinas (1225–1274), Italian philosopher and Christian saint, *Summa Theologica*

It says much about the Middle Ages that in the year 1550, after a thousand years of neglect, the roads built by the Romans were still the best on the continent.

—William Manchester (1922–), American historian, *A World Lit Only by Fire*

The abiding fact is that modern science grew out of the lovely Medieval idea of *Ordo Mundi*, the faith in a universal order, a religious feeling for the ultimate unity of all life.

—Thomas Goldstein, twentieth-century American historian, *Dawn of Modern Science*

In 1225 a baby is born in a castle in Italy into the upper-crust Aquino family. He is christened *Thomas*.

Like many well-born Italians of the Middle Ages, young Thomas Aquinas (uh-KWY-nuhs) is sent to a monastery as an oblate, a monk-in-training. He goes to the ancient Monte Cassino Abbey run by the Benedictine order of monks. It is a quiet, scholarly place. It also conveys status. Thomas is bright, and his family has the right connections. He should go far in the church. But the unexpected happens. The pope and the emperor are feuding; the monastery gets caught in the conflict; and the oblates are told to go home. (The emperor expels the monks because they are obedient to the pope, not to him.) That incident of politics changes Thomas Aquinas's life.

St. Thomas Aquinas, portrayed here in a Renaissance fresco by Fra Bartolommeo, became a Dominican friar against his mother's wishes. The sun on his chest is a symbol of the religious order.

The monastery where Thomas had spent nine years was cloistered, calm, and dedicated to contemplation and study. Aquinas now heads for the newly founded

University of Naples. That urban university is open and exciting. Scholars there are just discovering ancient Greek scientific works translated from Arabic. It leaves their minds unsettled.

Thomas Aquinas, caught in the fervor of the time, joins a new democratic order of monks—the Dominicans. They focus on active preaching and teaching rather than on isolation and personal prayer. And they are Mendicants. That means that Dominicans take vows of poverty and have to beg to exist. That idea doesn't please Thomas's aristocratic mother. She has her son kidnapped and tries to keep him locked in the family castle until he changes his mind. He won't. When she finally releases Thomas in 1245, he is 20 years old, and he heads off to do some more studying. This time he chooses the most prestigious university in Europe—the University of Paris.

In the thirteenth century, Paris is the place to be if you

WHO GETS THE UPPER CRUST?

In medieval times, bread was divided according to status. Workers got the burnt bottom of the loaf; the family got the middle; and special guests got the top, or "upper crust." Now, an upper-crust family is one with high status.

Have you ever heard the expressions "chew the fat" or "bring home the bacon"? They come from the Middle Ages, too, when most people were poor and having a ham or a side of pork was very special. You'd invite your friends to dinner and "chew the fat" together.

A painting (left) from a French history book (ca. 1400) offers a glimpse into a medieval classroom. A professor at the University of Paris lectures while his students are busy taking notes.

MONKS, UNCHAINED

In this Renaissance painting, Pope Honorius III approves the rules of the new Dominican order in 1216. The pope later called St. Dominic and his followers *pugiles fidei*, "champions of the faith," for defending and promoting Christianity.

New religious orders, such as the Dominicans, called their members friars—brothers—rather than monks. They weren't bound to a monastery. They could attach themselves to a university or other school, or they could travel and preach. Their mission was to be useful.

Mediterranean lands have light soil that is easily plowed. Not so in northern Europe. Slavs (peasants from eastern Europe) are said to have invented the moldboard plow in the sixth century, shown here in the foreground of a fresco called *The Month of April.* The plow does a triple whammy on tough soil, with a knife blade that cuts deep into the ground, a plowshare that cuts grass at ground level, and a curved moldboard that lifts and turns the soil. It opened new lands to farming, and food production soared.

TUMULT means "mayhem or uproar."
In a FEUDAL society, lords owned the land and its harvest. (That's where the word *landlord* comes from.) Serfs did the work but had little to show for it.

like tumult and activity. While most of Europe is still feudal, Paris is the center of an emerging market economy. Old ideas are being blown away.

Elsewhere, peasants are serfs—little better than slaves—working all their lives for wealthy lords who keep them as ignorant as possible. People who can escape those medieval traps flock to the cities. There they find trade guilds, busy markets, and the exhilaration and upsets of a society in the process of change.

In the countryside, farms are producing surplus crops. That extra food feeds city dwellers and helps make large cities and city markets possible. Among the new inventions making agriculture more efficient is the moldboard plow. The damp, heavy soils in northern Europe demand a powerful tool. The moldboard does the job. It leads to strip farming in open fields. Large farms are becoming more productive than the small farms found in scattered hamlets. So rural life is changing at the same time that cities are emerging.

Change is both energizing and upsetting. The feudal world was known. What a free, capitalist world might be like is unknown. It seems to offer little security or control. But there is no stopping the new forces.

In the monasteries, clerics are focused on saving their souls through prayer, study, and isolation. When it comes to science, they quote Pythagoras, Plato, and Augustine. That trio all concentrated, in one way or another, on ideal forms in nature, which often kept them from considering the real world.

But at the budding universities, new scholars, inspired by the rediscovery of Greek science, are interested in understanding the forces of nature. Those new scholars are fascinated by Aristotle. Aristotle looked at the world

A Natural Birth

Thomas Aquinas, Roger Bacon, and other medieval thinkers laid the foundations for the Renaissance ("rebirth" in French). The Renaissance is generally said to be the fifteenth and the sixteenth (into the seventeeth) centuries. It was a transitional time—between the Middle Ages and modern times—when the visual arts (painting, sculpture, and architecture) reached high levels. Artists began carefully examining and painting the natural world. That new vision helped prepare the way for the scientific explosion that would begin about the year 1600.

As an artist, Leonardo da Vinci (1452–1519) studied the world visually—by sketching horses in motion (above) or birds in flight. His artistic research contributed to many technical fields—among them anatomy, astronomy, mechanics, and engineering. His bird studies led to a design for a flying machine. The German painter Albrecht Dürer (1471–1528) likewise explored nature through art. Note the almost photographic quality in his study of a piece of turf (right).

about him and observed, made notes, and classified its inhabitants—plant and animal. He was down to earth. (Sorry, but pun intended.)

The medieval thinkers have been ignoring things around them in order to concentrate on a world-to-come. Slowly their views begin to change. You can see it in the paintings of the time. Idealized gilded backgrounds are giving way to realistic plants and animals.

A small number of Christian scholars are now taking the idea of a round Earth seriously. Their picture of the spherical Earth comes straight from Ptolemy. In that vision, the round Earth stands stolid and unmoving in the center of a fixed universe. The Sun circles the Earth. The medieval thinkers who believe this have arrived back where Aristotle was in the fifth century B.C.E. Still, that is too advanced for some.

In a CAPITALIST economic system, individuals or corporations own the products of their own work. They can sell what they produce and buy what they can afford in a free market.

A SUBJECT FOR DISCUSSION: FAITH AND SCIENCE—CAN THEY GET ALONG?

Have you ever had a strong opinion that was different from that of your parents? I'm not talking about going somewhere your parents think you shouldn't go. I'm talking basic opinions—ideas about life in general. It's a foolish question, because of course you have. No one intelligent grows up without doing independent thinking. Developmental experts will tell you that challenging the ideas of those who came before you is a natural part of growing up.

Little children believe devotedly in whatever parents, teachers, and Santa Claus say. They ask questions, but they almost always believe the answers adults tell them.

Teenagers? Do they believe anything adults say? You know the answer to that.

Maturity comes if you can put it all together—the learning and experience of past generations and the questioning insights of youth. Not everyone does that successfully. Lots of people get older; only a few grow wiser.

Societies are a lot like people. Generations build on one another. Some, acting like little kids, mindlessly accept and repeat the mistakes of the past. Others, like some teenagers, rebel and destroy without much thought. And a few stride ahead with profound insights.

Which brings us to those Dark and Middle Ages. They were complex times, fascinating to study. But for most people who lived through them or looked back on those years, they were like a bad adolescence—something to be blocked out, forgotten, gotten past, rejected.

It was a time when science (human reason) and religion (faith) got mixed together with the addition of a whole lot of superstition. The chance to be educated went only to a privileged few in out-of-the-way monasteries. Some historians say we've been rebelling against that situation ever since.

That rebellion set the tone for much of modern history. It began during the Renaissance. Learning and thinking—especially in the arts and sciences—started to break out of the straitjacket of religious control. That led to an astonishing burst of scientific creativity and to the modern science-based world. And it brought an incredible payoff in technological achievement.

But does reason need to be in conflict with faith? Have we gone too far and lost our shared values? Do spirituality and science have anything in common?

To a scientist, the brain is a collection of neurons (nerve cells) and chemicals. But we all know it is something more than that, too. What is that something else? Where can we find guidance to understand our dual nature? Pope John Paul II, speaking in 1999, said, "Deprived of reason, faith has stressed feeling and experience, and so runs the risk of no longer being a universal proposition."

Deprived of faith, reason and science often seem empty and shallow.

Today we live with the fear that science and technology may be out of control: manipulating human life, spoiling the environment, threatening destruction, while at the same time giving us an incredible gift of leisure and affluence. Can we have it all—the moorings of profound philosophical and religious thinking and the splendors of science and technology?

The Renaissance thinkers and most scientists who came after them didn't concern themselves with those questions. They were rebelling against the long nightmare that had kept most people ignorant of their environment and themselves. Like kids bursting out of a locked closet, the thinkers couldn't wait to discover the whole universe. They did an amazing job of it.

Painter Fra Angelico (1387–1455) commemorated the great Dominican scholar Albertus Magnus in a fresco in the Convent of San Marco, Florence. (*Fra* is short for "friar" or "brother.")

In Paris, Arabic-Aristotelian science is called sinful by church leaders who attempt to keep it from their students—which of course just makes it more enticing. (The old guard is not going to sit back and take change without fuss or fight.) Arabic numbers, when known, are shunned by the authorities. The scholars who read Fibonacci's book on Hindu-Arabic numbers do it surreptitiously.

SURREPTITIOUS (ser-uhp-TISH-uhs) is what I think of as a delicious word. It means sneaky, stealthy, clandestine, or secret. You can't write a detective novel without someone being surreptitious.

In 1210, Aristotle's scientific writings are banned at the University of Paris. The University of Toulouse is more daring. It announces in its catalog for 1229 the "teaching of the books on natural science that have been banned at Paris." That means Aristotle.

In Paris, in 1229, the ban on Aristotle still holds. Only Aristotle's writings on logic are studied officially. But unofficially, everyone—especially the students—seems to be reading and talking about Aristotle's science. Young Thomas Aquinas is entranced by Aristotle's ideas. He is lucky to study them with the leading scholar in Paris, Albertus Magnus, who is sneaking some of Aristotle into his lectures.

Roger Bacon (ca. 1214–ca. 1292) was born in England to a wealthy family and was well schooled at Oxford University and the University of Paris. He became a Franciscan friar, but in 1277, the Franciscans banned Bacon's work, and he was imprisoned. Some scholars think we have made too much of Roger Bacon. They say he wasn't quite the *wunderkind* (child genius) he's sometimes perceived to be. But I would have liked to have known him.

Aquinas hears the great English thinker Roger Bacon, who is known as *Doctor Mirabilis*. (In Latin that means "Marvelous Teacher.") Bacon is living in Paris and is writing commentaries on Aristotle's physics.

The enormously influential Bacon knows of Hero's inventions in Alexandria—the vehicles and moving statues—and sees no reason for them not to be part of his world. "Cars can be made so that without animals they will move with unbelievable rapidity," he writes. Bacon will spend 14 years in prison for what is called heresy. What he has actually done is attack the clergy for ignorance and vice. His books are accused of being "suspected novelties" and banned.

Bacon is inspired by Aristotle, but he refuses to be limited

Thomas Aquinas seems to have had a photographic memory. He remembered things he had barely seen. Memory, in the world before printed books, was a very valuable commodity. Thomas wrote out rules of memory that were learned by every schoolchild.

by anyone's logic. He insists that scientific truth is not something that you blindly accept from authorities—even from Aristotle. Rather, it is the fruit of observation and experimentation. Bacon gives up his professor's chair to devote himself to experimental science. Almost no one in his time understands that move, but he has been led there by Aristotle's thinking.

The controversy over Aristotle is enormously stimulating to students who can think—as Thomas Aquinas can.

Aquinas, who is a big, heavyset fellow, doesn't talk much. Some of his classmates call him Dumb Ox. But those who know him best realize that he has a rare mind. Albertus Magnus says, "This ox will one day fill the world with his bellowing." Magnus is prescient. Aquinas will bring those two forces-in-conflict—the Roman Catholic Church and Aristotelian science—into harmony. He believes that Christian revelation (the words of the New Testament) and human knowledge are part of a single truth and shouldn't be rivals.

PRESCIENT is one of those words with several acceptable pronunciations. I say PRESH-uhnt. It means being able to foretell actions or events before they happen. *Foresight* is a synonym of *prescience*.

Still, it isn't easy to get people to give up old feuds. Three years after his death in 1274, a high church body condemns 12 of Aquinas's works. Roman Catholics are not to read his writings. But new ideas, especially good ideas, can't be held hostage. A half century later, in 1323, Thomas Aquinas is officially made a saint by the Roman Catholic Church. (It is some 800 years after Augustine lived. Augustine too is now a Catholic saint.)

Brother Thomas raised new problems in his teaching, invented a new method, used new systems of proof. To hear him teach a new doctrine, with new arguments, one could not doubt that God...gave him the power to teach, by the spoken and written word, new opinions and new knowledge.

—William of Tocco, who was Aquinas's first biographer and knew him

St. Thomas Aquinas insisted that faith and reason are compatible (see page 232). In other words, Christians don't have to be obsessed with just the world-to-come. They can balance that with analysis of the world-of-here-and-now. Even more important for the future of science, Christians can learn things about the natural world from pagan philosophers (such as Aristotle)!

With Aquinas's help, Aristotle and Ptolemy go from

outcasts to scientific gods or icons or wisest of the wise. Before too long, almost everyone will think that if Aristotle said it, it must be true.

Aquinas's writings make Aristotle's ideas a foundation for much science in centuries to come. And, because Ptolemy built his theories on Aristotle's ideas, Ptolemy is soon revered too. It is a turnabout. Catholic scholars now become Aristotelians. That has mixed consequences. Aristotle was often wrong. But one thing is sure: Aristotle's science is better than the science of the medieval flat-Earth thinker Cosmas. Aristotle, who believed in the power of the human mind, helps prepare the way for the modern world, which is waiting to be born.

This 1471 painting by Benozzo Gozzoli is called *The Triumph of Saint Thomas Aquinas.* What triumph? Read these pages and then note the hierarchies in the painting. St. Thomas (seated) looms large in the center, flanked by Aristotle on the left and Plato on the right. The controversial Muslim philosopher Averroës (see page 211) is lying at his feet. The tiny pope (in white) and other church leaders are relegated to the bottom, while Christ, St. Paul, Moses, and the four evangelists appear above it all.

ROGER BACON PREDICTS...

Bacon predicts...
"Machines for navigation can be made without rowers so that the largest ships on rivers or seas will be moved by a single man in charge with greater velocity than if they were full of men."

Steam engines make oars obsolete in the early 1800s. George Catlin painted the steamboat *St. Louis* in 1831.

Like many inventors, Roger Bacon's imagination outpaced the technology of his time. He dreamed of all the machines on this page, and all but one of them became reality. (A flying machine that flaps its wings is tougher to build than it looks.) Bacon wrote, "These machines were made in antiquity and they have certainly been made in our times, except possibly the flying machine, which I have not seen or do I know anyone who has, but I know an expert who has thought out a way to make one."

So who is dreaming up tomorrow's technology today? What inventive minds are writing about and drawing designs for machines of the future? Will their machines work? There is no sure answer to that question. Your thoughts are as good as mine—probably better.

Bacon predicts ...
"Cars can be made so that without animals they will move with unbelievable rapidity."

Henry Ford drives his first car through the streets of Detroit in 1896.

Bacon predicts ...
"Flying machines are possible, so that a man may sit in the middle turning some device by which artificial wings may beat the air in the manner of a flying bird."

A German pioneer in flight makes a long glide near Berlin in 1896, but big flapping fliers have not yet been invented. Today's fliers all rely on fixed wings, rotating propellers, rising hot air, and other technology.

Bacon predicts ...
"Machines can be made for walking in the sea and rivers."

A diving bell with an air hose was invented in 1690. This modern battery-powered submersible, called *DeepWorker,* allows one person to dive 600 meters (about 2,000 feet), maneuvering with foot pedals.

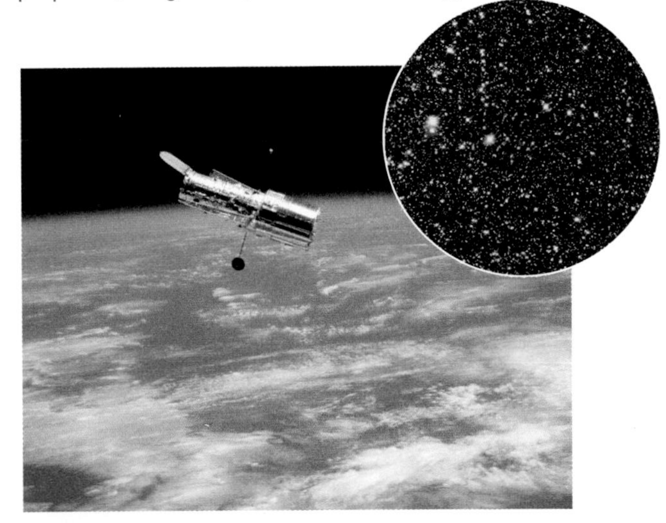

Bacon predicts ...
"Far distant things may appear very near and conversely.... So we may even make the sun, moon and stars descend lower in appearance."

Telescopes revolutionized astronomy in the seventeenth century. The Hubble space telescope (launched in 1990) brought us the distant Sagittarius star cloud, a treasure chest of glittering jewels.

27 Books Will Do It

It is a noble work to write out holy books, nor shall the scribe fail of his due reward. Writing books is better than planting vines, for he who plants a vine serves his belly, but he who writes a book serves his soul.

—Alcuin (ca. 735–804), English scholar and advisor to Emperor Charlemagne, as quoted in *The History of Education*, by Ellwood Cuberley

He who first shortened the labor of Copyists by the device of Movable Types was disbanding hired Armies and cashiering most Kings and Senates and creating a whole new Democratic world.

—Thomas Carlyle (1795–1881), English essayist and historian born in Scotland, *Sartor Resartus*

Johannes Gutenberg (ca. 1397–1468) developed movable type in the mid-fifteenth century, beginning a process that made books easy to produce. And that helped create "a whole new Democratic world" where almost everyone could learn to read. You've probably heard that before but there's more to the story.

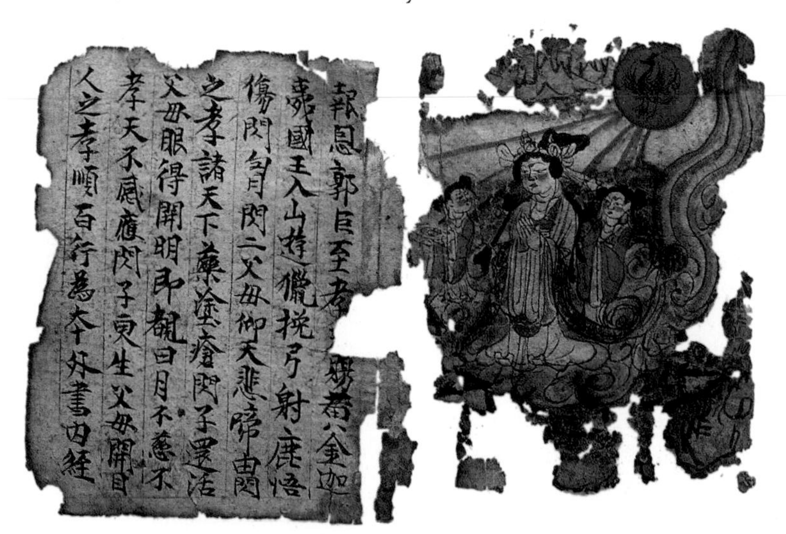

Sometime during the T'ang Dynasty (618 to 907), a Chinese writer put ink pen to paper to tell this illustrated story about Marici, the Japanese protector of warriors. The Chinese had been making paper for centuries, but their invention was just beginning to make its way West—to India, Arabia, and, finally, a few centuries later, Europe.

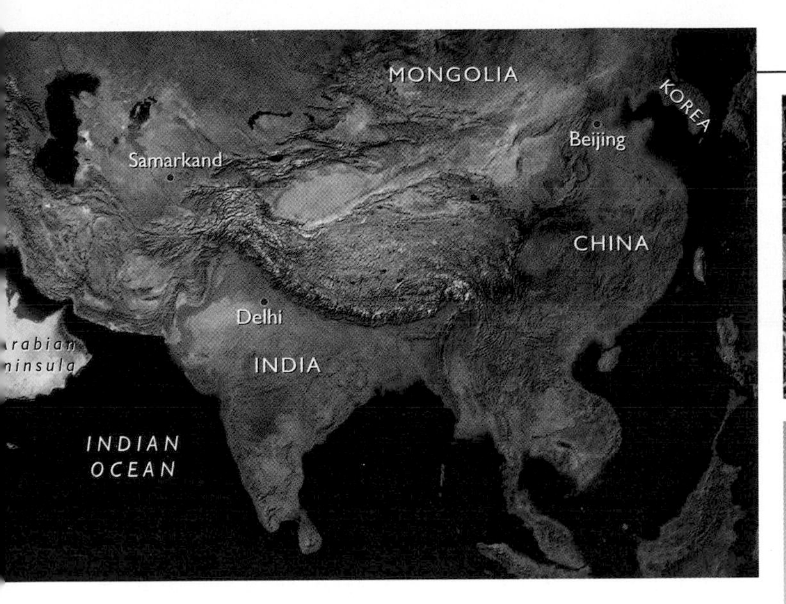

Samarkand (see map above left) was an ancient city destroyed by Alexander the Great in 329 B.C.E. Later, it became the center of the silk trade. It was destroyed again by Genghis Khan in 1220 and then rebuilt as the capital of Tamerlane's empire, which stretched from Delhi to the Black Sea. Still later, in the early twentieth century, Samarkand became part of the Soviet Union, a collection of Communist states that broke up in 1991. Samarkand is now in Uzbekistan. If you go there, don't miss the Shah-I-Zinda (directly above), an ensemble of tombs and mosques built around Tamerlane's time. The tilework (top) is brilliant, and the diverse architecture is well preserved.

Movable type, the printing press, and paper were all invented in China long before Gutenberg's time.

At first, the Chinese wrote their books on wood or bamboo and then on rolls of silk. But silk was expensive and bamboo was heavy. So, sometime before 100 B.C.E., they began to develop a method of making paper from tree bark, hemp, rags, and even old fishnets. Some authorities say that Ts'ai Lun actually invented paper in 105 C.E. More than six centuries later, in 751, when the Chinese had perfected the process, some papermakers were captured by Arabs and brought to Samarkand (now in Uzbekistan in Central Asia). They taught their jailers how to make paper. After that, the secret was out and the art of papermaking spread west.

WHERE TO PUT YOUR WORDS?

If you're going to write, you need something to write on. Clay is a possibility; **parchment** (made from the skin of goats or sheep) is another. But clay tablets are cumbersome; they're hard to put in your pocket. As to skins, they are expensive, and using them is unkind to animals. How about plants? It was the Chinese who discovered that any form of the very common organic compound **cellulose** will make paper. So bark, flax, cotton, hemp (a plant fiber from Asia), and wood pulp (all containing cellulose) are among the sources of paper.

TWO WAYS OF WISDOM

Taoism (DOW-iz-uhm) is a Chinese philosophy based on the teaching of Lao-tzu (born ca. 600 B.C.E.). He believed that the "way" to happiness is in finding harmony with the natural world. Taoists aspire to live a simple and natural life.

Buddhism originated in India but is now a world religion. The Buddhist path to enlightenment is through meditation, morality, and reason. It is based on the teachings of Siddhārtha Gautama (born ca. 563 B.C.E.), known as the Buddha. He said, "Look within, thou art the Buddha," declaring that each person can find God within himself or herself.

A Chinese scroll painting depicts Lao-tzu, one of the Three Pure Ones, referring to the founders of the three religious beliefs in China—Confucianism, Taoism, and Buddhism.

But paper was only part of the process. Books had to be handwritten by copyists or scribes. You can guess how long it might take to make a book, especially if it had painted illustrations. So books were mostly for the very rich. When was the first book printed using a reproducible method (instead of handwriting)? No one knows. But in 1900 a Taoist (DOW-ist) monk accidentally found an ancient hidden library sealed off in an old cave in China. In it was a book made of seven huge sheets of paper joined together to make a scroll about 5 meters (about 16 feet) long. Six of the sheets were Buddhist teachings; the seventh was a picture of the Buddha. It had been beautifully printed from seven wooden blocks—and we know exactly when. The book, a work of Buddhist scriptures called the Diamond Sutra, had an inscription that read, "Reverently made for universal free distribution by Wang Jie on behalf of his two parents on the 13th of the 4th moon of the 9th year of Xiantong." That date is May 11, 868. (Today, the British Library holds the Diamond Sutra.)

Block printing was used even earlier as a way to print designs on fabric. The Chinese

Venetian explorer Marco Polo left for China in 1271, at age 17. In 1275, he met Kublai Khan (1215–1294), the grandson of Genghis Khan and conqueror of much of China, in Khanbalik (now Beijing), the opulent capital. The young explorer won over the Mongol emperor. This French illuminated manuscript (ca. 1400) depicts Kublai Khan giving his golden seal to Marco Polo.

This page (left), showing the Buddhist god Vaishravana and attendants, is one of the earliest dated examples of woodblock printing. It is from 947, half a millennium before Gutenberg lived.

also used blocks to print money, which startled Marco Polo when he visited the court of Kublai Khan in 1275. "With these pieces of paper they can buy anything," wrote Polo in amazement.

But books are something else. Carving a wooden block full of words (as was done with the Diamond Sutra; see page 205) is a lot of work, and the block can't be used for any other book. There had to be a more efficient method—and there was. In the eleventh century, a Chinese artisan named Pi Sheng developed movable type. Since the traditional Chinese languages do not use an alphabet, Pi Sheng had to find a way to cut word symbols, called ideograms, onto pieces of wood and then line up those wood pieces in a frame that could be inked, used as a printing block, and later taken apart so the ideograms could be reused. It was more complicated than you might imagine. Suppose he wanted to include a picture in that frame? Remember: everything had to be level so it would print evenly. Pi Sheng did it.

In Korea, 200 years later, handsome books were produced with movable bronze type. Metal type doesn't wear out as quickly as wood type does, and it can be cast with delicacy and precision. Printing was being refined in the East.

In Korea, movable metal type was cast by pouring molten copper into a mold made of compressed sand.

At the same time in Europe, books were still printed from single carved wooden blocks, much as the Diamond Sutra had been printed.

So why does Gutenberg get so much credit?

Because he deserves it.

Printing—as Gutenberg developed it—was more sophisticated and efficient than anything done before anywhere.

Gutenberg didn't just make printed books. He produced

You probably wouldn't have wanted to be involved with printing in sixteenth-century Korea. Government regulations included this sentence: "The supervisor and compositor shall be flogged [whipped] thirty times for an error per chapter; the printer shall be flogged thirty times for bad impression, either too dark or too light, of one character per chapter."

A sixteenth-century Dutch engraving illustrates the printing process. The capped man in the foreground is setting type—arranging individual metal letters (taken from the boxes at the far left) into words and sentences. Both letters and sentences are set backward, in order to print forward. On the right, a man is operating the printing press—pressing the inked letters onto paper, one sheet at a time.

Someone who is LITIGIOUS likes to sue people, or is often sued, in a court of law. A MATRIX is a mold that gives rise to copies, or it can be a structure for holding pieces in place. It comes from the same root as *mother* does. If you've seen the popular movies, you can guess why they're named *The Matrix*.

gorgeous books—books as splendid (or close to it) as the hand-illuminated manuscripts that scribes and artists were creating in Europe. To do that took the work of a genius.

We don't know a lot about Gutenberg, but we do know he was a litigious (li-TIJ-us) figure. The first time he turns up in court records is when a woman sues him for breach of promise. She said he promised to marry her. He said he didn't. He won.

After that there are years of lawsuits relating to his printshop and the financing of his invention. What those lawsuits tell us is that this man is persistent—he doesn't give up. That persistence is a necessary part of being an inventor.

Gutenberg faces four problems in devising a method for mass-producing books. He needs to design letter type itself, as well as a matrix to hold the letters in place. He needs to find a new kind of ink that can be applied easily and will reproduce uniformly. He needs to find paper that will absorb the new ink. And he needs to design a mechanical method of printing to replace the old method of hand-pressing each individual sheet.

He does all those things. It is the work of many years.

When it comes to printing, Gutenberg has a big advantage over the Asians. The Latin alphabet has only 23 letters (*w*, *k*,

Johannes Gutenberg holds movable type in this fifteenth-century engraving.

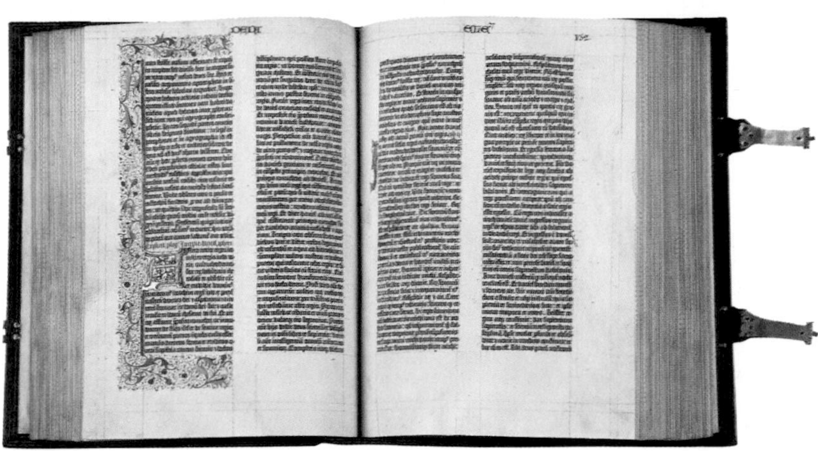

Here's the actual printing press (above) developed by Johannes Gutenberg, now in the Gutenberg Museum in Mainz, Germany. The printing blocks (foreground) belonged to Italian publisher Aldus Manutius (see next page).

In 1445, Gutenberg printed his first book, a Bible. On the left is one of his now-famous and very valuable Bibles, housed in the Morgan Library in New York.

Only the rich could afford to commission hand-copied books in Gutenberg's time. *Les Très Riches Heures du Duc de Berry* (*"The Very Rich Hours of the Duke of Berry"*) is one of the masterpieces among the so-called books of hours. This page, illuminated by the Limbourg brothers, is dedicated to the month of January and shows the duke's feast. All three brothers died of the plague in 1416, leaving the book unfinished.

and *y* are missing). When you add punctuation marks, capitals, and other symbols, it adds up to about 150 characters. In contrast, Chinese has about 30,000 characters. The challenge is different. Chinese ideograms are relatively large; Roman letters are usually small. Gutenberg has to make sure that every metal block with an *A*, for example, is interchangeable with every other *A* block. The letter *X* is larger than *I*—so how can they both fit in the same matrix? His skills as a goldsmith and metal caster (which is what he was before he became a printer) are invaluable.

One of Gutenberg's most important inventions is a mold for casting type quickly. It is a machine tool—an early step in the process that will soon allow machines to do more and more jobs of human workers.

Aldus Manutius (1450–1515) of Venice was one of the world's busiest printers. A sign outside his press said: "If you would speak to Aldus, hurry—time presses." Aldus produced small, cheap books, and he couldn't keep up with the demand. His workers were mostly Greek refugees, who helped translate many of the Greek classics for European readers.

No Chinese or European (as far as we know) had used a press for printing before Gutenberg. Presses were used for other purposes: They extracted juice from grapes to make wine. They were used to bind books. Gutenberg decides to design a press for printing. But the ink used by scribes won't do. He has to come up with a better type of ink. Flemish painters are experimenting with pigments that have linseed oil added. Gutenberg devises an oil-based ink. Then he designs paper that will absorb the new kind of ink. Other design hurdles loom. All this takes years and a lot of money. Finally, in 1445, he prints his first book, a Bible. It is gorgeous, the work of a perfectionist.

THE DANGER OF PRINTING

William Tyndale (1494–1536) translated the New Testament into English. It was originally written in Greek (with Jesus's words in Aramaic, a language he spoke). But a Latin translation became standard in much of the Christian world. Since few English people could read Latin they had to rely on the priests to tell them what was in the New Testament. Tyndale wanted them to be able to read for themselves, and with Gutenberg's invention, it should have been no problem. But the authorities claimed that people couldn't be trusted to make judgments themselves; left on their own they might have blasphemous (unholy) thoughts. There was more to it than that. Most churches sold indulgences; it was a major source of income. If you bought an indulgence, it was supposed to free you from sin, or it could buy a place in heaven for a dead relative. This was a kind of fraud, but as long as most people couldn't read the Bible for themselves, the fraud thrived. Tyndale wanted to open "the eyes of the blind." He believed that the true words of the Bible would set people free from what was institutional corruption in the medieval Catholic Church. That, not surprisingly, was a serious threat to church authorities.

Tyndale couldn't find a publisher for his New Testament in England. He did find one in Worms, Germany, where he also met Martin Luther. In 1525–1526, 3,000 copies of his book were printed and sent to England.

King Henry VIII and the English archbishop of the Catholic Church were furious. Tyndale had to run for his life—hiding out for the next few years. Finally, King Henry VIII found him. He had Tyndale killed. All the copies of his Bible that could be discovered were burned. A few years later, Henry VIII broke with the Catholic Church. Tyndale's translation—a very good one—became the basis for the King James Version of the New Testament. It was published during the lifetime of Henry VIII's great-grandnephew—King James I—the man who sent John Smith and some others to Virginia to build a settlement.

This 1563 woodcut is titled *The Martyrdom and Burning of Master William Tindall.* Tindall, Tyndale—spelling wasn't uniform, even for names.

Back in eighth-century Spain, a book cost about the same as two cows. (A cow was as valuable as a car is today.) By Gutenberg's time—the first half of the fifteenth century—hand-copied books are even more expensive, often costing several ounces of gold. That helps explain why most people are illiterate. Learning to read isn't worth the bother if you can never hold a book in your hands. Students—even college students—learn by listening to lectures and taking notes, not by reading.

Noah Webster (1758–1843) was an educator and writer who compiled the first American dictionary so that Americans didn't have to use British spellings (*colour*, for example) and could look up uniquely American words (*squash*, *skunk*, *hurricane*, etc.)

Nicholas of Cusa (1401–1464) was a Catholic bishop, mathematician, astronomer, and philosopher. Why isn't he as famous as Gutenberg, his German contemporary? Maybe because he wrote before printing presses really got churning. Or maybe because he was ahead of his time. He believed that the Earth moves around the Sun, that the stars are like our Sun (with inhabited worlds orbiting them), and that space is infinite. When he looked at the circle, he saw it filled with polygons (as Archimedes had) and said it was a metaphor for the search for truth—which never ended.

Gutenberg's new system spreads books like leaves in a windstorm. Printers are soon turning out inexpensive, mass-produced volumes. Aldus Manutius founds the first true publishing house and issues Greek and Latin classics, contemporary poetry, and reference works—all in small volumes at reasonable prices.

"Even before the end of the fifteenth century, unhappy scribes complained that their city [Venice] was already 'stuffed with books,'" writes Daniel J. Boorstin in *The Discoverers*. ("Unhappy scribes?" Changes in technology can leave people reeling.) By 1501, printing presses have turned out 7 editions of Ptolemy's *Geographia*; 33 other editions of that book are published in the next century. Ptolemy becomes to geography what Noah Webster later is to American dictionaries—a benchmark. Happily his central idea is a sound one—the idea of a global Earth.

Later, Charles Babbage, the nineteenth-century inventor of the first true calculating machine (the granddaddy of the computer), described the importance of easily available books when he wrote, "The modern world commences with the printing press." Do printed books have an impact on science? Here is twentieth-century science writer Isaac Asimov on that subject:

> *The base of scholarship broadened and . . . the views and discoveries of scholars could be made known quickly to other scholars. Scholars began to act as a team, instead of as isolated individuals. The realm of the unknown could more and more be assaulted by concerted blows. Scientists were no longer fists, but arms moving a battering ram."*

After Gutenberg, people with inquiring minds who aren't rich—like Christopher Columbus and Leonardo da Vinci—

can buy books and read for themselves both the works of the ancients and the ideas of their contemporaries.

Some authorities react with alarm. Bad ideas can be spread in books, they say. They are right. The printing press, ink, and paper make no judgments. Anything can be printed. People must make up their own minds about what they read. But can ordinary people be trusted to think for themselves? No one knows. There has never been a way to get book learning to the masses before this time.

Columbus reads both Ptolemy and Eratosthenes. He believes Ptolemy. So Columbus relies on his copy of *Almagest* when he sets out exploring. That means he thinks the Earth is smaller than it actually is, and Asia, larger. When he reaches the Caribbean Islands, he is sure he is near Asia (and he would have been, if Ptolemy had been right).

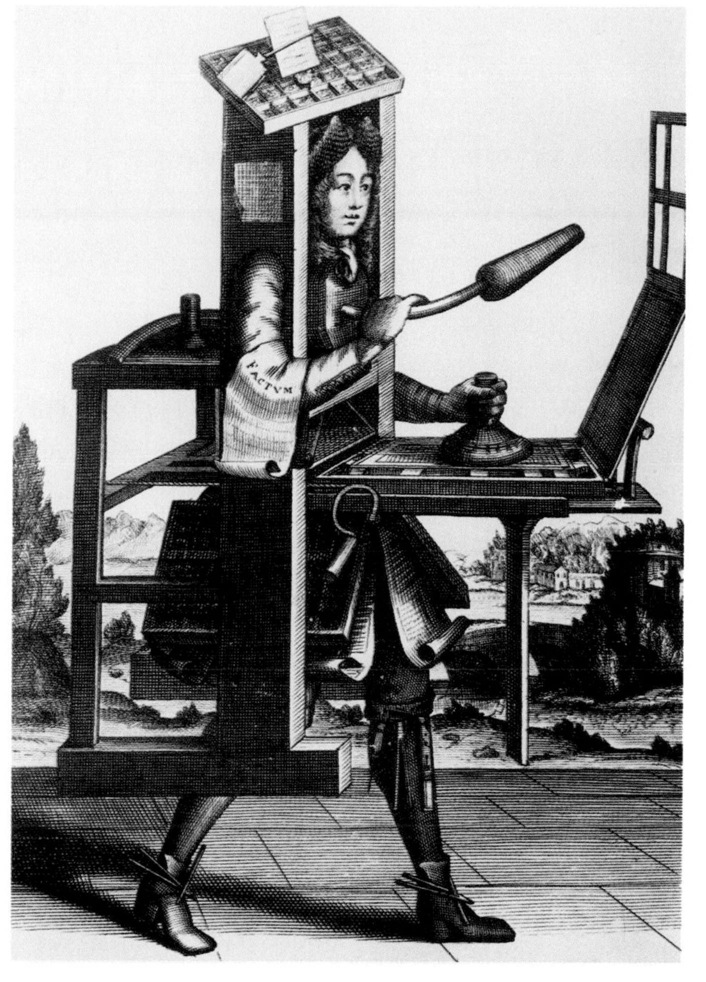

Reporters have always wanted to be where the action was. Publishers have always sought the latest news—"wonder births" of quintuplets, the bad omen of a comet sighting, a gala fair, the breakout of a revolution. This seventeenth-century copper engraving shows a mobile printer carrying his wooden press from town to town, where he acted as reporter, typesetter, publisher, and newsboy all in one.

In Portugal, mathematicians study Eratosthenes' calculations and believe he is right. They don't think Columbus has a chance of making it to Asia. It is too far away. So the Portuguese send Vasco da Gama around Africa.

Now consider this: If Columbus had paid attention to Eratosthenes—and if he had known how big the world really is—he might not have dared sail west. Where would we be now?

28

The Antipodes: Discovering Down Under

An age will come after many years when the Ocean will loose the chain of things, and a huge land lie revealed; and Tethys [a sea creature] shall discover new worlds, and Thule will no longer be the farthest land.

—Seneca (ca. 3 B.C.E.–65 C.E.), Roman philosopher and playwright, *Medea*

In the Middle Ages a pound of ginger was worth a sheep, while a similar weight of mace could buy three sheep or half a cow. Pepper, counted out berry by berry, was nearly priceless. It could pay taxes and rents, dowries and tributes.

—Charles Corn, twentieth-century American writer and editor, *The Scents of Eden: A Narrative of the Spice Trade*

History is not a random sequence of unrelated events. Everything affects, and is affected by, everything else. This is never clear in the present. Only time can sort out events. It is then, in perspective, that patterns emerge.

—William Manchester (1922–), American historian, biographer, and professor, *A World Lit Only by Fire*

On September 4, 1522, a battered, worm-infested ship is sighted heading for Seville on Spain's Guadalquivir (gwah-thuhl-kee-VEER) River. No one has expected it; it is as if a ghost ship has appeared.

The ship's crew—a pitiful band of 18 Europeans and 4 East Indians—are living skeletons. The 18 Europeans are all that remain of a hopeful expedition of 270 that set sail three years earlier. The families of the men think them dead.

The ship limps slowly up the river, but word of its appearance spreads rapidly. By September 6, when it reaches Seville, the whole city is consumed with curiosity. Where has this ship been?

Thule was the name of an island described by the Greek navigator Pytheas in 310 B.C.E. (Perhaps it was Iceland or Norway. No one knows.) Thule was said to be the most northerly land in the world and the farthest from Greece. The Latin phrase *ultima Thule*, from Seneca's *Medea*, is often used to describe someplace "at the end of the earth." Today Thule is also the name of a settlement in northern Greenland, founded by Danes in 1910 and now a Danish/U.S. science station.

That's the Guadalquivir River (left) and the busy port of Seville, Spain, in a sixteenth-century painting by Alonso Coello Sanchez.

The captain, Juan Sebastián del Caño, will soon be lionized across Europe and beyond. Actually, he is a traitor. Only later will history tell the truth about him.

The story begins on this same river with an armada of five ships. The ships are not particularly impressive, but the *capitán-general* is. His name is Ferdinand Magellan, and he has checked, rechecked, and strengthened every timber, every sail, every length of rope.

Magellan is preparing for a two-year expedition. His food supplies include several tons of pickled pork, almost as much honey, and 200 barrels of anchovies. In case they must fight, there are thousands of lances, shields, and helmets. To keep the ships in repair, they pack 40 loads of lumber, pitch, and tar. And there are mirrors, scissors, colored glass beads, and kerchiefs to trade with unknown natives. (Later he will discover that he has been cheated; much that he ordered and paid for is stolen on the docks and not loaded on the ships.)

Ferdinand Magellan is holding a map in this 1574 fresco by Giovanni Antonio da Varese. Below is a map of the Moluccas (the Spice Islands) made in 1519, the year Magellan set off from Spain to reach them—by sailing west, not east. (See the globe on page 251.)

Some people say Magellan is a fanatic or, at best, a dreamer. Actually, he is a bit of both, as achievers must be. Short and muscular with a bushy black beard and a limp—a battle wound—he has an idea, and it consumes him. He believes he can do what Columbus has not done—reach Asia and its fabled Moluccas (Spice Islands) by sailing west. He expects to find the

FINESSE is the social skill of handling situations in a subtle or smooth way—getting what you want from people without them realizing it, for example. COURTIERS, full of finesse, are like groupies who hang out with royals instead of rock bands. Through FLATTERY (fake praise) and FAWNING (brownnosing), they get on the good side of people with power and wealth. Perhaps, they hope, a little of both will rub off.

Moluccas, load his ships with cloves and other spices, and then turn around and sail back, thus pioneering a new trading route. Spices, especially cloves, nutmeg, mace, cinnamon, and ginger, are worth their weight in gold and sometimes more than gold.

European monarchs, merchants, adventurers, and scholars, in search of power, wealth, and knowledge, are trying to find new ways around the world. At the same time, they are bringing surprises to distant cultures. They are not alone. Other forces are changing peoples and lands that were once isolated.

In the seventh century, an inspired young camel driver named Muhammad married the widow of a wealthy Arab spice merchant. His followers, soon called Muslims, began spreading a new faith—Islam. Often they mixed missionary activities with business. Before long, Islamic Arabs traveling east controlled the overland route to Asia's market cities. They did everything they could to keep others from the rich spice trade. Europeans chafed. They wanted the wealth of the spice trade for themselves.

After Vasco da Gama sails around Africa, the sea route seems a possible way to trade with the Far East. Perhaps the long land trek on the Silk Road can be replaced by a ship

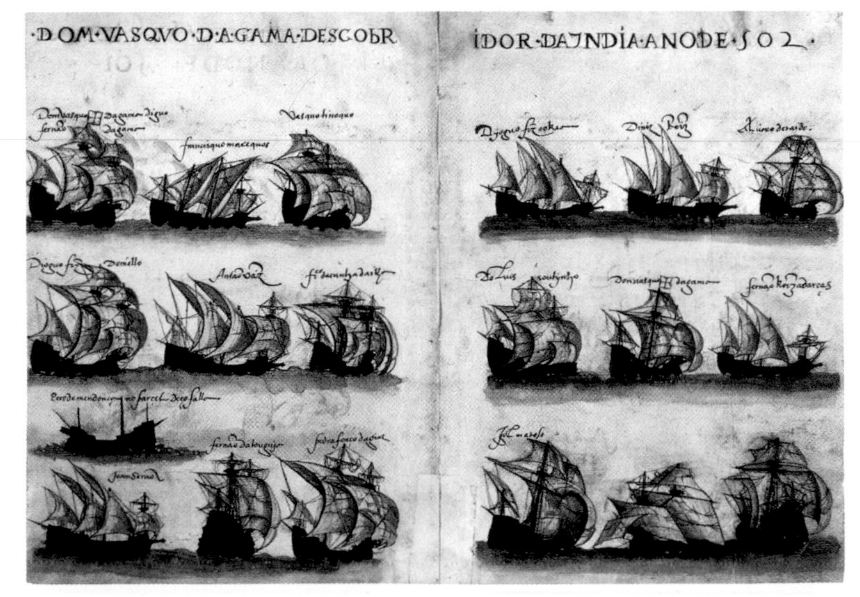

Vasco da Gama (that's his fleet at right, illustrated in a 1558 book) usually gets credit for being the first European to sail around Africa. But many historians believe that sailors from Phoenicia (now Lebanon and parts of Syria, Jordan, and Israel) did it back in 600 B.C.E. The Greek historian Herodotus wrote of their voyage, "It was not until the third year that they returned through the Pillars of Hercules [at Gibraltar], and made good their voyage home."

journey. But the voyage around Africa's Cape of Good Hope is tough and dangerous.

Is there a better way to go? Magellan is convinced that sailing west to reach the East will be faster and safer. If he is right, he will become rich and famous. And if he reaches his destination, he knows it will prove once and for all that the Earth is a round globe.

Although he is descended from Portuguese nobility, Magellan has little finesse. When he goes to Lisbon to the royal court and presents his plan, he does it without the flattery and fawning that King Manuel I expects from his subjects. The king, expressing his disdain, turns around and shows Magellan his back. It is an insult. The courtiers titter.

Carlos I (1500–1558) became king of Spain as a teenager. (Here, he's 12 years old, wearing a gift of armor from his grandfather.) The young ruler hoped to win sea rights from Spain to the Indonesian Spice Islands—rights that Portugal (with the pope's approval) had already claimed.

Achievers and heroes—as this man will prove to be—don't give up. He goes to Spain and presents himself to the 18-year-old monarch, King Carlos I, who will soon be Holy Roman Emperor Charles V. But he is Carlos when he agrees to sponsor the expedition. To get Spanish backing, Magellan must give up his Portuguese nationality and become Spanish. Before Magellan does that, he studies the guarded secret records in the Portuguese treasury. There he finds logs and

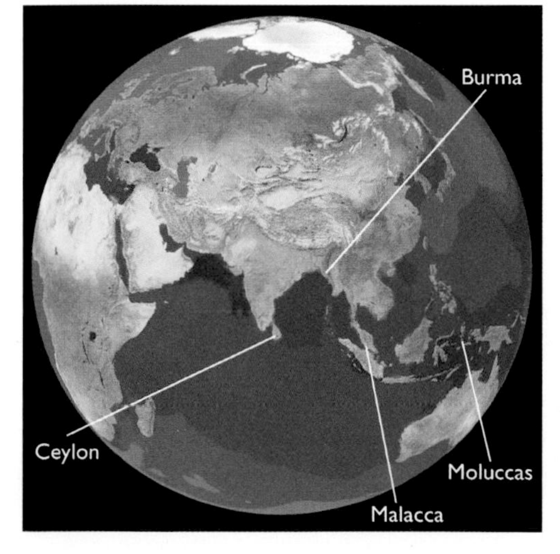

Ludovico di Varthema, an Italian adventurer, set out in 1502 on an overland trip to India. From there he traveled to Ceylon and Burma and, eventually, the Moluccas. (Don't confuse Malacca, a port on the Malay Peninsula, with the Moluccas, spice-rich islands near New Guinea, Indonesia.) Di Varthema was the first European to get there. His widely read journal is full of fascinating descriptions of Asian kings, Buddhist rites, and places he visited.

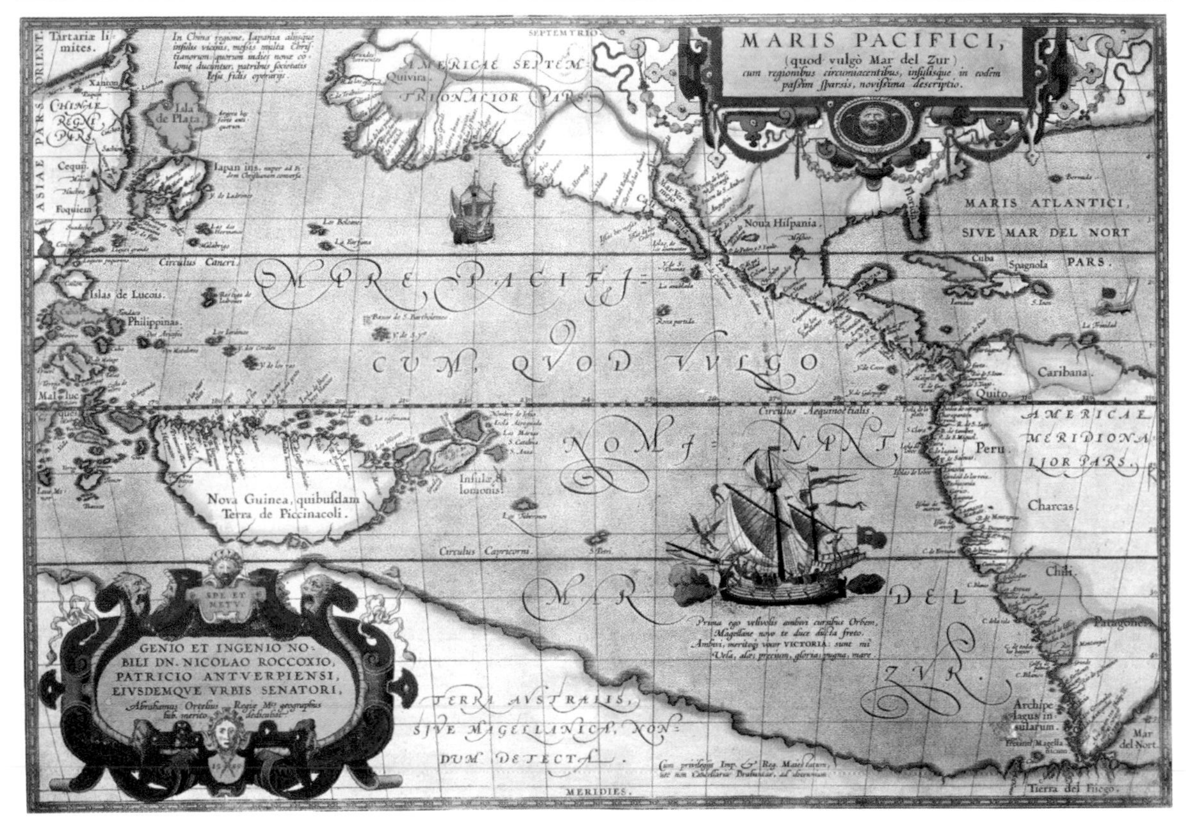

This 1589 map by Abraham Ortelius understates the vastness of the Pacific Ocean. (Compare it to the satellite map on the opposite page.) When Magellan's fleet emerged from the long, narrow passage now known as the Strait of Magellan (lower right corner, just below South America), his crew was facing an immense, unknown sea. Magellan thought he was near the Moluccas, but the crossing took months. The ship is the *Victoria*, the sole survivor of Magellan's fleet and the first vessel to sail around the world.

sailing accounts from those who have been to the Americas. He learns as much as he can from the voyagers who have sailed before him.

The trick will be to find a passageway through the American land. That land is like a giant jigsaw puzzle, and most of the pieces are still missing. Like other mariners, Magellan believes the land Columbus discovered is skinny. Perhaps it is a long island or two.

He knows that Vasco Núñez de Balboa climbed a peak in Central America in 1513 and looked out at a sea. No one realizes that Balboa stood at the narrow waist between two giant continents—the only place where the Atlantic and Pacific Oceans are relatively close. No one knows how big Earth actually is. And almost everyone believes there is only one ocean and that the new land is a minor interruption in the great Ocean Sea. Magellan studies everything that is known about the Americas.

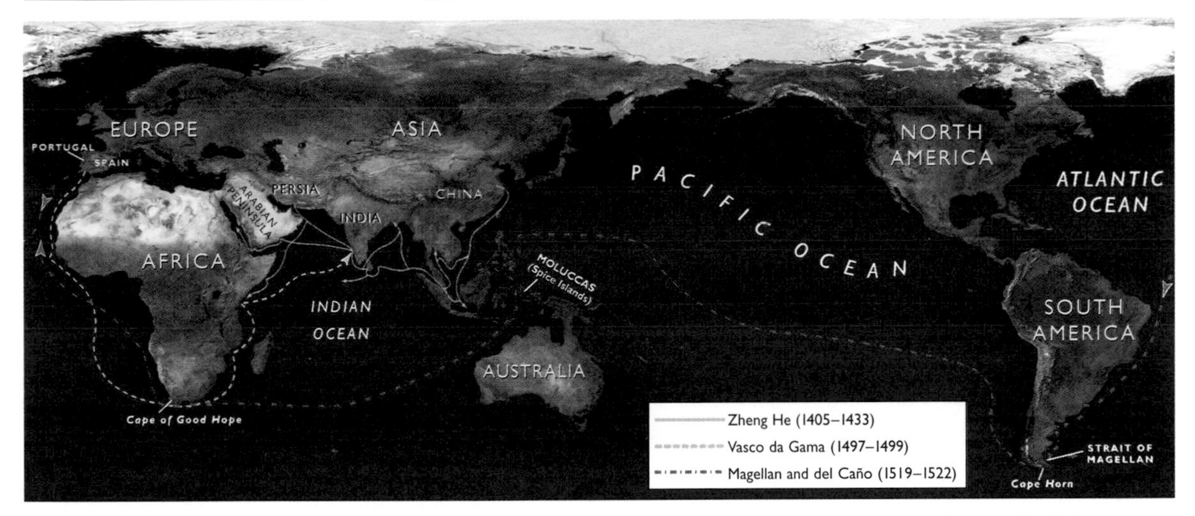

Magellan's crew had almost no provisions or freshwater when the ships set out from the tip of South America across the Pacific Ocean. For months, the men ate rats and boiled leather, and many of them died of scurvy (a disease caused by lack of vitamin C). The explorers had only one thing in their favor—the calm weather that gave the sea its name. (*Pacific* means "peaceful.")

A seasoned soldier and sailor who has served with distinction in Africa, India, Malaysia, and Mozambique, Magellan is one of a very few Europeans who has been to Malacca—the rich trading center for the lucrative (money-making) spices. In 1511, when a Portuguese fleet captured Malacca, Magellan and his friend Francisco Serrão were part of the invasion force. Afterward, Magellan went back to Europe around Africa's Cape of Good Hope. His friend Serrão stayed behind and set out to find the Spice Islands. He was shipwrecked in a bad storm but ended up in the Spice Islands anyway. Serrão fell in love with those islands, settled there (the first European to do so), and married the daughter of a local prince. He must have had quite a life. "I have found a New World," he wrote to Magellan in a letter full of enthusiasm. "I beg you to join me here, that you may sample for yourself the delights which surround me."

Serrão's detailed letters inform and inspire Magellan. Besides that, Magellan has brought a slave, Enrique, from the Malay Islands. Enrique knows the Spice Islands region, and he has a talent for languages. With him, Magellan is as well prepared to travel to the Spice Islands as any European in his time can be.

When the Portuguese learn of the expedition, they are alarmed and try to destroy it before it can get started. Magellan, who is now working for Spain, is called a traitor. Rumors are spread that he is incompetent. The rumors are

CHINA'S FLOATING CITY

In the early 1400s, a Chinese admiral named Zheng He (JUNG-HUH) sailed from China to Africa in a ship that was more than four times bigger than any of Magellan's ships. Magellan commanded three-masted caravels; Zheng He commanded a nine-masted giant. It was a floating city with layered decks, fancy cabins, vegetable gardens, and sophisticated navigational technology. He led more than 27,000 men in an armada of 300 ships, which was larger than anything the West would see for the next 500 years.

Except for the time of the Roman Empire, China had long been technologically more advanced than Europe. There is considerable evidence that Chinese ships visited the Americas, perhaps as early as the fifth century. In Zheng He's day, China's navy had 3,500 ships. (Today's United States Navy has about 295.) So what happened? Why didn't China lead the world in exploration? Why didn't Zheng He or other Chinese explorers sail around the world? Why, after years of sailing triumphs, did China suddenly stop all long-range explorations? Why, in 1525, did the Chinese government order the destruction of all its oceangoing ships?

No one knows. Some say the Chinese were content with what they had. They weren't greedy or particularly inquisitive. To the Chinese who knew about Europe, Europeans seemed barbaric. They saw no reason to go there. They were content with themselves.

But turning away from inquiry and exploration means turning away from learning and scientific progress. Chinese politicians, deciding to isolate their nation, made a poor decision. It happened all the time in Europe. (Consider the whole of the Dark Ages.) But Europe was made up of many nation-states, so when one state made a mistake, another often saw an opportunity. China was under unified control. When it looked inward, it missed a singular chance to lead the world.

Europeans were often both greedy and inquisitive. It made them eager to explore.

The primitive magnetic compass (right) from China is roughly 2,000 years old. The more sophisticated sundial and compass (above) is from the Ming dynasty (1368–1644).

false but effective. Magellan has a hard time finding a crew of seamen. When he leaves, it is with a ragtag lot of sailors who speak several languages, and a few Spanish noblemen (called dons) who are not pleased to have a Portuguese captain. Some of them, including del Caño are already planning a mutiny.

One man, Antonio Pigafetta from Venice, has been sent by business interests to report on this possible route to the spice trade. Pigafetta will keep a detailed diary that will tell future generations the truth of the expedition. Going on this voyage takes courage and daring (like later trips into unexplored space).

Perhaps you know the story: of the mutiny and Magellan's strength in thwarting it. Of the discovery of a strait at the tip of South America and of the terrors and the tortuous twists of its waters. (It takes 38 days to get through it.) Of the second mutiny, when those who sail the largest of the ships turn around and head back to Spain with most of the expedition's provisions. Of the awful voyage across the enormous Pacific—99 days without fresh food. Of Magellan's leadership and example during all that harrowing time. And, finally, of the landing on an island where Enrique talks to the natives and they answer him. He is speaking their language!

Magellan realizes that Enrique has arrived home. He is the first man to sail around the world. Magellan has found the East by sailing west. But the *capitán-general* doesn't have long to celebrate. They are in the Philippine Islands, and Magellan is about to be killed in a senseless battle.

What does all this have to do with science? A whole lot:

Magellan's will states clearly that when he dies, Enrique is to be freed and paid 10,000 maravedis. But the new commander refuses to do as the will says. So Enrique betrays the Spaniards to a local king. In the fight that ensues, 30 Europeans are killed.

THE SPICE OF LIFE

In the Middle Ages, Europeans wore coarse woolen garments. Wool is hot, heavy, and difficult to wash. After some wear without washing, woolen clothes are apt to be itchy and filled with vermin. Silk, from the East, was light, washable, and elegant.

Pearls and precious stones came from the East. And so did handsome woven carpets with gorgeous designs. If you were wealthy, you wanted them instead of threshes—the leftovers from a grain harvest, used to cover most floors—especially at the entry, or "threshold."

And, of course, most desired of all were the exotic spices that seemed to add to the taste of foods and preserve them (in those days before refrigerators). Spices also served as medicines, love potions, and perfumes. They were even used in religious ceremonies. The spice trade spawned a new kind of worldwide marketing system out of which, eventually, came free enterprise.

In seventeenth-century China, this square silk badge, with its white crane, could only be sewn onto the coat of a first-rank civil servant. Other birds symbolized eight lower ranks.

Before the age of exploration, Europeans thought they were at the center of the world. The voyages of Christopher Columbus, Ferdinand Magellan, and others shattered that illusion. This 1574 fresco paints a much clearer picture. Not only is the continent of Europe on the small side, all the landmasses together add up to only a fraction of the area of the oceans.

It's one thing to have a theory. It is something else to have a proof. Science depends on both.

Pythagoras believed the Earth was round. Two thousand years later, Magellan proved he was right. That knowledge electrifies the medieval world. After del Caño arrives in Seville on Magellan's worn-out, worm-eaten vessel (the only ship left from the expedition), couriers (messengers) race across Europe to carry the news to the pope. Soon, everyone knows of the voyage. Their picture of Earth changes. They learn that when Magellan and his crew sailed near the antipodes—what they thought was the bottom of Earth—they hadn't been upside down.

You may not think of geography as a science. But it is one, and it plays an important role in our understanding of the universe. Europeans had limited knowledge of world geography. They thought themselves at the center of the Earth. They believed that hell was the underside of Earth; they called it the Underworld. They believed that the Garden of Eden and Paradise were somewhere on the actual Earth. They thought their medieval Christian world, which surrounded the Mediterranean Sea, was the biggest and most important part of the world.

When Magellan discovered the Philippine Islands, he found people who had never heard of Europe and its ideas and values. And they seemed to be doing just fine. Before long, Europeans became aware of the immensity of Asia and the colossal size of the Americas. It upset their worldview.

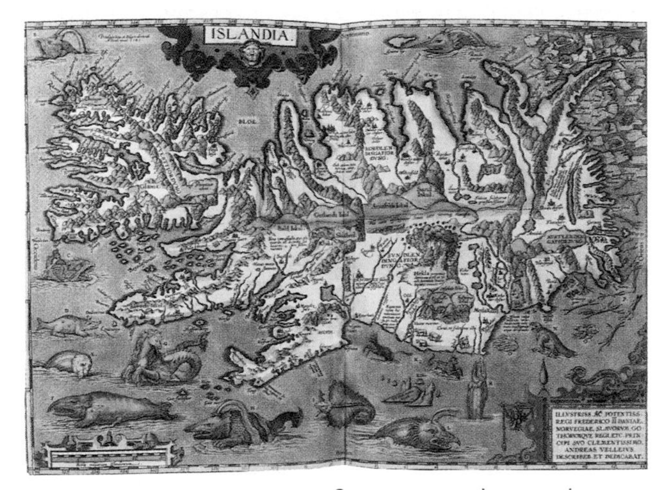

Sea monsters decorated maps even after Magellan's crew hadn't encountered any on their three-year voyage around the world. On this 1585 map of Iceland, enormous, threatening beasts lurk just offshore. Perhaps the monsters remained as symbols of the undeniable dangers of the sea.

Magellan's voyagers had sailed around the globe and had not seen any sea monsters. That was important knowledge. Could it be that Gaius Julius Solinus, the third-century Roman writer known as the Polyhistor, had made up the gryphons and the dogheaded, horse-footed men he described so vividly? Solinus's exotic creatures decorated most medieval maps. They had been believed. But Magellan's crew had seen nothing of them.

And then there was Magellan's discovery of the vastness of the Pacific Ocean. This was no little sea. If all the Earth's landmasses could be dumped into the Pacific, there would still be plenty of water left for swimming. And Europeans hadn't even known that ocean existed! Can you imagine how that knowledge stretched their minds?

There was still more. The voyage was a great technological feat. When Magellan's crew arrived back in Spain, according to Pigafetta, the voyage had covered 81,449 kilometers. That's 50,610 miles. Columbus sailed only about 4,100 kilometers (2,548 miles) on his first voyage. Like the first trip to the Moon, Magellan's voyage showed what human intelligence and daring can do. It energized a Western world that, after 1,000 years of semi-hibernation, was doing more than yawning. It was getting ready to wake up and run. This new information would help make that possible. For those who thought scientifically, Magellan changed everything.

"Manned by a cadaverous crew and a few ghostly captives, [Magellan's ship *Victoria*] wound its way up the Guadalquivir River laden with pungent cloves that were worth their weight in gold, and that had cost their weight in blood. The *Victoria*'s return conclusively demonstrated that the Far East could be reached by sailing west, that the Pacific Ocean was incomparably larger than imagined and that calendars would not work without an international date line."
—from W. Jeffrey Bolster's review of *Over the Edge of the World*, by Laurence Bergreen, a book described as a "first-rate historical page turner."

29

Cosmic Voyagers: Is It Fiction, or Could It Be True?

There is an infinite number of worlds, some like this world, others unlike it.... For the atoms out of which a world might arise, or by which a world might be formed, have not all been expended on one world or a finite number of worlds, whether like or unlike this one. Hence there will be nothing to hinder an infinity of worlds.... And further, we must not suppose that... in another sort of world there could not possibly be, the seeds out of which animals and plants arise and all the rest of the things we see.
—Epicurus (341–270 B.C.E.), Greek philosopher, "Letter to Herodotus"

And so I say again, again you must confess
That somewhere in the universe
Are other meetings of the atom stuff resembling this of ours.
—Lucretius (ca. 99–ca. 55 B.C.E.), Roman poet and philosopher, "There Are Many Worlds"

It is in the highest degree unlikely, of course, that Earth is the only world in the universe that has developed life. It seems quite possible than any world with physical and chemical characteristics of Earth may develop life, and there may conceivably be millions of such Earth-like planets in each of billions of galaxies.
—Isaac Asimov (1920–1992), American science and science-fiction writer, "Terrestrial Intellligence"

Far, far from Earth, on the planet Thule, in a galaxy named Aleeya (UH-lee-uh), creatures with curiosity—thinking beings—set out to solve the problem of intergalactic (galaxy-to-galaxy) travel. Like our Milky Way Galaxy, Aleeya is an average spiral galaxy, which means it has about 100 *billion* stars. Young stars, arranged a bit like a pinwheel, speed around a central bulge of old stars. If you

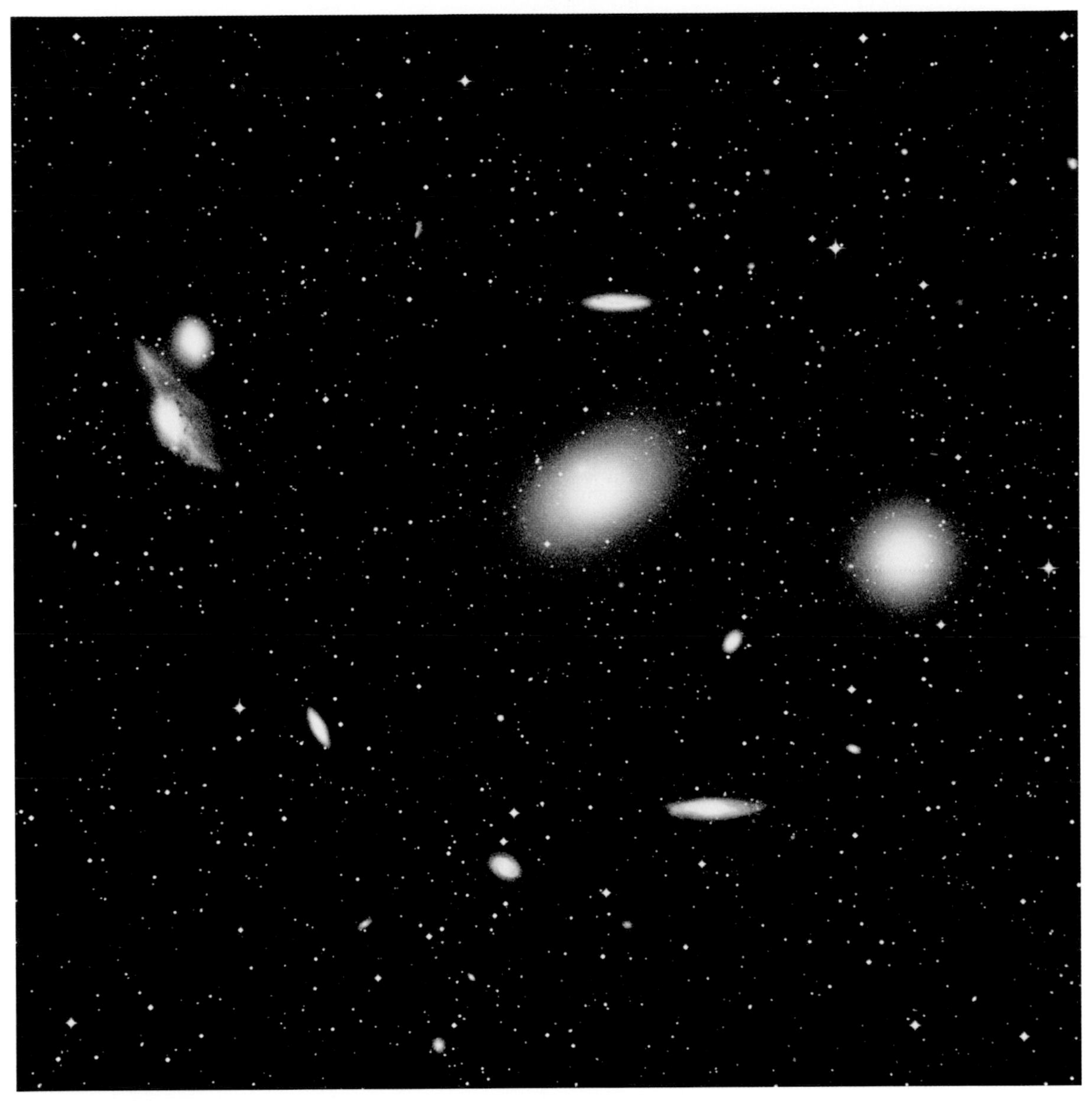

The nearest big group of galaxies is the Virgo Cluster. Yet even traveling at the speed of light, you'd need 60 million years to get there. This handful of galaxies (above) is just a tiny core. The whole Virgo Cluster contains a whopping 2,000 galaxies!

look at Aleeya from another galaxy, the spinning stars seem to form a flat disk. If you tried to count all the stars in the Aleeya Galaxy, counting one every second, it would take you more than 3,000 years. But the inhabitants of the planet Thule, circling a medium-sized star, have other things on their minds besides counting. They are trying to understand the universe and find out if there are other thinking creatures anywhere.

The most important upshot of the discovery of extraterrestrial life would be to restore to human beings something of the dignity of which science has robbed them. Far from exposing *Homo sapiens* as an inferior creature in the vast cosmos, the certain existence of alien beings would give us cause to believe that we, in our humble way, are part of a larger, majestic process of cosmic self-knowledge.
—Paul Davies (1946–), British physicist and philosopher, *Are We Alone?*

For a long time, the old guard scientists on Thule keep saying that nothing can go faster than the speed of light. And with the immense distances in space, even traveling as fast as light (which no one can actually do), it would take forever to get to distant planets. "Too bad," say the experts, "but the speed of light is a universal speed limit. We'll never get out of this galaxy."

A few renegades won't listen. They figure out that if you can harness the energy in black holes and then find wormholes in space to tunnel between galaxies, you can get around the speed-of-light trap. Which they finally do. They set out in their year 13700000000 to begin to explore the universe, which is less than a billion years ago. (The Thulean calendar begins with what they believe is the birth of the universe.)

Even with their astonishing technology, this is a daunting task. Given the hundreds of billions of galaxies and the far greater number of solar systems with asteroids and planets—well, where does one start? The Thulean goal is to see if there is life in other parts of the universe. By analyzing light rays, they can tell the composition of a planet, and that gives them clues as to where life might be possible. When their astronomers notice the blue glow of a small planet orbiting a medium-sized

VOYAGE OF A LIFETIME—AND THEN SOME

Light is faster than anything else in the known universe, but don't think of it in terms of speed. Think in terms of distance and time. (Speed is a ratio—distance divided by time.)

Where would you want to go in space, and how far is it? How much time will it take to get there? Here are a few travel tips for those of you imagining a speed-of-light voyage:

Moon: 1.3 seconds
Sun: 8.5 minutes
Neptune: 4 hours

Proxima Centauri (nearest star beyond our sun)**:** 4.24 *years*
Center of the Milky Way Galaxy: about 30,000 years
Sagittarius Dwarf Elliptical Galaxy (nearest galaxy)**:** 80,000 years
Andromeda Galaxy: 2.3 million years

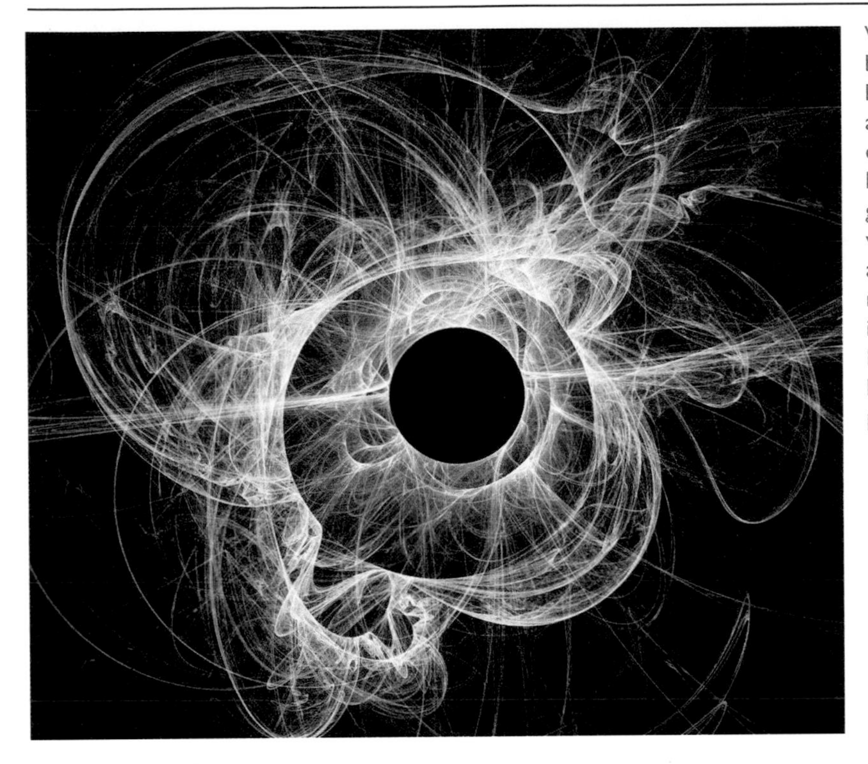

We can't see black holes—but we can see the effect they have on surrounding matter and energy. In this imaginative computer illustration, not even light can escape the powerful gravitation at the center—which is not really a hole, but a tiny dead star. The point of no return—the circle's edge—is called the event horizon. Beyond it, swirling clouds of hot dense matter are being pulled inside.

star on the outskirts of the Milky Way Galaxy, it looks promising. A space expedition is sent to check on the planet.

Early reports confirm a complexity of life-forms on that planet. Thule's cosmic explorers, able to fade into the dark matter that makes up so much of the universe, are invisible to the inhabitants of this planet, which is called Earth by its most evolved life-form, known as humans.

A BLACK HOLE is a star whose own massive gravity has caused it to collapse in on itself. The gravitational pull is so strong that not even light can escape. A WORMHOLE is a theoretical backdoor tunnel that connects two very distant points in space as if they were neighbors.

Planet Earth is deemed suitable for further study. Thuleans send a series of expeditions beginning in Earth's Jurassic period—some 100,000,000 years before the present. By earth year 1500 (about 500 YBP*, a wink in cosmic time) they have compiled a mountain of information (all annotated and analyzed in Thulean information processors implanted in Thulean brains). One report after another is filed and made available to the Thulean public. The reports of Earth's progress are as fascinating as a soap opera to the Thuleans. What will happen in the next installment, they ask? The evolution of the earthlings known as "mammals" is of particular interest.

*years before present

One report notes: "These mammals vary from tiny rodents to tall, long-necked animals called giraffes. Humans are a particularly fascinating form of mammal. They have a moderately developed form of intelligence and a disturbing tendency to kill each other." The report goes on: "Humans need to be studied carefully, as they may have the potential to eventually conquer space travel. If that happens, will they be a menace? Might they bring their warring ways to other planets? A detailed report on their ethics, beliefs, and ways of life is being prepared and will follow." For a few advance notes, see below and the next two pages. (Before you do, consider: It is earth year 1500. How would you describe life on Earth?)

How did earthlings see their world in that landmark year, 1500? This 1457 map from Genoa, Italy, is like a "before" picture. A giant landmass—Asia, Africa, and Europe—dominates the planet. People could only imagine what lurked beyond the watery edges. As the new century approached, a drastically different "after" picture was about to come to light.

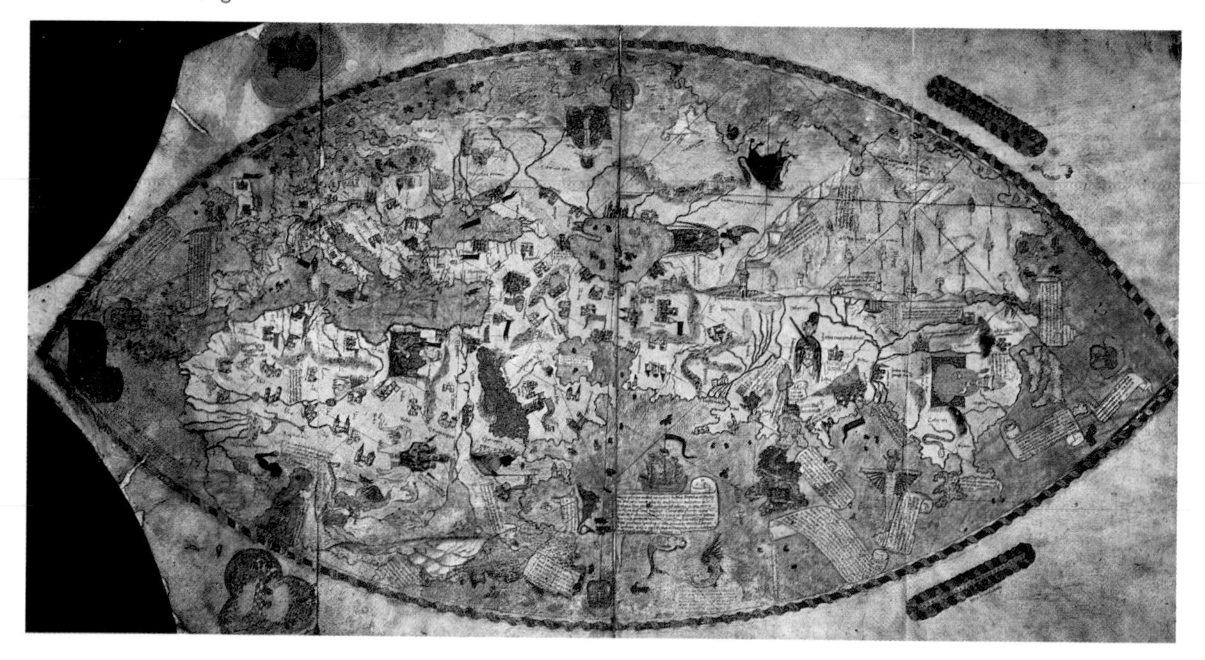

Overview of Report Filed in Earth Year 1500

by Thulean Intergalactic Mission #11631

Planet Earth has been a worthwhile subject for study with many surprises and twists and turns in its history. Earlier reports dealt with the development of life-forms on this planet—

especially an astonishing collection of animals that some call "dinosaurs." (We use Earth terminology in these reports.) In this paper we confine ourselves to a period of about 4,500 earth years when the creatures called "men" and "women" establish themselves in communities, learn to read and write, and make and lose some remarkable advances toward understanding the laws of the universe and of themselves.

Auxiliary reports, charts, and drawings give details on civilizations called Sumerian, Ming, Egyptian, Nubian, Mayan, European, Islamic, and American. We were particularly fascinated by the peoples of a small city named Athens. There, men and women seem to have understood the importance and beauty of the scientific method. In an astonishingly brief period of time, they made discoveries about the cosmos that (we are embarrassed to note) took us Thuleans much longer to find.

Perhaps their progress was speeded when they realized that science can only exist in an atmosphere where truth and freedom dominate. We Thuleans know all too well that lies don't work in the field of science. Scientific achievements must be verified again and again. False results won't stand before repeated tests. Science demands truthfulness. Thus, a society where science flourishes is a society where truth is cherished. The remarkable Athenians discovered that link between democracy, truth, freedom, and science. While the Athenian democracy lasted, progress in science was spectacular. Progress in human rights was also outstanding in Athens and in a republic that developed in nearby Rome (except for a nasty practice known as slavery).

Then things changed. (See details in sub-reports.) In their year of 1500, most Earth governments are dictatorial and repressive. Democracies no longer exist. Open debate is rare.

A few very privileged individuals hold power; most others have little control over their fate. Life for most humans is unpleasant and short. The learning of the Greeks, while not completely lost, is known to only a few priests and scholars. Science seems just about dead. But there are stirrings at a few thriving urban centers and a burst of creativity—a renaissance in painting and sculpture—that is making some humans open their eyes to the world around them. This will be worth watching. A new Earth technology for spreading information—the printed book—looks promising.

Otherwise, the backward steps of humans on this small planet should teach us Thuleans lessons. Is there hope for these earthlings? Will they recover the knowledge and questing spirit of the Greeks? Will they use their intelligence to go further? Will a society where truth and freedom are supported be developed? Reports on the next 500 years will be forwarded as compiled.

Our solar system is nestled in an outer arm of the spiral Milky Way Galaxy. Though a little lonely, it's a terrific spot for viewing the scenic part—the dense, colorful center that stretches across our night sky (below). The Milky Way contains hundreds of billions of stars, colorful nebulae, interstellar matter, and globular clusters—but mostly space. There's more nothing than something, but what there is—us included—circles together around an intriguing core.

WHERE DO YOU STAND ON THE UNIVERSE—AND IN IT?

Your celestial address—where you are in relation to the rest of the universe—is a matter of fact and opinion. The fact is, you're on Earth. Earth orbits the Sun, which is in a spiral arm of the Milky Way Galaxy, which is part of the Local Group of galaxies, which is in a particular part of a vast, seemingly endless universe. All that is true. But what's your opinion about where we stand in the universe? What's our role, our importance, our status?

Peter D. Ward, a paleontologist (fossil expert), and Donald C. Brownlee, an astronomer, suggest in their book *Rare Earth* that we might be very special—unique, even. They argue that the conditions necessary for the survival of complex life are so unusual that Earth may be the only place that life exists in the universe. Do you agree?

In *Are We Alone?* physicist Paul Davies takes another view. He concludes that "consciousness, far from being a trivial accident, is a fundamental feature of the universe, a natural product of the outworking of the laws of nature to which they are connected in a deep and still mysterious way." If this view is correct, if consciousness is a basic phenomenon that is part of the natural outworking of the laws of the universe, then we can expect it to have emerged elsewhere. The search for alien beings can therefore be seen as a test of the world view that we live in a universe that is progressive, not only in the way that life and consciousness emerge from primeval (beginning of life) chaos, but also in the way that mind plays a fundamental role.

One way or another, it's in our nature to find out. Freeman Dyson, a physicist and author, says: "I do not feel like an alien in this universe. The more I examine the universe and study the details of its architecture, the more evidence I find that the universe in some sense must have known we were coming."

30 Finally! How Science Works

Science is about what can be observed and measured or it is about nothing at all. In science, as in democracy, there is no hidden secret knowledge; all that counts is on the table, observable and falsifiable.
—Dennis Overbye (1944–), American science writer and editor, *The New York Times*

In science one must search for ideas. If there are no ideas, there is no science. A knowledge of facts is only valuable in so far as facts conceal ideas: facts without ideas are just the sweepings of the brain and the memory.
—Vissarion Grigoryevich Belinsky (1811–1848), Russian literary critic, *Collected Works*

The important thing is not to stop questioning. Curiosity has its own reason for existing.... Never lose a holy curiosity.
—Albert Einstein (1879–1955), German-American physicist, *Life* magazine

There is something I want to be sure you understand before you close this book. It's this: Science is not about certainty; it's about uncertainty. Does that sound weird? Well, it's true. Science is all about trying ideas, discarding those that don't work, and building on those that do. It never stops.

As you know, scientifically minded serious thinkers once "proved" that the Earth was flat. They walked around and observed carefully and made that judgment. It helped them plan their lives. Then someone came along and "proved" the Earth is a globe in the center of the universe. And that worked. And then someone else came along and "proved" that the Sun is in the center of the universe, and that worked, until someone else came along and "proved" that the universe has no center, and...

Well, you get the point. Those people in the past who had wrong ideas weren't dummies. They were doing the best they could, given the knowledge of their times. We do the same thing today. And you can be sure that people in the future will look back and wonder why we believe some of the things we do. Good scientists know that. They learn to be humble. Does that make science unimportant? After all, if some of our scientific theories are going to be proved false, why bother studying them?

Because uncertainty is the most interesting of all places to be. If you believe something is an absolute truth, you can just memorize it and get on with your life. There's nothing to discuss.

Science isn't like that at all. Scientists—the good ones—are always questioning and questing. Yes, there are scientific ideas, often called principles or theories, that have proved themselves over time. Then they get called facts, or laws. By using those laws, modern science has achieved incredible things. But nothing is beyond question to the scientific mind—even those laws. So no good scientist will laugh at you for asking questions—because sometimes the silliest-

When the Egyptian Pharaoh Seti I (who ruled ca. 1302–1290 B.C.E.) looked at the skies, did he wonder if there was life on other planets? Maybe. He had the cosmos portrayed on the walls of his burial vault (see more on page 13). In this ceiling detail, the red dots in the wondrous creatures represent stars in Northern constellations. You don't have to be a Pharaoh to have a gorgeous picture of the heavens on your ceiling. Orbiting telescopes beam us colorful photographs that look like works of art, asking to be framed and admired.

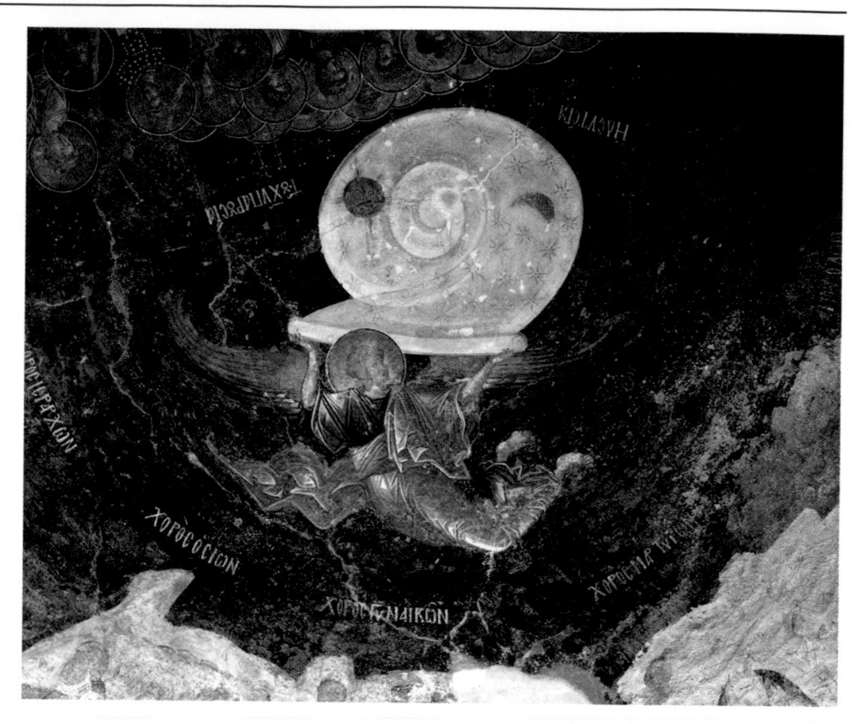

That's heaven in the form of a snail, with the Sun and Moon inside the shell. It's a fourteenth-century fresco from the ceiling of the Church of Chora in Istanbul, Turkey. The gorgeous mosaics in this Byzantine church (it's now a museum) were renowned in the Renaissance world of art. The artist didn't worry about laws of motion. He got an angel to push the heavens.

seeming questions have led to the most profound answers.

But we do know that some things certainly seem to be true. For instance, two modern theories called relativity and quantum theory have been tested again and again and again—and no experiment has been able to disprove them. (We usually call relativity a theory, but most scientists think of it as a law.)

Relativity describes the way nature works in the huge world of planets, galaxies, and space. Quantum theory describes the rules in the minuscule world inside atoms. Those theories have helped us understand the universe's dynamic 13.7-billion-year history.

Not too long ago, in earth time, no one thought the universe had a history. The experts in both religion and science believed the world had been created much as it is today and that we humans were acting on a fixed, permanent stage.

Those experts did their best, but they turned out to be wrong—really wrong. The question-askers, the scientifically minded, discovered that the universe is a changing place.

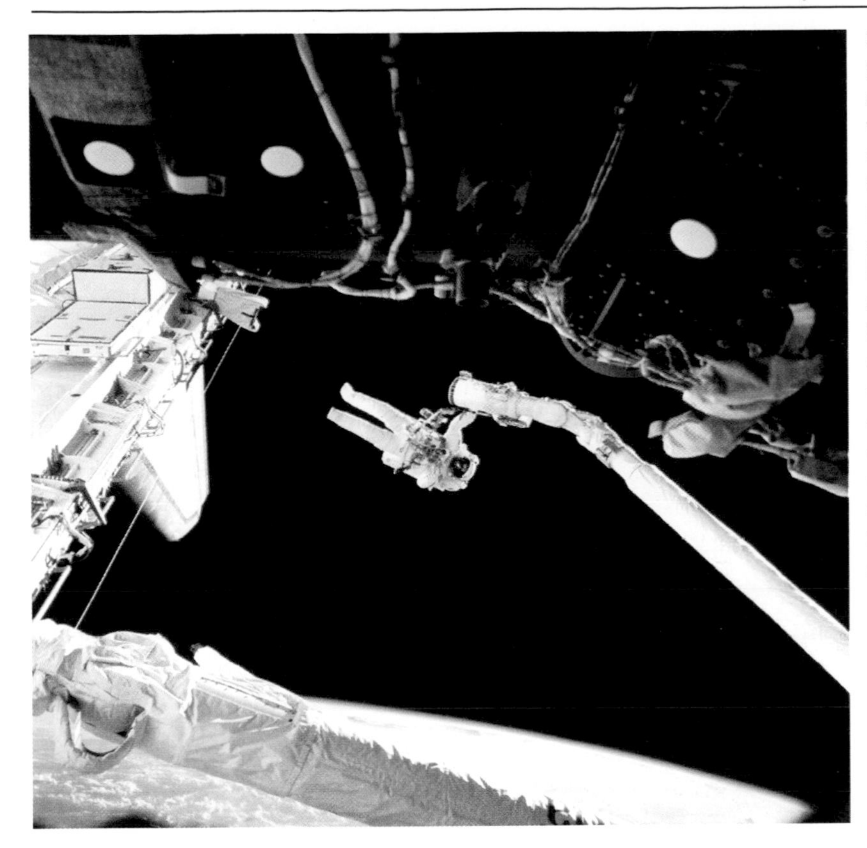

Some Greeks actually imagined people living in space. It took a few thousand years, but here we are, with a permanently staffed International Space Station (ISS) in orbit around Earth. In this October 2000 photo, astronaut William McArthur Jr. is holding the space shuttle *Discovery*'s Remote Manipulator System (RMS), a robotic arm used to build and maintain the structure. The ISS was a joint effort between NASA, 11 nations of the European Space Agency, the Russian Space Agency, Canada, Japan, and Brazil—the largest international science project in world history.

It has an incredible tale to tell. How did we find out about the Sun and the stars and about the atoms within us? That's what these books are about.

We began the story in this book with stargazers in ancient Sumer. In the books to come, we will head through time until we reach today's physicists and theories of superstrings (which, if they exist, are tiny vibrating bits of energy that lie beneath everything in the universe).

I hope you'll get hooked by all this and want to find out more about science on your own. But if you finish these books and say, "Whew, that's enough of that!" at least you won't be a total ignoramus when it comes to science. When someone mentions Aristotle or Isaac Newton or Albert Einstein, you can smile a savvy smile. You'll know who they are, and you'll also know they won't have the last word.

THE PRIME NUMBER SIEVE OF ERATOSTHENES

	2	3	~~4~~	5	~~6~~	7	~~8~~	~~9~~	~~10~~
11	~~12~~	13	~~14~~	~~15~~	~~16~~	17	~~18~~	19	~~20~~
~~21~~	~~22~~	23	~~24~~	~~25~~	~~26~~	~~27~~	~~28~~	29	~~30~~
31	~~32~~	~~33~~	~~34~~	~~35~~	~~36~~	37	~~38~~	~~39~~	~~40~~
41	~~42~~	43	~~44~~	~~45~~	~~46~~	47	~~48~~	~~49~~	~~50~~
~~51~~	~~52~~	53	~~54~~	~~55~~	~~56~~	~~57~~	~~58~~	59	~~60~~
61	~~62~~	~~63~~	~~64~~	~~65~~	~~66~~	67	~~68~~	~~69~~	~~70~~
71	~~72~~	73	~~74~~	~~75~~	~~76~~	~~77~~	~~78~~	79	~~80~~
~~81~~	~~82~~	83	~~84~~	~~85~~	~~86~~	~~87~~	~~88~~	89	~~90~~
~~91~~	~~92~~	~~93~~	~~94~~	~~95~~	~~96~~	97	~~98~~	~~99~~	~~100~~

All the circled numbers are prime. How can you tell? Start with 2, the lowest prime number. Cross out all its multiples (the even numbers). Then cross out all the multiples of 3 (in red) that are still hanging around. Next, which multiples of 5 (in green) are left? of 7 (in purple)? Cross those multiples out too—and so on.

Suggested Reading

Alexander the Great reading Homer (see page 120)

Bendick, Jeanne. *Archimedes and the Door of Science*. Warsaw, N.D.: Bethlehem Books, 1997 (first published, 1962). This is a delightful easy-to-read book. The author has a sense of humor, and the science is explained well.

Bragg, Melvyn. *On Giants' Shoulders: Great Scientists and Their Discoveries—from Archimedes to DNA*. New York: John Wiley & Sons, 1999. Melvyn Bragg brings together some modern scientists to talk informally about major figures from the past (from Archimedes to Crick and Watson). The conversation is uneven and quirky, but informative too.

Boorstin, Daniel J. *The Discoverers: A History of Man's Search to Know His World and Himself*. New York: Vintage Books, 1985. A lively, wide-ranging story of intellectual discoveries about time, geography, nature, and society.

Bronowski, Jacob, *The Ascent of Man*. Boston: Little, Brown and Company, 1974. A wonderful classic dealing with science, history, and ideas. In addition to an engaging narrative, it's beautifully illustrated.

Burke, James. *Connections*. Boston: Little, Brown and Company, 1995. You may have seen James Burke on TV. He's an imaginative Englishman who is good at showing how inventions and scientific discoveries are linked, from Ptolemy's astrolabe to the discovery of electricity.

Kuhn, Thomas S. *The Copernican Revolution: Planetary Astronomy in the Development of Western Thought*. Cambridge: Harvard University Press, 1992 (first published, 1957). A scholarly account of astronomical observation and theory from the Greeks to Copernicus, and Copernicus's contribution to the founding of modern scientific practice. Not easy reading, but worthwhile.

Rensberger, Boyce. *How the World Works: A Guide to Science's Greatest Discoveries*. New York: William Morrow, 1987. Simple explanations of 24 major scientific concepts, alphabetically arranged.

Tallack, Peter, ed. *The Science Book*. London: Weidenfield & Nicolson, 2003. A picturebook timetable of scientific discoveries from 3500 B.C.E. to 2000 C.E., with contributions by many experts.

NSTA Recommends

The National Science Teachers Association (NSTA) considers these books to be "outstanding"—clear, factual, and loaded with great science. Hungry for more? The NSTA lists several thousand highly recommended science books at this web address: www.nsta.org/ostbc.

Books for Kids

Ancient Communication: From Grunts to Graffiti; *Ancient Computing: From Counting to Calendars*; and *Ancient Construction: From Tents to Towers*, all by Michael Woods and Mary B. Woods (Lerner Publications, 2000). You'll be amazed at the creative technologies that ancient people used in their daily lives. These authors describe the mechanical ancestors of today's cell phones, computers, and high-rise buildings—and your human ancestors too.

Ancient Science, by Jim Wiese (John Wiley & Sons, 2003). Read about science in the Stone Age, the birth of civilization, the building of ancient pyramids, and Chinese fireworks. Then explore with easy activities you can do on your own.

Archaeology for Kids: Uncovering the Mysteries of Our Past, by Richard Panchyk (Chicago Review Press, 2001). The author takes the reader on a virtual journey from Neanderthal caves to the civilizations of Babylonia, Egypt, and China, to ancient Greece and the Roman Empire, and on to the Mayan and Aztec cultures of the New World. It also provides activities for would-be apprentices.

Aristotle: Philosopher and Scientist, Great Minds of Science series, by Margaret J. Anderson and Karen F. Stephenson. (Enslow Publishers, 2004). An all-too-brief summary of the life of a complex and influential thinker, this book is a good place to begin to trace Aristotle's ideas.

The Pantheon, Great Building Feats series, by Lesley A. DuTemple (Lerner Publications, 2003). Explore the architecture and technology of the Pantheon—ancient Rome's most famous structure—including the columns, walls, the distribution of weight of the flat, vaulted, and domed ceilings, and more. You can use this information to build a scaled-down Pantheon or to create your own structures.

Stretch a Little Further

Construct a Catapult, Science by Design series, by Lee Pulis (NSTA Press, 2000). In the Middle Ages, the ultimate weapon was the catapult. Build your own, as you explore history and technology with this book.

Echoes of the Ancient Skies: The Astronomy of Lost Civilizations, by E. C. Krupp (Dover Publications, 2003). An expert in ancient astronomy takes readers through time and space to look at early observatories from East and West and to explore the ways in which astronomical phenomena influenced the lives of past civilizations.

Lost Discoveries: The Ancient Roots of Modern Science—from the Babylonians to the Maya, by Dick Teresi (Simon & Schuster, 2003). Examine mathematics, astronomy, cosmology, physics, geology, chemistry, and technology from many ancient civilizations.

Picture Credits

All base maps were provided by Planetary Visions Limited and are used by permission. © 1996–2004 Planetary Visions Limited, and © 2004 Planetary Visions Limited/German Aerospace Center, unless otherwise noted.

Abbreviations for picture sources:
AR: Art Resource, New York
PR: Photo Researchers, Inc., New York
BAL: Bridgeman Art Library, London, Paris, New York, and Berlin

ii: © stephanecompoint.com; vi:Réunion des Musées Nationaux/AR; vii: IBM Research, Almaden Research Center

Chapter 1
Frontispiece: SPL/PR; 3: (full) Georg Gerster/PR; 4: Scala/AR; 6: Erich Lessing/AR; 7: (top) Werner Forman/AR, (bottom) Erich Lessing/AR; 8: Erich Lessing/AR

Chapter 2
10: Erich Lessing/AR; 11: (top) Erich Lessing/AR; 13: Werner Forman/AR; 14: (both) NASA; 15: NASA; 16: Or.8210/S 3326 section 11 Chinese star chart. By permission of the British Library; 17: AR; 18: Erich Lessing/AR; 19: Werner Forman/AR

Chapter 3
20: Erich Lessing/AR; 21: Alexander Marshack; 23: (top) Scala/AR; (bottom) Erich Lessing/AR; 24: The Metropolitan Museum of Art, Rogers Fund, 1948 (48.105.52), Photograph © 1979 The Metropolitan Museum of Art.; 25: Frank Zullo/PR; 26: D. Nunuk/PR; 27: NASA; 28: (top) Vanni Archive/CORBIS; (bottom) Jason Hawkes/CORBIS; 29: Werner Forman/AR; 30: Erich Lessing/AR; 31: Scala/AR; 32: (both) Erich Lessing/AR; 33: (top) Scala/AR; (inset) Réunion des Musées Nationaux/AR

Chapter 4
34: Erich Lessing/AR; 35: Fresco by Giulio Romano, Scala/AR; 36: Archivo Iconografico, S.A./CORBIS; 37: (both) Erich Lessing/AR; 41: (left) David Parker/PR; (right) Francois Gohier/PR; 42: Erich Lessing/AR

Chapter 5
44: Hulton Archive/Getty Images; 45: Serif Yenen, Meander Image Bank, Istanbul; 47: Scala/AR; 48: (top) Scala/AR; (bottom) Watercolor by George Ledwell Taylor, Victoria & Albert Museum, London/AR; 49: (top) Francois Gohier/PR; (inset) Frank Zullo/PR; (bottom) Hulton Archive/Getty Images; 50: Erich Lessing/AR; 51: Scala/AR; 52: The British Museum, London, U.K./BAL; 53: AR

Chapter 6
55: Historical Picture Archive/CORBIS; 56: (top) Mark A. Schneider/PR; (inset) Michael W. Davidson at Florida State University; (left center) Biophoto Associates/PR; (right center) Astrid and Hanns-Frieder Michler/PR; 57: AFP/CORBIS

Chapter 7
58: Werner Forman/AR; 60: (top) Bettmann/CORBIS; (center) Erich Lessing/AR; 61: (top) Oliver Strewe/Lonely Planet Images; (bottom) The Twelve Apostles in Victoria, Australia, Phillip Hayson/PR; 62: John Chumack/PR

Chapter 8
64: Réunion des Musées Nationaux/AR; 65: (top) SEF/AR; (inset) Delphi, Phokis, Greece/Index/BAL; 66: G. Hellner/E. Feiler, DAI, Athens; 67: (right center) Linda Braatz-Brown. Reprinted by permission of The Math Forum @ Drexel, an online community for mathematics education, http://mathforum.org/. © 1994–2004, The Math Forum @ Drexel; (bottom left) Victoria & Albert Museum, London/AR; 68: Clore Collection, London/AR; 69: Staatliche Museen, Berlin, Germany/BAL; 70: Réunion des Musées Nationaux/AR

Chapter 9
73: Scala/AR; 76: The Granger Collection, New York; 77: Chandra X-Ray Observatory/NASA/PR; 78: Werner Forman/AR (photo: E. Strouhal); 79: (left) Paul Klee, *Garden of the Chateau*, 1919, Bridgeman-Giraudon/AR; 81: Stocktrek/CORBIS; 83: David Muench/CORBIS; 84: (left) Erich Lessing/AR; (center) Michael Patrick O'Neill/PR; 85: (top) Charles O'Rear/CORBIS; (bottom) Scala/AR

Chapter 10
86: Scala/AR; 89: Yann Arthus-Bertrand/CORBIS; 90: © IBM Research, Almaden Research Center; 91: Bridgeman-Giraudon/AR; 93: © Biblioteca Apostolica Vaticana

Chapter 11
94: Sculpture by Charles De George, Bridgeman-Giraudon/AR; 95: Dr. Fred Espenak; 97: Scala/AR; 99: Royalty-free/CORBIS; 100: (both) Erich Lessing/AR; 101: © Jona Lendering, Livius.org; 102: Bibliothèque Nationale, Paris, France/Archives Charmet/BAL; 103: NASA; 104: E. R. Degginger/PR; 105: Réunion des Musées Nationaux/AR

Chapter 12
107: Edward Owen/AR; 108: Scala/AR; 109: Erich Lessing/AR; 110–11: Réunion des Musées Nationaux/AR; 113: Tunç Tezel, Istanbul

Chapter 13
115: Victoria & Albert Museum, London/AR; 117: Jerry Schad/PR

Chapter 14
120: Painting by Ciro Ferri, Scala/AR; 121: (top) Scala/AR; (center) Réunion des Musées Nationaux/AR; 123: (left and center) Erich Lessing/AR; (right) Werner Forman/AR; (bottom) Erich Lessing/AR; 124–25: Thomas Hartwell/CORBIS SABA; 126: (left) based on drawing in *Pharos* by Hermann Thiersch, 1909; (right) Bettmann/CORBIS; 127: (all) © stephanecompoint.com

Chapter 15
128: Scala/AR; 129: Erich Lessing/AR; 131: Bridgeman-Giraudon/AR; 133: (top) CORBIS

Chapter 16
136: Nicolo Orsi Battaglini/AR; 137: Erich Lessing/AR; 138: (top) Joseph Sohm; Visions of America/CORBIS; 139: AR; 140: The Menil Collection, Houston, and © 2004 Artists Rights Society, New York/ ADAGP, Paris; 142: Michael T. Sedam/CORBIS; 143–44: Richard E. Schwartz

Chapter 17
146: Pushkin Museum, Moscow, Russia/BAL; 152: (left) Erich Lessing/AR; (right) Image Select/AR; 155: (bottom) Bettmann/CORBIS; 157: Erich Lessing/AR; 159: Redrawn after illustration in *Ancient Inventions* by P. James and N. Thorpe, 1994

Chapter 18
161: © 2003 Bob Sacha; 165: NASA

Chapter 19
166: Bridgeman-Giraudon/AR; 167: Erich Lessing/AR; 168: Bettmann/CORBIS; 169: Scala/AR; 171: (bottom) Keren Su/CORBIS; (inset) Erich Lessing/AR; 173: Erich Lessing/AR

Chapter 20
175: Sheila Terry/PR; 176: Bettmann/CORBIS; 180: akg-images; 181: Bettmann/CORBIS; 182: Quadrant by John Giamin, ca. 1550, Museo di Storia della Scienza, Florence, Bettmann/CORBIS; 183: Archivo Iconografico, S.A./CORBIS

Chapter 21
184: Erich Lessing/AR; 185: © National Maritime Museum, London; 186: Courtesy of the Bodleian Library, University of Oxford. MS. Pococke 375 folios 3v-4r.; 187: Scala/AR; 188: Erich Lessing/AR; 189: Jean Loup Charmet/PR

Chapter 22
191: (top) Scala/AR; (bottom) Erich Lessing/AR; 192: Erich Lessing/AR; 193: Ancient Art and Architecture Collection Ltd./BAL; 194: Scala/AR; 195: Scala/AR; 196: Giraudon/BAL; 197: Cameraphoto Arte, Venice/AR; 198: The Triped was created by Damon Knight for his short story "Rule Golden"; this image was painted by Wayne Barlowe for his book, *Barlowe's Guide to Extraterrestrials*. Used by permission of the Estate of Damon Knight and by permission of Wayne Barlowe; 199: (top) Bettmann/CORBIS; (bottom) Scala/AR

Chapter 23
201: Scala/AR; 202: Gianni Dagli Orti/CORBIS; 204: Erich Lessing/AR; 205: Werner Forman/AR

Chapter 24
206: Reuters NewMedia Inc./CORBIS; 207: Bridgeman-Giraudon/AR; 208: Scala/AR; 209: (top) Ken Welsh/BAL; (inset) Erich Lessing/AR; 211: Mishneh Torah, ca. 1351, National Library, Jerusalem, Lauros/Giraudon/BAL; 212: (top center) Fresco by the Master of Saint Francis, S. Francesco, Assisi, Scala/AR; (top right) Icon, St. Catherine Monastery, Mount Sinai, Erich Lessing/AR; (bottom) Jonathan Blair/CORBIS; (inset) Franz-Marc Frei/CORBIS; 213: The Dean and Chapter of Hereford and the Hereford Mappa Mundi Trust; 214: Bridgeman-Giraudon/AR; 216: Snark/AR; 217: (top) Christophe Loviny/CORBIS; (bottom) Bibliothèque Nationale, Paris, France/BAL

Chapter 25
218: Réunion des Musées Nationaux/AR; 219: Borromeo/AR; 220: Courtesy of the Bodleian Library, University of Oxford. MS. Sansk d.14.; 221: Abilio Lope/CORBIS; 222: Snark/AR; 223: Roger Wood/CORBIS; 225: (bottom) Painting by Joseph Heintz the Younger, Alinari/AR; 226: (bottom) Syd Greenberg/PR; 227: (top) Scott Camazine and Sue Trainor/PR; (bottom) Thomas R. Taylor/PR

Chapter 26
228: Nicolo Orsi Battaglini/AR; 229: (top) Bridgeman-Giraudon/AR; (bottom) Painting by Leandro Bassano, Cameraphoto Arte, Venice/AR; 230: Scala/AR; 231: (left) Réunion des Musées Nationaux/AR; (right) Erich Lessing/AR; 233: (top) Scala/AR; (bottom) AR; 235: Réunion des Musées Nationaux/AR; 236: Smithsonian American Art Museum, Washington, D.C./Art Resource, N.Y.; 237: (top left) Underwood & Underwood/CORBIS; (top right) Otto Lilienthal in 1896, Underwood & Underwood/CORBIS; (bottom left) *DeepWorker*, Sustainable Seas Expeditions, 2003; (bottom right and inset) NASA

Chapter 27
238: Réunion des Musées Nationaux/AR; 239: (top right) Gérard Degeorge/CORBIS; (bottom right) Brian A. Vikander/CORBIS; 240: (top) Foto Marburg/AR; (bottom) Bibliothèque Nationale, Paris, France/BAL; 241: (top) HIP/Scala/AR; 242: (top) Private Collection/The Stapleton Collection/BAL; (bottom) Private Collection/BAL; 243: (top) Erich Lessing/AR; (bottom) The Pierpont Morgan Library/AR; 244: (top) Réunion des Musées Nationaux/AR; (bottom) Bibliothèque Nationale, Paris, France/BAL; 245: Private Collection/BAL; 246: Archivo Iconografico, S.A./CORBIS; 247: Bettmann/CORBIS

Chapter 28
249: (top and center) Bridgeman-Giraudon/AR; (bottom) Bibliothèque Nationale, Paris, France/BAL; 250: The Pierpont Morgan Library/AR; 251: (top) Erich Lessing/AR; (bottom) Geosphere/Planetary Visions/PR; 252: Private Collection/BAL; 254: Keren Su/CORBIS; 255: Werner Forman/AR; 256: Sala del Mappamondo, Palazzo Farnese, Viterbo, Scala/AR; 257: Royal Geographical Society, London, U.K./BAL

Chapter 29
259: Celestial Image Co./PR; 261: Mehau Kulyk/PR; 262: Scala/AR; 264–65: John Chumack/PR

Chapter 30
267: Archivo Iconografico, S.A./CORBIS; 268: Erich Lessing/AR; 269: NASA/PR

Suggested Reading
271: Scala/AR

Permissions

Excerpts on the following pages are reprinted by permission of the publishers and copyright holders.

Page 9: From George Johnson, *Fire in the Mind*
(New York: Alfred A. Knopf, 1995)

Page 20: From Gwendolyn Brooks, *In the Mecca*
(New York: Harper & Row, 1968)

Page 21: From Isaac Asimov, *Asimov on Numbers*
(Garden City, N.Y.: Doubleday, 1977)

Page 43: From Ernest Zebrowski Jr., *History of the Circle*
(Piscataway, N.J.: Rutgers University Press, 1999)

Page 54: From Robert Frost, "Fire and Ice," in *The Collected Poems*
(Henry Holt, 1969)

Page 72: From Alfred North Whitehead, *Science and the Modern World*, 1926.
Reprinted with the permission of Cambridge University Press.

Page 73: From Norton Juster, *The Phantom Tollbooth*
(New York: Random House, 1961)

Page 77 and 87: From *The God Particle* by Leon Lederman with Dick Teresi.
Copyright © 1993 by Leon Lederman and Dick Teresi.
Reprinted by permission of Houghton Mifflin Company. All rights reserved.

Page 106 and 184: From Stephen Hawking, *A Brief History of Time*
(New York: Bantam Books, 1988)

Page 117: From Isaac Asimov, *Asimov's Biographical Encyclopedia of Science & Technology*
(Garden City, N.Y.: Doubleday, 1964)

Page 120: From Morris Kline, *Mathematics in Western Culture*
(New York: Oxford University Press, 1953)

Page 123: From Carl Sagan, *Cosmos*
(New York: Random House, 1980)

Page 213: From Joseph and Frances Gies, *Life in a Medieval City*
(New York: Harper Perennial, 1981)

Page 219: From Jan Gullberg, *Mathematics: From the Birth of Numbers*
Copyright © 1997 by Jan Gullberg.
Used by permission of W. W. Norton & Company, Inc.

Page 257: From W. Jeffrey Bolster, review of *Over the Edge of the World* by Laurence Bergreen,
The New York Times Sunday Book Review section, 7 December 2003

Page 260: From Paul Davies, *Are We Alone?*
Copyright © 1995 by Orion Productions.
Reprinted by permission of Basic Books, a member of Perseus Books, L.L.C.

Index

D

E

F

G

H

I

J

K

L

M

N

O

P

Q

R

S